Biohydrometallurgical Recycling of Metals from Industrial Wastes

T0221377

Biohydrometallurgical Recycling of Metals from Industrial Wastes

Hong Hocheng, Mital Chakankar, and
Umesh Jadhav

CRC Press
Taylor & Francis Group
Boca Raton London New York

CRC Press is an imprint of the
Taylor & Francis Group, an **informa** business

Dedication

This book is dedicated to our family and students.

Contents

Preface

Microbial processing of metals from ores is not new and is known for centuries. However, as years go by, tremendous use of natural resources and depletion in ore deposits is resulting in the scarcity of metals occurring naturally. The situation can be overcome by the use of a secondary source, which in itself has a greater amount of metals than the natural source: metal waste from industries. The only concern is the method for the recovery of those metals in pure form and at the same time to not harm nature. Many methods in use, although simple and fast, are also quite expensive, and primarily cause environmental pollution. The most obvious natural method is biological recovery of metals, that is, biohydrometallurgical processing of metals.

The use of biological means will not only help conserve the dwindling ore resources but will also fulfill the need for an unambiguous method to extract metals in a nonpolluting, low-energy, low-cost, and natural manner. Also, it would lead to a more sustainable world to live in.

This book will cover almost all the aspects of biohydrometallurgy and its application in the recovery of metals from secondary sources like wastes. The ultimate objective is to provide the researchers with as much information as possible with respect to different wastes for metal recovery and biological treatment methods, which are both environmentally friendly and economically viable, under one cover.

The authors owe a special gratitude to the Ministry of Science and Technology and National Tsing Hua University in Taiwan for their support. The help and support of our lab mates, friends, and family are deeply acknowledged.

Authors

Professor Hong Hocheng is currently the chair professor at National Tsing Hua University, Taiwan, R.O.C. He received his PhD from the University of California, Berkeley (1998). He completed his BSc in mechanical engineering from the National Taiwan University, Taiwan, R.O.C., and Diplom-Ingenieur in manufacturing engineering from Rheinisch-Westfälische Technische Hochschule (Aachen), Germany. He is an ASME and AMME fellow. His fields of interest are innovative manufacturing processes, nontraditional machining, micro-/nano-manufacturing, fatigue of MEMS, and machining of composites. He has around 160 research papers published and 30 international invited paper/review article/keynote lecture/plenary speech/awardee's lecture/book chapters. He is a member of the editorial boards of 15 international scientific journals.

Dr. Mital Chakankar received her PhD in microbiology from the Department of Microbiology, Shivaji University, Kolhapur, India. She carried out research in the field of microbial production of biosurfactants, microbial degradation of pollutants, and biosensors for the detection of microorganisms and biomolecules. She is currently a postdoctoral research fellow at the Department of Power Mechanical Engineering, National Tsing Hua University, Taiwan. Her current research interests include bioleaching of metals from industrial wastes. She has published 12 research papers in international journals.

Dr. Umesh Jadhav received his PhD in environmental biotechnology from the Department of Biochemistry, Shivaji University, Kolhapur, India, in 2009. He carried out research in the field of microbial degradation of pollutants and bioleaching of metals. Presently, he is an assistant professor at the Department of Microbiology, Savitribai Phule Pune University, Pune, India. He is involved in the application of microorganisms for geotechnical applications. He has published 31 research papers. Also, he has four book chapters to his credit.

1 Introduction

1.1 BRIEF HISTORY OF INDUSTRIAL PROGRESS AND GENERATION OF INDUSTRIAL WASTES

Industries are the cornerstones of social and economic development. They have dynamically contributed to the growth and standard of human civilization. High-production activities in manufacturing structurally altered the economies from traditional to modern economies (Naudé and Szirmai 2012). Historically, Britain and the United States underwent a rapid industrial and economic development in the mid-eighteenth century. The industrialization which started in Britain in the eighteenth century swept through Europe and the United States and reached Japan and Russia by the end of the nineteenth century. However, by 1900, this wave had subsided (Pollard 1990). This was followed in the twentieth century by a new wave in Japan, the East Asia, and most recently China. In this industrial revolution, production technology diffused across the globe, which has fundamentally affected the nature of global production. Today, developing countries are trying to catch up through manufacturing development, while developed countries are advancing their technologies to maintain their competitiveness. However, the murky side of this technological roar is the accumulation of industrial wastes.

Industrial growth has a positive impact on humans by enhancing the quality of resources and life, but at the same time, it has a negative impact due to environmental pollution and depletion of natural resources. At first, when industrial activity was less, this impact was localized, but with the industrial progress and global expansion, this impact increased exponentially, leading to the generation of gigantic amounts of wastes; consequently, pollution is inflating at alarming rates and causing serious environmental issues globally.

Increasing population, rapid urbanization, and escalating industrialization have significantly accelerated global waste generation and thus led to unprecedented dissemination of toxic components in the environment. This is a serious global environmental issue due to its impact and stress on the environment and living beings. Moreover, continuous exploitation of natural resources has resulted in progressive depletion of nonrenewable sources (Sharholy et al. 2007, Minghua et al. 2009, Seng et al. 2011, Guerrero et al. 2013).

According to the recent World Bank report, global solid waste generation is projected to be doubled by 2025, from 3.5 million tons to 6 million tons per day, and will keep elevating so on and so forth in the foreseeable future. Furthermore, the amount of trash will be three times of today's waste by 2100, as depicted in Figure 1.1. The global solid waste volume was about 11 billion tons per year (using 2.5 ton trucks can turn 300 circles around the equator) in 2011, and the

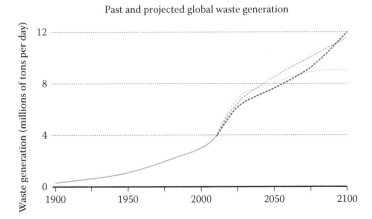

FIGURE 1.1   A scenario that assumes that current demographic and per capita waste production trends will continue (blue line) projects waste to peak sometime after 2100, as does a scenario with even greater population growth (red line). Only a scenario with a smaller, wealthier world population and more environmentally friendly consumption behaviors (gray line) enables peak garbage to occur in this century. (Reprinted from Hoornweg D. et al., *Nature*, 502, 615, 2013. With permission.)

global per capita solid waste generation was approximately 1.74 tons per year (Hoornweg et al. 2013, Song et al. 2015).

On the other hand, extensive amounts of natural resources are depleting every day due to the high demand for new products (Menikpura et al. 2013, Plagányi et al. 2013). Globally, 120–130 billion tons of natural resources are consumed every year and produce around 3.4–4 billion tons of municipal solid waste (Giljum et al. 2008, Chalmin and Gaillochet 2009). The generation of wastes not only exhausts natural resources and consumes large amounts of energy and water, but also exerts pressure by polluting the land and environment, along with additional expenses on waste management.

Furthermore, disposal of this waste is more troublesome. Due to a huge generation of industrial waste, not all the waste can be collected and treated; therefore, enormous amounts of industrial wastes are dumped illegally and also exported to other countries (transboundary movement), which threatens the environment and human health (Chartsbin 2013).

Increased waste generation over the past two centuries has severely impacted the environment and also substantially elevated the rate at which trace metals are released into the environment. This can be clearly seen by the progressive pollution of rivers such as the Meuse, Kibe, and Rhine (Middelkoop 2000); the Los Angeles smog; the proclaimed "death" of Lake Erie; and chemical poisoning by mercury in Minamata, Japan, and many more. Lake Pepin, a natural lake on the upper Mississippi River (the United States), is accumulating trace metals from both natural and industrial sources (Balogh et al. 2009). The Ganges of India has also been found to have high heavy metals in sediments and fish (Gupta et al. 2008).

1.2   THREATS OF METAL-CONTAINING WASTES
TO THE ENVIRONMENT

Metals are an indispensable part of human life and a "foundation stone" of industrial civilization. The use of metals by humans dates back to ages where rocks and minerals were melted to obtain metals. Today, various pyro-, hydro-, and biohydrometallurgical methods are used in metal production industries. Among these methods, pyro- and hydrometallurgical methods produce huge quantities of toxic and hazardous by-products as waste, which is composed of heavy metals and is released in the surroundings, polluting the air, land, and water.

Nevertheless, organic compounds from these wastes get degraded; inorganic compounds remain undegraded and accumulate, which imparts toxicity toward living beings. Heavy metals are one such component of industrial wastes which remain unblemished. These heavy metals from the soil are further mobilized to water due to various natural and anthropogenic ways.

Numerous industries, for example, electroplating, metal finishing, electronic, steel and nonferrous processes, petrochemical and pharmaceutical, and used electronic/household goods, discharge large quantities of waste by-products in the form of wastewater, residues, sludge, dust, etc., comprising a variety of toxic heavy metals such as Zn, Cd, Cr, Cu, Ni, Pb, V, Mo, Co, etc., and thus threatening the public health and the environment if managed improperly (Sörme and Lagerkvist 2002, Lee and Pandey 2012).

Industrial waste generation and composition vary depending on the type of industry and processes/technologies implemented and also on the waste management system of the country.

Hazardous wastes from industries are of particular concern, as they entail serious environmental risk if managed improperly. Gigantic amounts of waste are generated every year globally. Nevertheless, part of this enormous waste is recycled. However, a lot of remaining waste is landfilled or dumped, creating health and environmental hazards. Over the time, the hazardous materials in these dumped wastes leach out from the landfills into the groundwater and streams, contaminating the soil, water, and air.

The industrial wastes which are dumped in rivers and oceans cause water pollution. As a result, groundwater and surface water get contaminated, affecting our ecosystem. The use of this water for agriculture deteriorates the food quality.

Heavy metals are found throughout the earth's crust and occur naturally in the soil due to biogeochemical cycles. Their concentration is within defined limits; however, it increases above the defined values due to anthropogenic activities such as mining and smelting operations, industrial production and use, and domestic and agricultural use of metals and metal-containing compounds. Thus, heavy metals turn into contaminants and cause risks to human health, plants, animals, ecosystems, or other media (Shallari et al. 1998, Herawati et al. 2000, Goyer 2001, D'Amore et al. 2005, He et al. 2005).

Environmental contamination can also occur through metal corrosion, atmospheric deposition, soil erosion of metal ions and leaching of heavy metals, sediment resuspension, and metal evaporation from water resources to the soil and groundwater (Nriagu 1989). Industrial sources include metal processing in refineries, coal burning in power plants, petroleum combustion, nuclear power stations and high-tension lines, plastics, textiles, microelectronics, wood preservation, and paper processing plants (Pacyna 1996, Arruti and Fern 2010, Sträter et al. 2010).

Soil acts as the major sink for heavy metals that are released into the environment due to industrial pollution and remain persistent for a long time as they are nonbiodegradable (Adriano 2003, Kirpichtchikova et al. 2006). This may also affect the biodegradation of other organic contaminants (Maslin and Maier 2000). Upon entering the soil in large amounts, heavy metals primarily affect its biological characteristics: the total content of microorganism changes, the microbial diversity reduces, and the intensity of basic microbiological processes and the activity of soil enzymes decrease. In addition, heavy metals also change the humus content, structure, and pH of soils (Levin et al. 1989). These processes ultimately lead to the partial or, in some cases, complete loss of soil fertility and its structure. This may create an ecological imbalance and renders the land unsuitable for agriculture.

The increase in contamination may also affect crop productivity adversely and could lead to a significant long-term loss in production and income (Wuana and Okieimen 2011). The connection between soil and water contamination and metal uptake by plants is determined by many chemical and physical soil factors as well as by the physiological properties of crops. These factors include climate, irrigation, atmospheric deposition, the nature of the soils on which the plant is grown, and the time of harvesting.

The accumulation of heavy metals in plant tissues eventually leads to toxicity and changes in the plant community (John et al. 2009, Kim and McBride 2009, Gimmler et al. 2017). The toxic metals in soils are reported to inhibit root and shoot growth, affect nutrient uptake and homeostasis, and are frequently accumulated by agriculturally important crops. Thereafter, they enter the food chain with a significant potential to impair animal and/or human health (McLaughlin et al. 2000a,b, Ling et al. 2007). The reduction in biomass of plants growing on metal-contaminated soils has been found to be due to the direct consequence of chlorophyll synthesis and photosynthesis inhibition (Dong et al. 2005, Shamsi et al. 2007), carotenoids inhibition (John et al. 2009), inhibition of various enzyme activities, and induction of oxidative stress, including alterations of enzymes involved in the antioxidant defense system (Dazy et al. 2009, Kachout et al. 2009).

The emissions from industries contain gaseous contaminants such as carbon dioxide (CO_2), methane, sulfur, oxides of nitrogen, etc. which result in health and environmental hazards such as acid rains, smog, respiratory disorders in humans, etc. Air pollution has a direct impact on human health and causes various illnesses. Moreover, the increasing pollution is destroying the forests and habitats of wildlife, affecting various species, many of which are now highly endangered and some may even become extinct in the near future. Also, major industrial disasters such as oil spills and leakage of radioactive materials are hard to clean up, which may take up to several years and have a high impact in a shorter time span. Further, this pollution is the major cause of global warming. Emission of greenhouse gases, smoke, etc. results in air pollution, which in turn increases global warming. Industrial processes release various greenhouse gases such as CO_2 and methane in the atmosphere. These gases absorb thermal radiation from the sun, thereby increasing the temperature of the earth, leading to global warming. The effects of global warming include various natural disasters such as melting of glaciers, extinction of species, floods, tsunamis, hurricanes, etc.

Because of their high solubility in aquatic environments, heavy metals can be absorbed by living organisms. Once they enter the food chain, large concentrations of heavy metals may accumulate in the human body. If the metals are ingested beyond the permitted concentration, they can cause serious health disorders (Babel and Kurniawan 2004, Barakat 2011). Therefore, it is necessary to treat metal-contaminated wastewater prior to its discharge into the environment.

Table 1.1 summarizes the maximum contaminant level (MCL) standards of some metals established by the U.S. Environmental Protection Agency (USEPA) (Babel and Kurniawan 2004).

Heavy metals such as copper, iron, zinc, manganese, cobalt, selenium, and chromium are required for normal physiological functions of mammals in very low concentrations and are referred to as "essential trace elements" or "micronutrients." Their deficiencies can lead to diseases and even death of the plant or animal. However, a high concentration of one or more of these results in altered physiological functions in plant/animal cells. The most important heavy metals with regard to potential hazard and occurrence in contaminated soils are Pb, Cr, As, Zn, Cd, Cu, and Hg (Alloway 1995, World Health Organization 1996). Soil contamination with these metals decreases crop production due to bioaccumulation and biomagnification in the food chain, along with superficial and groundwater contamination (USEPA 1996). Exposure to these pollutants through dietary intake of plant-derived food and beverages, drinking water, or air can have long-term effects on human health (Chaffei et al. 2004, Godt et al. 2006, Järup and Åkesson 2009). Other metals such as aluminum (Al), antinomy (Sb), arsenic (As), barium (Ba), beryllium (Be), bismuth (Bi), cadmium (Cd), gallium (Ga), germanium (Ge), gold (Au),

TABLE 1.1
MCL Standards for the Most Hazardous Heavy Metals (Babel and Kurniawan 2004)

Heavy Metal	Toxicities	MCL (mg/L)
Arsenic	Skin manifestations, visceral cancers, vascular disease	0.050
Cadmium	Kidney damage, renal disorder, human carcinogen	0.01
Chromium	Headache, diarrhea, nausea, vomiting, carcinogenic	0.05
Copper	Liver damage, Wilson's disease, insomnia	0.25
Nickel	Dermatitis, nausea, chronic asthma, coughing, human carcinogen	0.20
Zinc	Depression, lethargy, neurological signs, and increased thirst	0.80
Lead	Fetal brain damage and diseases of the kidneys, circulatory system, and nervous system	0.006
Mercury	Rheumatoid arthritis and diseases of the kidneys, circulatory system, and nervous system	0.00003

Source: Reprinted from Barakat, M.A., *J. Chem.*, 4, 361, 2011.

indium (In), lead (Pb), lithium (Li), mercury (Hg), nickel (Ni), platinum (Pt), silver (Ag), strontium (Sr), tellurium (Te), thallium (Tl), tin (Sn), titanium (Ti), vanadium (V), and uranium (U) have no established biological functions and are considered as nonessential metals and cause toxicity above a certain tolerance level (Chang et al. 1996).

Knowledge of the basic chemistry and environmental and associated health effects of these heavy metals is necessary for understanding their speciation, bioavailability, and remedial options. The general effects of these heavy metals on biological systems include the effects on cellular organelles and components such as cell membrane, mitochondria, lysosome, endoplasmic reticulum, nuclei, and some enzymes involved in metabolism, detoxification, and damage repair (Wang and Shi 2001). Metal ions are also known to interact with cell components such as DNA and nuclear proteins, causing DNA damage and conformational changes that may lead to cell cycle modulation, carcinogenesis, or apoptosis (Chang et al. 1996, Wang and Shi 2001, Beyersmann and Hartwig 2008).

The health effects of some potentially hazardous heavy metals are presented next.

1.2.1 LEAD

The concentration of Pb in soil averages around 32 mg/kg and ranges from 10 to 67 mg/kg worldwide (Kabata-Pendias and Pendias 2001). Lead is released into the soil and water in the form of ionic lead, Pb(II), lead oxides and hydroxides, and lead–metal oxyanion complexes. Toxicity of lead is well known and can cause serious injury to the brain, nervous system, red blood cells, and kidneys (Baldwin and Marshall 1999). Biological effects depend on the dose level and exposure duration. Routes of lead exposure include ingestion of lead-contaminated food, water, and paints and inhalation of lead-contaminated dust particles or aerosols, and both have same effects (ATSDR 1992, 1999). Pb is known to accumulate in organs such as the brain, leading to poisoning (plumbism) or even death. It also causes inhibition of the synthesis of hemoglobin; dysfunction of the kidneys, joints, reproductive system, and cardiovascular system; and acute and chronic damage to the central nervous system (CNS) and peripheral nervous system (PNS). Other effects include damage to the gastrointestinal tract (GIT) and urinary tract, resulting in bloody urine, neurological disorders, and severe and permanent brain damage. While inorganic forms of lead typically affect the CNS, PNS, GIT, and other biosystems, organic forms predominantly affect the CNS (Lenntech Water Treatment and Air Purification 2004).

Children, especially under 6 years of age, are at considerable risk, and exposure may lead to the poor development of the gray matter of the brain and, consequently, to impaired development, lower intelligence quotient (IQ) (Udedi 2003), shortened attention span, hyperactivity, mental deterioration, and antisocial and diligent behaviors (Amodio-Cocchieri 1996, Factor-Litvak et al. 1998, Kaul et al. 1999). Lead also influences the nervous system, slowing down the nerve response, thereby affecting learning abilities and behavior. Symptoms in adults post lead exposure include decreased reaction time, loss of memory, nausea, insomnia, anorexia, and weakening of the joints (NSC 2009). Its absorption in the body is enhanced by Ca and Zn deficiencies. The symptoms of acute lead poisoning are skin damage, headache, irritability, nausea, abdominal pain, gastric and duodenal ulcers, and various symptoms related to the nervous system (Monika 2010).

Lead encephalopathy is characterized by sleeplessness and restlessness. Acute and chronic effects of lead result in psychosis, confusion, and reduced consciousness. Lead also induces brain damage, kidney damage, and gastrointestinal (GI) diseases, while chronic exposure may adversely affect the blood, central nervous system, blood pressure, kidneys, and vitamin D metabolism (ATSDR 1992, 1999, Amodio-Cocchieri 1996, Apostoli et al. 1998, World Health Organization 1996, Factor-Litvak et al. 1998, Kaul et al. 1999, Hertz-Picciotto 2000). In adult populations, reproductive effects such as decreased sperm count in men and spontaneous abortions in women have been associated with high lead exposure (Apostoli et al. 1998, Hertz-Picciotto 2000, Barbosa et al. 2005, Poon 2008, Chen et al. 2011).

1.2.2 CHROMIUM

Chromium is a less common element and naturally occurs in compounds, instead of in elemental form. It is commonly found in the chromium(IV) form in contaminated sites, which is known to be the more toxic form. Chromium is associated with allergic dermatitis in humans (Scragg 2006).

Inhalation of high levels of chromium(VI) causes irritation to the nose lining and nose ulcers. The main health problems seen in animals following ingestion of chromium(VI) compounds are irritation and ulcers in the stomach and small intestine, anemia, sperm damage, and male reproductive system damage. Chromium(III) compounds are much less toxic and do not appear to cause these problems. Some individuals are extremely sensitive to chromium(VI) or chromium(III) and thus have allergic reactions consisting of severe redness and swelling of the skin. An increase in stomach tumors was observed in humans and animals exposed to chromium(VI) in drinking water. Accidental or intentional ingestion of extremely high doses of chromium(VI) compounds by humans has resulted in severe respiratory, cardiovascular, GI, hematological, hepatic, renal, and neurological effects as part of the sequelae leading to death or in patients who survived because of medical treatment (ATSDR 2008). It is also known to cause cancer in humans and terrestrial mammals; however, the mechanism is unknown (Li Chen et al. 2009).

1.2.3 ARSENIC

Many arsenic compounds adsorb strongly to soils and are therefore transported only over short distances in groundwater and surface water. Exposure occurs via the oral route (ingestion), inhalation, dermal contact, and the parenteral route, to some extent (NRCC 1978, Tchounwou et al. 1999, ATSDR 2000). Arsenic is associated with skin damage, cardiovascular and peripheral vascular disease, developmental anomalies, neurological and neurobehavioral disorders, diabetes, hearing loss, portal fibrosis, hematological disorders (anemia, leukopenia, and eosinophilia), and carcinoma (ATSDR 2000, National Research Council 2001, Tchounwou et al. 2004, Centeno et al. 2005, Scragg 2006). Inorganic arsenic is acutely toxic, and intake of large quantities leads to GI symptoms, severe disturbances of the cardiovascular and central nervous systems, and eventually death. In survivors, bone marrow depression, hemolysis, hepatomegaly, melanosis, polyneuropathy, and encephalopathy may be observed (Järup 2003).

Arsenic exposure affects virtually all organ systems, including the cardiovascular, dermatologic, nervous, hepatobiliary, renal, GI, and respiratory systems (Tchounwou et al. 2003b, Jomova et al. 2011). Arsenic toxicity also presents a disorder, which is similar to, and often confused with, the Guillain–Barre syndrome, an anti-immune disorder that occurs when the body's immune system mistakenly attacks part of the PNS, resulting in nerve inflammation, which in turn causes muscle weakness (Kantor 2006).

1.2.4 ZINC

The average concentration of Zinc in soil is 70 mg/kg. Zinc is an essential trace element for human health and its shortage may result in birth defects. However, increased Zn concentration may result in Zn accumulation in fish, which is further biomagnified in the food chain. Zn can interrupt the activity in soils, as it negatively influences the activity of microorganisms and earthworms, thus retarding the breakdown of organic matter (Greany 2005). The clinical signs of zinc toxicosis have been reported as vomiting, diarrhea, bloody urine, icterus (yellow mucus membrane), liver failure, kidney failure, and anemia.

1.2.5 CADMIUM

Cadmium is a known heavy metal poison. The main routes of exposure are via inhalation or cigarette smoke and ingestion of food. Skin absorption is rare. Due to its chemical similarity with Zn, it gets substituted with Zn and results in the malfunctioning of metabolic pathways (Campbell 2006). Although Cd has less toxicity, it is more biopersistent and thus remains resident for many years after absorption. Cd is known to affect many biological enzymes, and the major threat to human health is chronic accumulation in the kidneys, affecting the enzymes responsible for the reabsorption of proteins in kidney tubules, which may cause renal damage, leading to kidney dysfunction, characterized by tubular proteinuria (Manahan 2003). Cadmium also reduces the activity of delta-aminolevulinic acid synthetase, arylsulfatase, alcohol dehydrogenase, and lipoamide dehydrogenase, whereas it enhances the activity of delta-aminolevulinic acid dehydratase, pyruvate dehydrogenase, and pyruvate decarboxylase (Manahan 2003). Chronic exposure to cadmium has a depressive effect on levels of norepinephrine, serotonin, and acetylcholine (Singhal et al. 1976).

The most well-known cadmium poisoning occurred in the Jintsu River Valley, near Fuchu, Japan, resulting in the *itai-itai* disease. Contamination of rice with Cd due to an upstream mine producing Pb, Zn, and Cd, resulted in Cd poisoning. The affected people had a painful osteomalacia (bone disease) combined with kidney malfunction (Rai and Pal 2002). Cd can also affect the neurodevelopment of the fetus and young children (Chen et al. 2011).

High exposure can lead to obstructive lung disease, cadmium pneumonitis, resulting from inhaled dust and fumes, characterized by chest pain, cough with foamy and bloody sputum, and death of the lining of the lung tissues because of excessive accumulation of watery fluids. It is also associated with bone defects such as osteomalacia, osteoporosis and spontaneous fractures, increased blood pressure, and myocardial dysfunction.

Cadmium is a severe GI irritant, which can be fatal if inhaled or ingested. After acute ingestion, symptoms such as nausea, vomiting, salivation, abdominal cramps, burning sensation, dyspnea, muscle cramps, vertigo, shock, loss of consciousness, and convulsions usually appear within 15–30 min (Baselt and Cravey 1995). Acute cadmium ingestion can also cause GIT erosion, pulmonary, hepatic or renal injury, and coma, depending on the route of poisoning (Baselt and Cravey 1995, Baselt 2000). Severe exposure may result in pulmonary edema and death (European Commission 2002).

1.2.6 COPPER

Copper is an essential micronutrient required for growth in plants, animals, and humans. In humans, it is required for hemoglobin production, while in plants, it plays an important role in seed production, disease resistance, and water regulation. However, exposure to higher doses can be harmful. Long-term exposure to high concentrations of copper causes irritation in the nose, mouth, and eyes, and causes headache, dizziness, nausea, vomiting, stomach cramps, and diarrhea (ATSDR 2004).

1.2.7 MERCURY

Mercury released in the environment exists in different forms such as the mercuric (Hg^{2+}), mercurous (Hg_2^{2+}), elemental (Hg_o), or alkylated form (methyl/ethyl mercury). It is most toxic in alkylated forms, which are soluble in water and volatile in air (Smith et al. 1995). Mercury enters the water naturally as well as through industrial pollution (Dopp et al. 2010). Algae and bacteria methylate the mercury entering the waterways. Methylmercury then makes its way through the food chain into fish, shellfish, and eventually humans (Sanfeliu et al. 2003).

Acute inorganic mercury exposure may give rise to lung damage. Chronic poisoning is characterized by neurological and psychological symptoms, such as tremor, changes in personality, restlessness, anxiety, sleep disturbance, and depression. Mercury is also known to cause spontaneous abortions, congenital malformations, and GI disorders (such as corrosive esophagitis and hematochezia). Metallic mercury may cause kidney damage, which is reversible after exposure has stopped. Also, it is an allergen which may cause contact eczema, and mercury from amalgam fillings may give rise to oral lichen. Poisoning by organic forms such as monomethyl and dimethylmercury results in erethism (an abnormal irritation or sensitivity of an organ or body part to stimulation), acrodynia (pink disease, which is characterized by rashes and desquamation of the hands and feet), gingivitis, stomatitis, neurological disorders, and total damage to the brain and CNS, and is also associated with congenital malformations (Lenntech Water Treatment and Air Purification 2004).

Methylmercury poisoning has a latency of 1 month or longer after acute exposure and the main symptoms relate to nervous system damage (Weiss et al. 2002), including earlier symptoms such as paresthesias and numbness in the hands and feet. High doses may lead to death, usually 2–4 weeks after the onset of symptoms. The Minamata catastrophe in Japan in the 1950s was caused by methylmercury poisoning from fish contaminated by mercury discharges into the surrounding sea (Järup 2003).

The elemental vapor form is highly lipophilic and is effectively absorbed through the lungs and tissues lining the mouth. After entering the blood, it passes through cell membranes, including both the blood–brain barrier and the placental barrier (Guzzi and La Porta 2008). Once it gains entry into the cell, Hg_0 is oxidized and becomes highly reactive Hg^{2+}. Methylmercury derived from eating fish is readily absorbed in the GIT, and because of its lipid solubility, it can easily cross both the placental and blood–brain barriers. Once absorbed, it has a very low excretion rate and accumulates in the kidneys, neurological tissue, and liver. All forms of mercury are toxic, and their effects include GI toxicity, neurotoxicity, and nephrotoxicity (Tchounwou et al. 2003, Scragg 2006).

1.2.8 NICKEL

Nickel is present in the environment at very low levels and is required in very small doses. It causes various kinds of cancer in different sites within the bodies of animals. Microorganisms also suffer from growth decline due to the presence of Ni, but they develop resistance to Ni after a while. Nickel is not known to accumulate in plants or animals, and as a result, it has not been found to biomagnify up the food chain. Nickel can cause skin allergies and cancer (Padiyar et al. 2011).

Along with the earlier-mentioned health effects, heavy metals such as arsenic, cadmium, lead, and mercury disrupt the proper functioning of endocrine systems. All of them have potential enough to cause reproductive problems by decreasing the sperm count, by increasing the number of testicular germ cells, and by causing cryptorchidism, hypospadias, miscarriages, endometriosis, impaired fertility, and infertility (Balabanič et al. 2011, Pant et al. 2012).

1.3 ADVANTAGES OF RECYCLING

Rapid economic growth, urbanization, and increasing population have caused (materially intensive) resource consumption to increase, and consequently the release of large amounts of wastes into the environment. From a global perspective, current waste and resource management strategies lack a holistic approach covering the whole chain of product design, raw material extraction, production, consumption, recycling, and waste management.

The generation of huge amounts of industrial wastes presents a massive disposal problem due to the scarcity of available landfills and the potentially hazardous heavy metal content of the waste. The landfilling of these wastes results in various health and environmental hazards, as discussed earlier. However, the one way out from this muddle is recycling of these wastes.

Industries are extensively using nonrenewable sources of mineral deposits, which will exhaust soon due to the alarming and uncontrolled exploitation and use of these metal resources and metals, respectively. Moreover, with booming industrialization, the demand for metals is also escalating.

Recycling is the key to reducing the large volume of waste generated and will also benefit the environment. Furthermore, the increasing demand for metals can also be compensated by recycling and reuse of metals from secondary sources.

In general, the treatment of solid wastes can confer at least two beneficial aspects: (1) recovery of valuable metals from the wastes and (2) detoxification of the contaminated wastes.

Recovery of metals from these "renewable resources" is considered important for prolonging the life of metallic elements reserve, while the removal of toxic substances from contaminated wastes could minimize environmental pollution, along with decreasing the cost of waste disposal.

Recycling will significantly reduce the loss of heavy metals to the environment and, at the same time, avoid the entry of new metals into circulation, thereby conserving energy resources and reducing the mining and beneficiation activities that disturb ecosystems. Aside from reducing greenhouse gas emissions, which contribute to global warming, recycling also reduces air and water pollution associated with making new products from raw materials (Needhidasan et al. 2014).

However, the concept of recycling covers an array of very different activities. Dedicated recycling of heavy metals may be carried out rather efficiently, with very small losses to the environment and residues, provided that an efficient environmentally friendly technique is applied. If such a sustainable technique of metal extraction from waste is developed, it would help conserve the depleting high-grade metallic ores, provide the extracted metals to industries, and conserve the biotic and abiotic components of the ecosystem from the hazardous effects of waste.

Apart from base metals, various precious and rare metals are employed in the manufacturing of products which are predominantly obtained from their respective primary production routes. The extraction process of heavy metals through mining has numerous negative aspects such as extensive land, energy, and water usage, gaseous emissions such as SO_2 and CO_2, and the generation of large amounts of secondary solid and liquid wastes. The environmental footprint of primary production of these metals is high, especially when the concentration of valuable components is quite low and requires the removal of large volumes of material, involving several steps of treatment. Primary production is also associated with high levels of greenhouse gas emissions.

Hence, sustainable recovery of precious metals from waste will significantly impact the environment as well as the economy (Cayumil et al. 2016). In addition to this, various other advantages of metal recovery from waste include a reduction in waste landfilling, resource utilization, minimization of gaseous emissions and generation of toxic compounds, along with high savings in energy consumption. For example, recycling of gold can reduce energy usage by up to 65% as compared with primary production; corresponding energy savings for Pd were estimated to be 14% (Wang and Gaustad 2012, Cayumil et al. 2016).

End-use product recycling is one approach to waste management that can reduce the burden on the environment, such as a reduction in the landfill space. Recycling also reduces the consumption of refined materials and energy used (saving resource usage) for new material extraction (CoCusi Coque 2004). It thus helps improve the environmental performance over the life cycle of products and also generates market values for selling recovered materials (Apisitpuvakul et al. 2008).

Metals can be recycled nearly indefinitely. Nevertheless, the economic recovery of metals after us, that is, from secondary sources, primarily depends on their initial use and their chemical reactivity. Thus, the success of secondary metals markets depends on the cost of retrieving and processing metals embedded in abandoned structures, discarded products, and other waste streams and its relation to primary metal prices.

Even if the incentive for recovery of precious metals is high, to attain the desired purity of the retrieved method is a challenging task, which will consequently result in the demand for the new metal. However, the recycling approach will provide incentives, along with the alternative framework to avoid excess use of metals, improving the technology for recovery of metals from secondary sources according to the changes in industrial and customer products, and will stimulate research on industrial processes for recovering metals.

One of the most striking environmental benefits of secondary metals production is the reduction in energy needed to produce a ton of metal. The primary reason for this phenomenon is that melting metals requires less energy than that needed for reducing naturally occurring oxides and sulfides (Chapman and Roberts 1983).

In the case of toxic metals, recovery is essential not because of incentives, but to ensure that these metals do not become biologically available again and pose risk to the environment. This also includes the reduction of human exposure through efficient recovery or by sequestration in products, landfills, or some vitrified form, need to primarily consider the environmental transport mechanisms.

As resource depletion, environmental concerns, and other factors drive primary production costs up, the relative importance of recycling in supplying the material needs of society will grow. Thus, recycling will help in sustainable recovery of heavy metals from secondary sources, thus decreasing the waste volume, and in coping with the increasing metal demand.

1.4   CONCLUSIONS

In recent decades, the world has witnessed a technological and industrial development fueled by a continuous demand for metal resources. This revolution will not decline, but rise exponentially, resulting in increased waste generation, a major global issue. Management of waste is thus an intimidating task and needs to be addressed systematically. Metals are an integral part of an industry that supports modern economics. Due to industrialization, the amounts of heavy metals that are being deposited onto the surface of the earth are many times greater than arising from natural background sources. Recycling of waste for the extraction of base and precious metals is of main concern not only due to the presence of high concentrations of these metals but also due to their hazardous nature, which threatens human health and the environment. Moreover, considering the elevating demand for metals and rare earth elements, due to the increasing global population, it is imperative to exploit secondary metal sources for metal recovery. This will help retrieve valuable metals and fill the gap between the demand and supply of metals if proper recovery techniques are applied. Because of the centrality of metals in industries, improving resource efficiency and reducing losses in the metals sector will help usher in positive changes, both industrially and environmentally.

REFERENCES

Adriano, D. C. 2003. *Trace Elements in Terrestrial Environments: Biogeochemistry, Bioavailability and Risks of Metals*, 2nd edn. New York: Springer.

Agency for Toxic Substances and Disease Registry (ATSDR). 1992. *Case Studies in Environmental Medicine—Lead Toxicity*. Atlanta, GA: Public Health Service, U.S. Department of Health and Human Services.

Agency for Toxic Substances and Disease Registry (ATSDR). 1999. *Toxicological Profile for Lead*. Atlanta, GA: Public Health Service, U.S. Department of Health and Human Services.

Agency for Toxic Substances and Disease Registry (ATSDR). 2000. *Toxicological Profile for Arsenic TP-92/09*. Atlanta, GA: Center for Disease Control.

Agency for Toxic Substances and Disease Registry (ATSDR). 2004. *Toxicological Profile for Copper*. Atlanta, GA: Center for Disease Control.

Agency for Toxic Substances and Disease Registry (ATSDR). 2008. *Toxicological Profile for Chromium*. Atlanta, GA: Public Health Service, U.S. Department of Health and Human Services.

Alloway, B. J. 1995. *Heavy Metals in Soils*, 2nd edn. Glasgow, Scotland: Blackie Academic and Professional.

Amodio-Cocchieri, R. 1996. Lead in human blood from children living in Campania, Italy. *Journal of Toxicology and Environmental Health* 47(4): 311–320. doi: 10.1080/009841096161663.

Apisitpuvakul, W., P. Piumsomboon, D. J. Watts, and W. Koetsinchai. 2008. LCA of spent fluorescent lamps in Thailand at various rates of recycling. *Journal of Cleaner Production* 16(10): 1046–1061. doi: 10.1016/j.jclepro.2007.06.015.

Apostoli, P., P. Kiss, S. Porru, J. P. Bonde, P. Apostoli, P. Kiss, S. Porru, J. P. Bonde, and M. Vanhoorne. 1998. Male reproductive toxicity of lead in animals and humans. *Occupational and Environmental Medicine* 55(6): 364–374.

Arruti, A. and I. Fern. 2010. Evaluation of the contribution of local sources to trace metals levels in urban PM2.5 and PM10 in the Cantabria Region (Northern Spain). *Journal of Environmental Monitoring* 12(7): 1451–1458. doi: 10.1039/b926740a.

Babel, S. and T. A. Kurniawan. 2004. Cr(VI) removal from synthetic wastewater using coconut shell charcoal and commercial activated carbon modified with oxidizing agents and/or chitosan. *Chemosphere* 54(7): 951–967. doi: 10.1016/j.chemosphere.2003.10.001.

Balabanič, D., M. Rupnik, and A. K. Klemenčič. 2011. Negative impact of endocrine-disrupting compounds on human reproductive health. *Reproduction, Fertility and Development* 23(3): 403–416. doi: 10.1071/RD09300.

Baldwin, D. R. and W. J. Marshall. 1999. Heavy metal poisoning and its laboratory investigation. *Annals of Clinical Biochemistry* 36(Pt 3): 267–300. doi: 10.1177/000456329903600301.

Balogh, S. J., D. R. Engstrom, J. E. Almendinger et al. 2009. A sediment record of trace metal loadings in the upper Mississippi River. *Journal of Paleolimnology* 41: 623–639.

Barakat, M. A. 2011. New trends in removing heavy metals from industrial wastewater. *Arabian Journal of Chemistry* 4(4): 361–377. doi: 10.1016/j.arabjc.2010.07.019.

Barbosa Jr., F., J. E. Tanus-Santos, R. F. Gerlach, and P. J. Parsons. 2005. A critical review of biomarkers used for monitoring human exposure to lead: Advantages, limitations, and future needs. *Environmental Health Perspectives* 113(12): 1669–1674. doi: 10.1289/ehp.7917.

Baselt, R. C. 2000. *Disposition of Toxic Drugs and Chemicals in Man*, 5th edn. Foster City, CA: Chemical Toxicology Institute.

Baselt, R. C. and R. H. Cravey. 1995. *Disposition of Toxic Drugs and Chemicals in Man*, 4th edn. Chicago, IL: Year Book Medical Publishers.

Beyersmann, D. and A. Hartwig. 2008. Carcinogenic metal compounds: Recent insight into molecular and cellular mechanisms. *Archives of Toxicology* 82(8): 493–512. doi: 10.1007/s00204-008-0313-y.

Campbell, P. G. C. 2006. Cadmium—A priority pollutant. *Environmental Chemistry* 3(6): 387–388. doi: 10.1071/EN06075.

Cayumil, R., R. Khanna, R. Rajarao, P. S. Mukherjee, and V. Sahajwalla. 2016. Concentration of precious metals during their recovery from electronic waste. *Waste Management* 57: 121–130. doi: 10.1016/j.wasman.2015.12.004.

Centeno, J. A., P. B. Tchounwou, A. K. Patlolla et al. 2005. Environmental pathology and health effects of arsenic poisoning: A critical review. In *Managing Arsenic in the Environment: From Soil to Human Health*, R. Naidu, E. Smith, J. Smith, and P. Bhattacharya (Eds.). Adelaide, South Australia, Australia: CSIRO Publishing Corp.

Chaffei, C., K. Pageau, A. Suzuki, H. Gouia, M. H. Ghorbel, and C. Masclaux-Daubresse. 2004. Cadmium toxicity induced changes in nitrogen management in *Lycopersicon esculentum* leading to a metabolic safeguard through an amino acid storage strategy. *Plant & Cell Physiology* 45(11): 1681–1693. doi: 10.1093/pcp/pch192.

Chalmin, P. and C. Gaillochet. 2009. From waste to resource: An abstract of world waste survey 2009. http://www.veolia-environmentalservices.com/veolia/ressources/files/1/927,753, Abstract_2009_GB-1.pdf (accessed February 17, 2017).

Chang, L.W., L. Magos, and T. Suzuki. 1996. *Toxicology of Metals*. Boca Raton, FL: CRC Press.

Chapman, P. F. and F. Roberts. 1983. *Metal Resources and Energy*, p. 138. Boston, MA: Butterworths.

Chartsbin. 2013. Global illegal waste dumping by country. Available at http://chartsbin.com/view/576 (accessed February 20, 2017).

Chen, A., K. N. Dietrich, X. Huo, and S. M. Ho. 2011. Developmental neurotoxicants in E-waste: An emerging health concern. *Environmental Health Perspectives* 119(4): 431–438. doi: 10.1289/ehp.1002452.

CoCusi Coque (Thailand) Co., Ltd. 2004. Seminar document of pilot scale project to recycling spent fluorescent lamps in Thailand.

D'Amore, J. J., S. R. Al-Abed, K. G. Scheckel, and J. A. Ryan. 2005. Methods for speciation of metals in soils: A review. *Journal of Environmental Quality* 34(5): 1707–1745. doi: 10.2134/jeq2004.0014.

Dazy, M., J.-F. Masfaraud, and J.-F. Férard. 2009. Induction of oxidative stress biomarkers associated with heavy metal stress in *Fontinalis antipyretica* Hedw. *Chemosphere* 75(3): 297–302. doi: 10.1016/j.chemosphere.2008.12.045.

Dong, J., F.-B. Wu, and G.-P. Zhang. 2005. Effect of cadmium on growth and photosynthesis of tomato seedlings. *Journal of Zhejiang University: Science B* 6(10): 974–980. doi: 10.1631/jzus.2005.B0974.

Dopp, E., L. M. Hartmann, A.-M. Florea, A. W. Rettenmeier, and A. V. Hirner. 2010. Environmental distribution, analysis, and toxicity of organometal(loid) compounds. *Critical Reviews in Toxicology* 34(3): 301–333. doi: 10.1080/10408440490270160.

European Commission. 2002. Heavy metals in wastes. European Commission on Environment, Denmark. http://ec.europa.eu/environment/waste/studies/pdf/heavy_metalsreport.pdf (accessed February 27, 2017).

Factor-Litvak, P., V. Slavkovich, X. Liu et al. 1998. Hyperproduction of erythropoietin in non-anemic lead-exposed children. *Environmental Health Perspectives* 106(6): 361–364. doi: 10.1289/ehp.98106361.

Giljum, S., C. Lutz, A. Jungnitz, M. Bruckner, and F. Hinterberger. 2008. Global dimensions of European natural resource use—First results from the Global Resource Accounting Model (GRAM). Sustainable Europe Research Institute (SERI) and Institute for Economic Structures Research (GWS), Paris, France. Available at http://www.petre.org.uk/pdf/Giljum%20et%20al_GRAMresults_petrE.pdf (accessed February 17, 2017).

Gimmler, H., J. Carandang, A. Boots, E. Reisberg, and M. Woitke. 2017. Heavy metal content and distribution within a woody plant during and after seven years continuous growth on municipal solid waste (MSW) bottom slag rich in heavy metals. *Journal of Applied Botany* 76(5–6): 203–217. http://cat.inist.fr/?aModele=afficheN&cpsidt=14386181#. WLJYUuRSXxY.mendeley (accessed February 26).

Godt, J., F. Scheidig, C. Grosse-Siestrup, V. Esche, P. Brandenburg, A. Reich, and D. A. Groneberg. 2006. The toxicity of cadmium and resulting hazards for human health. *Journal of Occupational Medicine and Toxicology (London, England)* 1: 22. doi: 10.1186/1745-6673-1-22.

Goyer, R. A. 2001. Toxic effects of metals. In *Cassarett and Doull's Toxicology: The Basic Science of Poisons*, C. D. Klaassen (Ed.), pp. 811–867. New York: McGraw-Hill Publisher.

Greany, K. M. 2005. An assessment of heavy metal contamination in the marine sediments of Las Perlas Archipelago, Gulf of Panama. MS thesis, School of Life Sciences Heriot-Watt University, Edinburgh, Scotland.

Guerrero, L. A., G. Maas, and W. Hogland. 2013. Solid waste management challenges for cities in developing countries. *Waste Management* 33(1): 220–232. doi: 10.1016/j.wasman.2012.09.008.

Gupta, A., D. K. Rai, R. S. Pandey, and B. Sharma. 2008. Analysis of some heavy metals in the riverine water, sediments and fish from river Ganges at Allahabad. *Environmental Monitoring Assessment* 157: 449–458.

Guzzi, G. and C. A. M. La Porta. 2008. Molecular mechanisms triggered by mercury. *Toxicology* 244(1): 1–12. doi: 10.1016/j.tox.2007.11.002.

He, Z. L., X. E. Yang, and P. J. Stoffella. 2005. Trace elements in agroecosystems and impacts on the environment. *Journal of Trace Elements in Medicine and Biology* 19(2): 125–140. doi: 10.1016/j.jtemb.2005.02.010.

Herawati, N., S. Suzuki, K. Hayashi, I. F. Rivai, and H. Koyoma. 2000. Cadmium, copper and zinc levels in rice and soil of Japan, Indonesia and China by soil type. *Bulletin of Environmental Contamination and Toxicology* 64: 33–39.

Hertz-picciotto, I. 2000. The evidence that lead increases the risk for spontaneous abortion. *American Journal of Industrial Medicine* 309: 300–309.

Hoornweg, D., P. Bhada-Tata, and C. Kennedy. 2013. Environment: Waste production must peak this century. *Nature* 502: 615–617. doi: 10.1038/502615a.

Järup, L. 2003. Hazards of heavy metal contamination. *British Medical Bulletin* 68: 167–182. doi: 10.1093/bmb/ldg032.

Järup, L. and A. Åkesson. 2009. Current status of cadmium as an environmental health problem. *Toxicology and Applied Pharmacology* 238(3): 201–208. doi: 10.1016/j.taap.2009.04.020.

John, R., P. Ahmad, K. Gadgil, and S. Sharma. 2009. Heavy metal toxicity: Effect on plant growth, biochemical parameters and metal accumulation by *Brassica juncea* L. *International Journal of Plant Production* 3: 65–76.

Jomova, K., Z. Jenisova, M. Feszterova, S. Baros, J. Liska, D. Hudecova, C. J. Rhodes, and M. Valko. 2011. Arsenic: Toxicity, oxidative stress and human disease. *Journal of Applied Toxicology* 31(2): 95–107. doi: 10.1002/jat.1649.

Kabata-Pendias, A. and H. Pendias. 2001. *Trace Metals in Soils and Plants*, 2nd edn. Boca Raton, FL: CRC Press.

Kachout, S. S., A. Ben Mansoura, J. C. Leclerc, R. Mechergui, M. N. Rejeb, and Z. Ouerghi. 2009. Effects of heavy metals on antioxidant activities of *Atriplex hortensis* and *A. rosea*. *Journal of Food, Agriculture and Environment* 7(3&4): 938–945.

Kantor, D. 2006. Guillain-Barre syndrome, the medical encyclopedia. Bethesda, MD: National Library of Medicine and National Institute of Health. http://www.nlm.nih.gov/medlineplus/ (accessed February 19, 2017).

Kaul, B., R. S. Sandhu, C. Depratt, and F. Reyes. 1999. Follow-up screening of lead-poisoned children near an auto battery recycling plant, Haina, Dominican Republic. *Environmental Health Perspectives* 107(11): 917–920. doi: 10.2307/3454481.

Kim, B. and M. B. McBride. 2009. Phytotoxic effects of Cu and Zn on soybeans grown in field-aged soils: Their additive and interactive actions. *Journal of Environmental Quality* 38: 2253–2259. doi: 10.2134/jeq2009.0038.

Kirpichtchikova, T. A., A. Manceau, L. Spadini, F. Panfili, M. A. Marcus, and T. Jacquet. 2006. Speciation and solubility of heavy metals in contaminated soil using x-ray micro-fluorescence, EXAFS spectroscopy, chemical extraction, and thermodynamic modeling. *Geochimica et Cosmochimica Acta* 70(9): 2163–2190. doi: 10.1016/j.gca.2006.02.006.

Lee, J.-C. and B. D. Pandey. 2012. Bio-processing of solid wastes and secondary resources for metal extraction—A review. *Waste Management* 32(1): 3–18. doi: 10.1016/j.wasman.2011.08.010.

Lenntech Water Treatment and Air Purification. 2004. Water treatment. Lenntech, Rotterdamseweg, the Netherlands. www.excelwater.com/thp/filters/Water-Purification.htm (accessed February 22, 2017).

Levin, S. V., V. S. Guzev, I. V. Aseeva, I. P. Bab'eva, O. E. Marfenina, M. M. Umarov. 1989. Heavy metals as a factor of anthropogenic impact on the soil microbiota. *Mosk Gos Univ*, Moscow, Russia, pp. 5–46.

Li Chen, T., S. S. Wise, S. Kraus, F. Shaffiey, K. M. Levine, W. D. Thompson, T. Romano, T. O'Hara, and J. P. Wise. 2009. Particulate hexavalent chromium is cytotoxic and geno-toxic to the North Atlantic right whale (*Eubalaena glacialis*) lung and skin fibroblasts. *Environmental and Molecular Mutagenesis* 50(5): 387–393. doi: 10.1002/em.20471.

Ling, W., Q. Shen, Y. Gao, X. Gu, and Z. Yang. 2007. Use of bentonite to control the release of copper from contaminated soils. *Australian Journal of Soil Research* 45(8): 618–623.

Manahan, S. E. 2003. *Toxicological Chemistry and Biochemistry*, 3rd edn. CRC Press.

Maslin, P. and R. M. Maier. 2000. Rhamnolipid-enhanced mineralization of phenanthrene in organic-metal co-contaminated soils. *Bioremediation Journal* 4(4): 295–308. doi: 10.1080/10889860091114266.

McLaughlin, M. J., R. E. Hamon, R. G. McLaren, T. W. Speir, and S. L. Rogers. 2000b. Review: A bioavailability-based rationale for controlling metal and metalloid contami-nation of agricultural land in Australia and New Zealand. *Australian Journal of Soil Research* 38(6): 1037–1086. doi: https://doi.org/10.1071/SR99128.

McLaughlin, M. J., B. A. Zarcinas, D. P. Stevens, and N. Cook. 2000a. Soil testing for heavy metals. *Communications in Soil Science and Plant Analysis* 31(11–14): 1661–1700. doi: 10.1080/00103620009370531.

Menikpura, S. N. M., J. Sang-Arun, and M. Bengtsson. 2013. Integrated solid waste man-agement: An approach for enhancing climate co-benefits through resource recovery. *Journal of Cleaner Production* 58: 34–42. doi: 10.1016/j.jclepro.2013.03.012.

Middelkoop, H. 2000. Heavy-metal pollution of the river rhine and meuse floodplains in the Netherlands. *Netherlands Journal of Geosciences/Geologie En Mijnbouw*, Igitur. https://dspace.library.uu.nl/handle/1874/28766.

Minghua, Z., F. Xiumin, A. Rovetta, H. Qichang, F. Vicentini, L. Bingkai, A. Giusti, and L. Yi. 2009. Municipal solid waste management in Pudong New Area, China. *Waste Management* 29(3): 1227–1233, doi: 10.1016/j.wasman.2008.07.016.

Monika, J. K. 2010. E-waste management: As a challenge to public health in India. *Indian Journal of Community Medicine* 35(3): 382–385. doi: 10.4103/0970-0218.69251.

National Research Council. 2001. *Arsenic in Drinking Water*. Update. 2001. Online at: http://www.nap.edu/books/0309076293/html/.

National Research Council Canada (NRCC). 1978. *Effects of Arsenic in the Environment*. National Research Council of Canada, pp. 1–349.

Naudé, W. and A. Szirmai. 2012. The importance of manufacturing in economic develop-ment: Past, present and future perspectives. UNU-MERIT Working Papers, Maastricht Economic and Social Research Institute on Innovation and Technology, Maastricht, the Netherlands. https://www.merit.unu.edu/publications/wppdf/2012/wp2012-041.pdf (accessed February 22, 2017).

Needhidasan, S., M. Samuel, and R. Chidambaram. 2014. Electronic waste—An emerging threat to the environment of Urban India. *Journal of Environmental Health Science & Engineering* 12(1): 36. doi: 10.1186/2052-336X-12-36.

Nriagu, J. O. 1989. A global assessment of natural sources of atmospheric trace metals. *Nature* 338: 47–49.

NSC. 2009. Lead poisoning. National Safety Council, Washington. http://www.nsc.org/ NSCDocuments_Advocacy/Fact%20Sheets/Lead-Poisoning-Fact-Sheet.pdf (accessed February 19, 2017).

Pacyna, J. M. 1996. Monitoring and assessment of metal contaminants in the air. In *Toxicology of Metals*, L. W. Chang, L. Magos, and T. Suzuli (Eds.), pp. 9–28. Boca Raton, FL: CRC Press.

Padiyar, N., P. Tandon, and S. Agarwal. 2011. Nickel allergy—Is it a cause of concern in everyday dental practice? *International Journal of Contemporary Dentist* 12(1): 80–81.

Pant, D., D. Joshi, M. K. Upreti, and R. K. Kotnala. 2012. Chemical and biological extraction of metals present in E waste: A hybrid technology. *Waste Management* 32(5): 979–990. doi: 10.1016/j.wasman.2011.12.002.

Plagányi, É. E., I. van Putten, T. Hutton, R. A. Deng, D. Dennis, S. Pascoe, T. Skewes, and R. A. Campbell. 2013. Integrating indigenous livelihood and lifestyle objectives in man-aging a natural resource. *Proceedings of the National Academy of Sciences of the United States of America* 110(9): 3639–3644. doi: 10.1073/pnas.1217822110.

Pollard, S. 1990. *Typology of Industrialization Processes in the Nineteenth Century*. Harwood Academic Publishers.

Poon, C. S. 2008. Management of CRT glass from discarded computer monitors and TV sets. *Waste Management* 28: 1499. doi: 10.1016/j.wasman.2008.06.001.

Rai, U. N. and A. Pal. 2002. Health hazards of heavy metals. *International Society of Environmental Botanist* 8(1).

Sanfeliu, C., J. Sebastià, R. Cristòfol, and E. Rodríguez-farré. 2003. Neurotoxicity of organo-mercurial compounds. *Neurotoxicity Research* 5: 283–305.

Scragg, A. 2006. *Environmental Biotechnology*, 2nd edn. Oxford, U.K.: Oxford University Press.

Seng, B., H. Kaneko, K. Hirayama, and K. Katayama-Hirayama. 2011. Municipal solid waste management in Phnom Penh, Capital City of Cambodia. *Waste Management & Research: The Journal of the International Solid Wastes and Public Cleansing Association, ISWA* 29(5): 491–500. doi: 10.1177/0734242X10380994.

Shallari, S., C. Schwartz, A. Hasko, and J. L. Morel. 1998. Heavy metals in soils and plants of serpentine and industrial sites of Albania. *Science of the Total Environment* 209(2–3): 133–142. doi: 10.1016/S0048-9697(98)80104-6.

Shamsi, I. H., K. Wei, G. Jilani, and G. P. Zhang. 2007. Interactions of cadmium and aluminum toxicity in their effect on growth and physiological parameters in soybean. *Journal of Zhejiang University Science B* 8: 181–188.

Sharholy, M., K. Ahmad, R. C. Vaishya, and R. D. Gupta. 2007. Municipal solid waste charac-teristics and management in Allahabad, India. *Waste Management* 27(4): 490–496. doi: 10.1016/j.wasman.2006.03.001.

Singhal, R. L., Z. Merali, and P. D. Hrdina. 1976. Aspects of the biochemical toxicology of cadmium. *Federation Proceedings* 35(1): 75–80.

Smith, L. A., J. L. Means, A. Chen et al. 1995. *Remedial Options for Metals-Contaminated Sites*. Boca Raton, FL: Lewis Publishers.

Song, Q., J. Li, and X. Zeng. 2015. Minimizing the increasing solid waste through zero waste strategy. *Journal of Cleaner Production* 104: 199–210. doi: 10.1016/j.jclepro. 2014.08.027.

Sörme, L. and R. Lagerkvist. 2002. Sources of heavy metals in urban wastewater in Stockholm. *Science of the Total Environment* 298(1): 131–145. doi: 10.1016/S0048-9697(02)00197-3.

Sträter, E., A. Westbeld, and O. Klemm. 2010. Pollution in coastal fog at Alto Patache, Northern Chile. *Environmental Science and Pollution Research* 17(9): 1563–1573. doi: 10.1007/s11356-010-0343-x.

Tchounwou, P. B., W. K. Ayensu, N. Ninashvili, and D. Sutton. 2003a. Environmental exposure to mercury and its toxicopathologic implications for public health. *Environmental Toxicology* 18(3): 149–175. doi: 10.1002/tox.10116.

Tchounwou, P. B., J. A. Centeno, and A. K. Patlolla. 2004. Arsenic toxicity, mutagenesis, and carcinogenesis—A health risk assessment and management approach. *Molecular and Cellular Biochemistry* 255: 47–55.

Tchounwou, P. B., A. K. Patlolla, and J. A. Centeno. 2003b. Invited reviews: Carcinogenic and systemic health effects associated with arsenic exposure—A critical review. *Toxicologic Pathology* 31(6): 575–588. doi: 10.1080/01926230390242007.

Tchounwou, P. B., B. Wilson, and A. Ishaque. 1999. Important considerations in the development of public health advisories for arsenic and arsenic-containing compounds in drinking water. *Reviews on Environmental Health* 14(4): 211–229.

Udedi, S. S. 2003. From guinea worm scourge to metal toxicity in Ebonyi State. *Chemistry in Nigeria as the New Millennium Unfolds* 2: 13–14.

USEPA. 1996. Report: Recent developments for in situ treatment of metals contaminated soils. U.S. Environmental Protection Agency, Office of Solid Waste and Emergency Response. https://www.epa.gov/sites/production/files/2015-08/documents/recent_dev_metals_542-r-97-004.pdf (accessed February 20, 2017).

Wang, S. and X. Shi. 2001. Molecular mechanisms of metal toxicity and carcinogenesis. *Molecular and Cellular Biochemistry* 222: 3–9.

Wang, X. and G. Gaustad. 2012. Prioritizing material recovery for end-of-life printed circuit boards. *Waste Management* 32(10): 1903–1913. doi: 10.1016/j.wasman.2012.05.005.

Weiss, B., T. W. Clarkson, and W. Simon. 2002. Silent latency periods in methylmercury poisoning and in neurodegenerative disease. *Environmental Health Perspectives* 110(Suppl. 5): 851–854. doi: 10.1289/ehp.02110s5851.

World Health Organization. 1996. *Trace Elements in Human Nutrition and Health*. Geneva, Switzerland: WHO/FAO/IAEA.

Wuana, R. A. and F. E. Okieimen. 2011. Heavy metals in contaminated soils: A review of sources, chemistry, risks and best available strategies for remediation. *ISRN Ecology* 2011: 1–20. doi: 10.5402/2011/402647.

2 Generation and Composition of Various Metal-Containing Industrial Wastes

2.1 INTRODUCTION

A substantial amount of wastes is generated as by-products of different processing industries. The best option for dealing with these wastes is to reduce their volume by recycling and reuse. Recovery of various metals, glass, plastics, and other materials from these wastes can help conserve energy and also protect the environment.

Metals are indispensable parts of these wastes and can turn out to be hazardous to the environment and human health, but if managed properly, these wastes can also act as secondary sources or artificial ores for many precious and valuable metals, as many times, the concentration of metals in these wastes is as much as or more than that found in natural ores.

This chapter summarizes the generation and composition of various metal-containing industrial wastes. According to the origins and characteristics, wastes can be classified into different types, such as electronic, energy storage, metal production, petrochemical, automobile wastes, etc.

2.2 ELECTRONIC WASTES

The development and miniaturization of more efficient technologies have resulted in a tremendous increase in the use of electronic equipment. Electrical and electronic equipment that have reached their end-of-life phase are discarded by the consumer and termed as e-waste. Electronic wastes comprise a wide range of electrical powered appliances such as cellular phones, refrigerators, computers, air conditioners, etc. E-waste generation is increasing at a rate that is three times more than the rate at which municipal waste is generated globally per year (Lepawsky and McNabb 2010). Figure 2.1 shows the global generation of e-waste. Every year, 20–50 million tons of waste electrical and electronic equipment (WEEE) are generated worldwide, and this figure is growing by about 45 million tons per year (Ogunseitan 2013). E-waste occupies a large amount of land due to land filling; in addition, the occurrence of toxic chemicals in e-waste can pose a risk to human and environmental health if improperly managed (Townsend 2011). E-waste is classified as hazardous if elements such as lead, mercury, arsenic, cadmium, selenium, and hexavalent chromium, and flame retardants are present beyond permissible quantities (Pant et al. 2012).

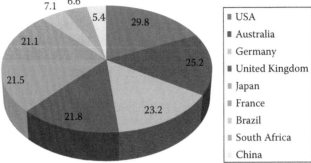

FIGURE 2.1   Global e-waste generation in kg/capita. (With kind permission from Springer Science + Business Media: *Int. J. Environ. Sci. Technol.*, Efficient management of e-wastes, 14(1), 2017, 211–222, Chatterjee, A. and Abraham, J.)

2.2.1 PRINTED CIRCUIT BOARDS

The crucial basic component of electrical and electronic equipment is the printed circuit board (PCB). It is often used in telecommunications, information technology, automation, entertainment equipment, automotive industries, etc. A nonconductive substrate, the printed conductive circuits, and electrical components together make a PCB. It comprises both metallic and nonmetallic materials. A PCB consists of 30% polymers, 30% refractory oxides (mainly silica, alumina, rare earth oxides), and 40% metals (copper, iron, tin, nickel, lead, aluminum, gold, and silver, among others) approximately (Cui and Forssberg 2003, Kasper et al. 2011). Besides these inorganic elements, the following important organic compunds are found in circuit boards: isocyanates and phosgene from polyurethanes, acrylic and phenolic resins, epoxides, and phenols, such as chip glues (Ludwig et al. 2003) which constitute about 3% of the total weight of electronic scrap (Li et al. 2004).

Depending on the operating need of the electronic system of the product, each PCB is designed differently. Technologically, it can be classified as single sided (with conductive circuits on only one side of the substrate), double sided (with conductive circuits on both sides of the substrate), and multilayer (with circuits between the substrate layers, which may vary from 4 to 16 layer) (Ladou 2006). It can also be classified based on the material of its substrate, where it is commercially termed as FR-1, FR-2, and FR-4 (where FR indicates the presence of flame retardants) and CEM-x. FR-1 and FR-2 boards are made of cellulose paper impregnated with phenolic resin, while fiberglass and epoxy resin make the FR-4 type. CEM-x is made of composites produced from fiberglass and cellulose paper impregnated with epoxy resin. FR-4 boards are generally used in computers, mobiles phones, and other communication equipment, whereas FR-2 boards are used in televisions and home appliances (Hall and Williams 2007).

Usually, copper is used for conductive circuits in most of the appliances; however, aluminum, nickel, chrome, or other metals may also be used in some cases

(Zeng et al. 2012). Chips, semiconductors, connectors, capacitors, etc. of electric components include metals such as iron, tin, nickel, lead, and aluminum, and precious metals (gold, silver, palladium, etc.). In addition, other elements, such as tantalum (Ta) in capacitors; gallium (Ga), indium (In), titanium (Ti), silicon (Si), germanium (Ge), arsenic (As), selenium (Se), and tellurium (Te) in chips and other units; tin (Sn) and cadmium (Cd) in solders; and gallium (Ga), silicon (Si), selenium (Se), and germanium (Ge) in semiconductors, are also present (Kasper et al. 2011).

A major portion of PCB metal content is composed of copper (nearly 10%–30%) (Chauhan and Upadhyay 2015), along with other toxic components such as lead, cadmium, mercury and polychlorinated biphenyls, and etched chemicals (Guidelines for Environmentally Sound Management of E-waste 2008). The release of 70% of toxic chemicals in the environment is due to the disposal of e-wastes (E-Waste Fact Sheet 2009). Polybrominated diphenyl ethers and polychlorinated dibenzo-p-dioxins and dibenzofurans are found in soils collected from crude e-waste processing areas (Leung et al. 2007). Every 500 million computers consist of 6.32 billion pounds of plastic, with 3 billion pounds of cadmium, 1.9 million pounds of chromium, 1.58 billion pounds of lead, and 632,000 pounds of mercury (Secretariat 2011). About 0.36 g of gold and 1.4 g of silver are present in one desktop computer (Kahhat et al. 2008, Chatterjee and Abraham 2017).

It has been reported that the concentration of precious metals in PCBs is 10 times higher than that of rich-content minerals (Li et al. 2007). Ores contain nearly 10 g/t of gold and palladium, whereas computer PCBs may contain over 250 g/t of gold and 110 g/t of palladium. Hence, is aptly termed as "urban mine" (Davis and Herat 2008). However, these quantities vary greatly with the equipment and its intended functions (Norgate and Haque 2012).

Typically, a PCB from a personal computer contains 7% Fe, 5% Al, 20% Cu, 1.5% Pb, 1% Ni, 3% Sn, and 25% organic compounds, together with (in parts per million, ppm) 250 Au, 1000 Ag, and 100 Pd. In addition, traces of As, Sb, Ba, Br, and Bi are present (Hagelüken 2006). A PCB of a mobile phone may contain, on average, 20%–30% copper, 6%–8% ferrous metals, about 5% nickel, 2.5% tin, 1.1% zinc; 250 mg of silver, 24 mg of gold, 9 mg of palladium, and other metals. The percentage of materials depends on the model and year of manufacture (Kasper et al. 2011, Polák and Drápalová 2012). Table 2.1 shows the amount of main metals present in PCBs of mobile phones (Petter et al. 2014). Schluep et al. (2009) gives a rough estimate of precious metals present in mobile phones as follows: One ton of mobile phones without batteries contains nearly 130 kg of Cu, 3.5 kg of Ag, 340 g of Au, and 140 g of Pd.

2.2.2 SOLDERS

Solders are an essential part of electronic devices as they are used as interconnecting materials in electronic packaging and provide electrical communication as well as mechanical support in modules (Ma and Suhling 2009). Lead-bearing/tin–lead (Sn–Pb) solders are predominantly used due to their magnificent solderability and reliability (Pan et al. 2005, Yoo et al. 2012). However, because of the known toxicity of lead, they have been substituted by lead-free solders, such as the Sn–Ag–Cu series, Sn–Ag–Cu–Bi series, Sn–Ag–Cu–Sb series, Sn–Ag series, Sn–Ag–Bi series,

TABLE 2.1
Concentration of Metals (%wt.) in Printed Circuit Boards Reported in Different Studies

%Wt.	Sum (2005)	Guo et al. (2009)	Yang et al. (2009)	Park and Fray (2009)	Yamane et al. (2011)	Tuncuk et al. (2012)
Gold	0.1	0.008	—	0.025	0.00	0.035
Silver	0.2	0.33	—	0.100	0.21	0.138
Copper	20	26.8	25.06	16.0	34.49	13
Nickel	2	0.47	0.0024	1	2.63	0.1
Tin	4	1.0	—	3.0	3.39	0.5

Source: Reprinted from *Waste Manage.*, 34(2), Petter, P., Veit, H.M., and Bernardes, A.M., Evaluation of gold and silver leaching from printed circuit board of cellphones, 475–482, doi:10.1016/j.wasman.2013.10.032. Copyright 2014, with permission from Elsevier.

Sn–Cu series, and Sn–Zn–Bi series of solders, comprising corresponding metals. Turbini et al. (2004) suggest that the Sn–Ag–Cu alloy is more harmful to the environment than the Sn–Pb solder. When such solders are discarded, these heavy metals leach out, posing risk to the environment and human health.

2.3   ENERGY STORAGE WASTES

Energy storage devices such as batteries, button cells, etc. are produced in a way so as to provide a wide range of power, energy, size, weight, safety, and cost requirements.

2.3.1   BATTERIES

Batteries are increasingly being extensively consumed as electrochemical power sources in modern-life equipment around the world, which consequently has resulted in their elevated waste disposal (Xu et al. 2008). For instance, in the United States and Europe, the consumption of batteries is estimated to be 8 billion units per year (Bernardes et al. 2004). In Japan, around 6 billion batteries were produced in 2004, while almost 1 billion units are consumed every year in Brazil (Salgado et al. 2003). Worldwide, 15 billion primary batteries are thrown away every year, all of which end up in landfill sites. Currently, only a very small percentage of about 2% of consumer disposable batteries are recycled (Sathaiyan 2014). To meet the demand for various requirements of power, energy, safety, size, weight, and cost and be applied to a wide range of electronic equipment, batteries must have different chemistries and materials (Worrell and Reuter 2014).

There are two types of batteries, nonrechargeable primary batteries and rechargeable secondary batteries. Primary batteries, which produce an electric current by means of an irreversible chemical reaction, are composed of Zn–Mn blends, C–Zn blends, or are primary lithium or mercury batteries. Among these, the Zn–Mn battery, developed after the C–Zn battery, is the most frequently used and accounts for nearly

80% of the spent batteries generated in Korea (Shin et al. 2007). Rechargeable batteries such as Ni–Cd, Ni–metal hydride (Ni–MH), and lithium-ion batteries (LIBs) are used in many industrial applications, especially in portable electronic products. For example, Ni–Cd batteries are most frequently used in wireless communication and wireless computing devices. All of these batteries are a rich repository of valuable metals, including Zn, Mn, Ni, Cd, Co, and Al (Vassura et al. 2009).

2.3.1.1   Zinc–Manganese Dioxide Batteries

Alkaline and Zn–C (Leclanché cell) batteries, along with Zn–Cl batteries, are primary batteries and usually used in devices that require low voltages between 1.7 and 0.8 V. Zn–C and Zn–Cl batteries are also called dry batteries. Zn–C batteries are the cheapest batteries due to their simple assembly and hence comprise about 90% of the total annual sales, but also run out rapidly and hence are thrown away, consequently threatening human health and the environment (Kierkegaard 2007). Nowadays, the annual consumption of zinc-manganese dioxide batteries is estimated to have reached 40 billion units worldwide (Xin et al. 2012). Zn–Mn batteries, including alkaline and Zn–C batteries, are used in radios, recorders, toys, remote controls, watches, calculators, cameras, and in many other objects where small quantities of power are required (Sayilgan et al. 2009).

The anode is composed of an alloy of zinc (16%–20%), with lead (Pb) for improved malleability; cadmium (Cd) and mercury (Hg) as corrosion inhibitors; a cathode made of manganese dioxide (28%–30%), mixed with carbon (7%–13%) to increase conductivity and hold moisture; and with ammonium chloride and zinc chloride (6%–10%) as electrolytes, forming a thick paste in contact with a carbon conductor (Shin et al. 2009).

Alkaline batteries contain an alkaline electrolyte, usually potassium hydroxide (5%–9%), which ensures corrosion protection for zinc (11%–16%), and a cathode composed of pure manganese dioxide (32%–38%). They also contain Hg and Cd as corrosion inhibitors and Pb for improved malleability (Shin et al. 2009, Schumm 2013). The assembly of these batteries using zinc powder ensures a higher surface area between the anode and the cathode, allowing a higher bulk density among the zinc–manganese dioxide systems. Alkaline and Zn–C batteries typically contain about 12%–28% and 26%–45% of Zn and Mn, out of the total powder mass, respectively; the remaining components mainly include graphite, K and Fe, which are cheap, abundant, and nontoxic (Sayilgan et al. 2009).

The electrolytically produced manganese dioxide is used in alkaline batteries instead of either chemical MnO_2 or natural ore because of its higher manganese content, its increased reactivity, and greater purity. The electrolyte KOH affords greater conductivity and a reduced hydrogen gassing rate. Powdered zinc (high purity ~99.85%–99.00%, produced by electrowinning or distilling) provides a large surface area for high-rate capability and the distribution of solid and liquid phases more homogeneously (Frenay and Feron 1990, de Souza et al. 2001, Scarr et al. 2001). The cell is totally enclosed in a high-density steel can, with both edges covered with plated steel. A nonwoven fabric is used inside to separate the anode and the cathode from the electrolyte solution and an asphalt insulator is added to prevent any leakage from the battery, along with an adhesive plastic coat for finishing operations (Rayovac Corp. 1999).

2.3.1.2   Nickel–Cadmium Batteries

Reports state the average composition of nickel–cadmium (Ni–Cd) batteries as Fe (40%), Ni (22%), Cd (15%), plastic (5%), KOH (2%), and others (16%), along with some Co (Vassura et al. 2009). Ni–Cd batteries have metallic Cd as cathode and NiOOH (also Co) as anode, along with an absorbent polymer sheet impregnated with KOH (Nogueira and Margarido 2007). As compared with other batteries, this system has a faster charging rate and high resistance to overcharge and overdischarge cycles. Also, it can be manufactured in several different shapes and sizes and is highly economic in terms of cost per cycle (Buchmann 2013). Ni–Cd batteries are most frequently used in wireless communication and wireless computing devices. Among all kinds of batteries, Ni–Cd batteries are represented as the most toxic ones and classified as hazardous waste due to the presence of Cd (Ijadi Bajestani et al. 2014). Ni–Cd batteries are classified as hazardous waste because nickel and cadmium are heavy metals and suspected carcinogens (Shapek 1995).

2.3.1.3   Nickel–Metal Hydride Batteries

Nickel–metal hydride (Ni–MH) batteries were developed so as to fulfill the need of light, compact, and durable rechargeable batteries to be used in electronics and mobile phones. They have twice the energy density of Ni–Cd batteries with a similar operating voltage, and further, their constituents are less toxic to the environment (Ruetschi et al. 1995). However, their cost of production is higher as compared with Ni–Cd batteries (Putois 1995). Having the basic structure similar to Ni–Cd batteries, Ni–MH batteries have layers containing nickel hydroxide, acting as the positive electrode, and an alloy with hydrogen-absorbing properties as the negative electrode (Zhang et al. 1998).

The hydrogen-absorbing alloy has an absorption capacity equivalent to a thousand times its own volume, generating metal hydrides. This alloy combines metal A, which releases heat during the formation of hydrides, with metal B, whose hydrides absorb heat. Depending on how metals A and B are combined, their alloys are classified into the following types: AB (TiFe, etc.), AB_2 ($TiNi_2$, $ZnMn_2$, etc.), AB_5 ($LaNi_5$, etc.), and A_2B (Mg_2Ni, etc.) (Jossen et al. 2004). These alloys contain approximately 36%–42% Ni, 22%–25% Fe, 8%–10% rare earth metals (REM), and 3%–4% Co.

For economic reasons, the La present in AB_5 is often replaced by a rare earth alloy known as Mischmetal (Baddour-Hadjean et al. 2003). Mischmetal is a mixture of rare earth elements, usually composed mainly of Ce, associated with La, Nd, Pr, and others, in proportions in which they occur naturally in minerals (Gschneidner et al. 1990). Regarding the charge and discharge efficiency and durability of the different alloys cited, only AB_5 and AB_2 have practical applications.

Ni–MH batteries are assembled in different formats, such as the cylindrical, button, or prismatic type. Cell phones are powered by prismatic or cylindrical batteries connected in series and arranged inside the protective packaging in different spatial arrangements according to the manufacturer. The Ni–MH battery assembly coated with a metal or plastic housing comprises different layers wrapped in a spiral or a plate format. The two electrodes are separated by a polymer membrane (polypropylene or polyamide), amid an alkaline electrolyte, usually 30% KOH, which allows the movement of electric charges. In cylindrical batteries, the positive and negative plates are coiled spirally, and in prismatic batteries, they are assembled in layers. The system has a safety valve which starts working in case of high pressure to avoid an explosion (Ruetschi et al. 1995, Baumann and Muth 1997).

2.3.1.4 Lithium-Ion Batteries

In 2003, primary lithium batteries and lithium-ion secondary rechargeable batteries (LIBs) represented about 28% of the rechargeable battery world market (Lee and Rhee 2002), while the global consumption of LIBs reached 4.49×10^9 units in 2011 (Niu et al. 2014). The dominance of rechargeable lithium batteries in the market of high-power storage systems is attributed to properties such as a high-energy density (120 Wh/kg), high battery voltage (up to 3–6 V), longevity (500–1000 cycles), and wide temperature range (20°C–60°C) (Nishi 2001, Peng and Chen 2009, Al-Thyabat et al. 2013, Zhang et al. 2013). LIBs are widely used in laptops, mobile phones, video cameras, and other portable electronics because of their function update and cost decrease. They consist of heavy metals, organic chemicals, and plastics in the proportion of 5%–20% cobalt, 5%–10% nickel, 5%–7% lithium, 15% organic chemicals, and 7% plastics, the composition varying slightly with different manufacturers (Shin et al. 2005). The typical metal composition of LIBs based on $LiCoO_2$ species in the cathode is shown in Table 2.2, including or not the weight of the steel protective case.

Of the two types of LIBs, primary (nonrechargeable) batteries use metallic lithium as a cathode and contain no toxic metals, but there is the possibility of fire if metallic lithium is exposed to moisture while the cells are corroding; whereas secondary LIBs

TABLE 2.2
Metal Composition of Lithium-Ion Batteries based on $LiCoO_2$ in the Cathode (%wt.)

Metal	Excluding Steel Case (Mantuano et al. 2006)	Including Steel Case (Rydh and Svärd 2003)
Al	6.5–10.0	4.6–24.0
Cd	0.01–0.03	—
Co	30.8–42.9	12–20
Cu	13	5–10
Fe	0.03–0.10	4.7–25.0
Li	2.45–8.88	1.5–5.5
Mn	<0.01	—
Ni	0.02	—
Zn	≤0.01	—

Source: Reprinted from *J. Power Sources*, 187(1), Ferreira, D.A., Prados, L.M.Z., Majuste, D., and Mansur, M.B., Hydrometallurgical separation of aluminium, cobalt, copper and lithium from spent Li-ion batteries, 238–246, doi:10.1016/j.jpowsour.2008.10.077, Copyright 2009, with permission from Elsevier.

do not contain metallic lithium and use a material such as $LiXMA_2$ ($LiCoO_2$, $LiNiO_2$, and $LiMn_2O_4$) at the positive electrode (Xu et al. 2008). Recently, LIBs are being used in electrical and plug-in electrical vehicles, not only because they provide high-energy densities, but also because they deliver high-power density, so as to replace conventional gasoline engines (Huang and Ikuhara 2012, Mukherjee et al. 2012).

A secondary LIB consists of a cathode and an anode, a separator, and an organic electrolyte (Wang et al. 2009, Al-Thyabat et al. 2013). Cathode materials can be divided into three groups, namely (1) lithium-based metal oxides, such as $LiCoO_2$; (2) transition metal phosphates, such as $Li_3V_2(PO_4)_3$ and $LiFeO_4$; and (3) spinels such as $LiMnO_4$ (Whittingham 2004, Huang and Ikuhara 2012, Mukherjee et al. 2012). Carbon is used as an anode (Mukherjee et al. 2012). The cathode is linked with aluminum foil, while the anode is linked with copper foil, and both electrodes are adhered to a substrate by a polymeric binder, such as polyvinylidene fluoride (PVDF).

The PVDF binder does not react with most acids and bases, and most oxidants and halogens, and does not dissolve in organic reagents such as fatty hydrocarbons, aromatic hydrocarbons, aldehydes, and alcohol at room temperature, but gets dissolved only partly in some special ketones and ethers (Xu et al. 2008, Al-Thyabat et al. 2013). The electrodes are immersed in electrolytes having high ionic conductivity such as lithium salts in aprotic solvents, for example, $LiPF_6$ in organic solutions, such as dimethyl carbonate (DMC), ethylene carbonate (EC), and diethyl carbonate (DEC), while the separator is made of microporous polypropylene (PE) or polypropylene (PP) (Mukherjee et al. 2012, Al-Thyabat et al. 2013, Li et al. 2013, Gratz et al. 2014). The cathode and the anode are folded into a prismatic or cylindrical form and the assembled battery is fitted into a metal shell (Ferreira et al. 2009).

2.3.2 BUTTON CELL BATTERIES

The use of button cell batteries in electric watches, portable medical monitors, calculators, hearing aids, digital thermometers, toys, etc. has increased in recent years due to their small size, long operating life, and high capacity per unit mass (Sathaiyan et al. 2006, Aktas 2010). In silver oxide–zinc primary cells, the negative plate (cathode) is made of sintered fine silver oxide (Ag_2O) powder, while the positive plate (anode) is composed of activated zinc, both of which are separated by a semipermeable ion exchange membrane and the electrolyte used is sodium hydroxide. Additives such as mercury are included to avoid corrosion. After completion of their lifespan, these batteries can be recycled to obtain the valuable metals.

2.4 METAL PRODUCTION WASTES

2.4.1 SLAG

Metal manufacturers generate substantial quantities of waste sludge, filter dust, etc. through activities such as casting, machining and welding, parts assembly, surface finishing, quality control, and packaging. These wastes contain a relatively high amount of heavy metals, which can prove harmful if released into the environment (Brandl and Faramarzi 2006).

Metallurgical or incineration processes of various ores produce different slags or residues, which, depending on their origin and characteristics, are categorized as ferrous, nonferrous, and incineration slags. Various heavy metals such as Fe, Cr, Cu, Al, Pb, Zn, Co, Ni, Nb, Ta, Au, and Ag are present in these slags. Thus, their recovery and utilization are important from the viewpoint of metal resources management and environment protection, and hence, these slags should be considered as a secondary source of metals rather than an end-waste.

Ores obtained from mining are initially crushed, ground, and concentrated using gravitational or floatation methods to separate the minerals from gangue. After this, through the roasting process, sulfide minerals are oxidized to remove S in the form of SO_2. Further, in the smelting step, SiO_2, limestone, or iron sources (e.g., ironstone, iron silicates, or iron oxides) are added as fluxes and carbon (coke, charcoal, coal, or wood) is added as a fuel and reductant. Then, the metal phase is separated from the slag and sent to the converter, where it is further desulfurized and oxygen and other fluxes are added to aid removal of impurities (Federal Highway Administration (FHWA) 1997). Obtained free metal is combined with other elements or compounds to produce the desired alloy. Thus, slag is produced during smelting, converting, and refining steps.

Ferrous slags incorporate iron slag (blast furnace slag), steel slag, alloy steel slag, and ferroalloy slag. Alloy steel slag and ferroalloy slag contain high amounts of alloy elements such as Cr, Ni, Mn, Ti, V, and Mo. Ferrous slags are produced during iron production from natural ores or recycling of materials for Fe or steel production. Iron oxides are reduced to molten Fe in the furnace by adding a flux such as limestone or dolomite and a fuel and reductant such as coke. The properties of slag vary depending on the rate and method used for its cooling. Nearly 220–370 kg of slag is generated per ton of iron produced in a blast furnace (Proctor et al. 2000).

Sponge iron (direct reduced iron) is produced by the reduction of iron ore using noncoking coal with a small quantity of dolomite in a rotary kiln. An electrostatic precipitator (ESP), which is connected to a rotary kiln, produces a metal-containing dust, which accounts for nearly 40% of solid waste generated during the process (Jena et al. 2012).

Nonferrous slag is produced during the recovery of nonferrous metals from natural ores and includes copper slag, salt slag, tin slag, etc. Incineration slags include metals such as Fe, Al, Cu, Pb, Zn, and Ag. Nonferrous ores are smelted in various types of furnaces such as blast, reverberatory, electric arc, or oxygen flash furnaces. Compressed oxygenated air is used for the combustion of the furnace charge in an oxygen flash furnace. In a reverberatory furnace, the fuel and materials are mixed together and the furnace separates the material being processed from the hot gases, but not from the combustion gases. Ores such as Cu, Ni, Pb, and Sn are processed by a reverberatory furnace, while a blast furnace is used to smelt Cu, Pb, Sn, and Zn ores and an oxygen blast furnace is used for Cu ores (Piatak et al. 2015).

During the processing of zinc ore in process plants, calcine rich in ZnO is first produced from sulfide or oxide–carbonate concentrates, which are then leached with a hot sulfuric acid solution. After the separation of liquid/solid phases, metallic zinc is purified and electrowon from the solution, leaving behind a hazardous residue containing a notable amount of zinc, lead, and cadmium. This residue contains concentrated lead in the form of lead sulfate, albeit the huge amount of lost zinc is also present in the form of zinc ferrite ($ZnO \cdot Fe_2O_3$) (Turan et al. 2004).

Further, zinc ash, dross, flux skimming, and blowing secondaries are generated during the different operations involved in the galvanizing of metal sheets or tubes (Subramanian 1995). Zinc ash is produced by the oxidation of molten zinc during the dry galvanizing process on top of the bath and floats on the surface of the molten zinc. It comprises zinc oxides, along with a small amount of free metallic zinc and other impurities (%wt.) such as Zn (60–85), Pb (0.3–2.0), Al (0–0.3), Fe (0.2–1.5), and Cl (2–12). A flux layer used to reduce the rate of oxidation of molten zinc in the wet galvanizing process consists of metallic Zn (5.6%), $ZnCl_2$ (48.1%), ZnO (27.4%), Al chlorides (3.1%), and also chlorides and oxides of Fe, Cd, Al, etc.

Galvanizers' dross is produced by the reaction of molten zinc with the steel kettle wall and the steel article and contains an alloy of Fe (4%) and Zn (96%) which settles at the bottom of the zinc bath (Jha et al. 2001). Blowing waste obtained in the tube galvanizing process during surface cleaning is composed of 81% Zn metal, 0.3% Fe, 16% ZnO, 0.3% Pb, and 0.1% Cd.

Steel slag produced during steel making is categorized based on the type of furnace used in the creation, such as basic oxygen furnace (BOF slag) and electric arc furnace (EAF slag). It mainly comprises gangue contents (CaO, SiO_2, Al_2O_3, MgO) and some metal oxides (Goldrig and Jukes 1997, Shen et al. 2004, Perrine et al. 2006, Das et al. 2007, Tossavainen et al. 2007). BOF uses oxygen to oxidize the charge, whereas EAF uses electric current to produce the necessary heat required for melting recycled steel scrap (Szekely 1996, Leclerc et al. 2003). As per the literature, over 400 kg of waste per ton of steel output is produced by these plants (Thakur 2000, Yadav et al. 2001).

Salt slag is generated during the smelting process of recycled aluminum scrap in a reverberatory or rotary furnace in the secondary aluminum industry. Fluxes consisting of NaCl and KCl are added in the molten bath to prevent metal oxidation and remove impurities from the charge, during which the salt slag is formed. Approximately 0.5 tons of salt slag is formed when 1 ton of secondary aluminum is produced (Kirchner 1991, López et al. 1994).

Copper extraction from sulfide minerals is usually carried out using a pyrometallurgical process that entails obtaining a Cu concentrate by froth flotation, followed by smelting it to a molten high-Cu matte, which is converted into molten blister copper, which is finally electrorefined to ultrapure copper (Davenport et al. 2002). An oxide-rich phase and copper–matte (sulfide) phase are produced during smelting. It consists of a matrix of iron silicates, with droplets of matte ($Cu_2S.FeS$), white metal (Cu_2S), and blister copper (Cu_0) (Davenport et al. 2002, Gorai et al. 2003). It is also composed of FeO, Fe_2O_3, and SiO_2, with small amounts of Al_2O_3, CaO, and MgO as well as metals such as Cu, Co, and Ni in metallic or oxide/sulfide form (Gorai et al. 2003, Deng and Ling 2004, Alp et al. 2008).

The concentration of Cu in slag from smelting is typically in the range of 0.5%–2%, while converter slags can contain 2%–15% Cu. This is a considerable amount of copper, very much higher than those found in many ores (Baghalha et al. 2007). About 2.2 tons of slag is generated for every ton of copper metal produced (Panda et al. 2015).

Pb/Zn smelting slag has high concentrations of Fe, Zn, Cu, Pb, and As, which can cause serious damage to the environment and human health (Guo et al. 2007).

2.4.2 DUST

Dust, scale, and sludge produced by the metal industries can either be recycled back to the process (blast furnace, BOF, or EAF) or used externally as raw materials in other industries (known as industrial ecology). Concentrations of metals such as iron, zinc, chromium, nickel, and molybdenum in these wastes vary significantly depending on the process from which the waste is generated (Bayat et al. 2009). Gas and dust produced during smelting of ores in furnaces contain a substantial amount of metals. Hence, they are recycled in furnaces, which not only hampers their efficiency but also increases the amount of energy required for smelting. Moreover, it damages the refractory bricks and imposes a circular load on furnaces (Bakhtiari et al. 2008).

Nearly 1%–2% of scrap being used in smelter furnaces enters into the exhaust gases and is converted into dust (Zunkel 1996). They contain a high amount of iron oxides, along with Cr and Ni; however, the amount of gangues present is less as compared with slags (Li and Tsai 1993b, Laforest and Duchesne 2006). Some of the dust is also collected from flue gas pipelines because of the violent agitation of molten steel in the furnace during the stainless steel production processes. Blast furnace flue dusts (Asadi Zeydabadi et al. 1997) are a mixture of oxides discharged from the top of the blast furnace, comprising mainly iron oxides and zinc, silicon, magnesium, and other minor element oxides in lesser amounts, along with some volatile elements such as zinc, cadmium, chromium, and arsenic (Jha et al. 2001).

Zinc dust from the smelting operation is generated in EAF, where different types of scrap are charged for the production of steel. These scraps contain different non-ferrous metals. During smelting of these materials, zinc metal, along with other metals, get evaporated and condensed during cooling. Lime and silica charged in the furnace are also collected with the zinc dust. A part of the dust is again charged into the furnace till the concentration of nonferrous metals reaches a permissible limit in the smelting operation for the production of steel. After this, this dust is removed from the system as waste and disposed of in landfills. The chemical composition of flue dusts varies depending on their origin, but mainly consists of zinc oxide, its ferrite, and other metal oxides: 19.4% Zn, 24.6% Fe, 4.5% Pb, 0.42% Cu, 0.1% Cd, 2.2% Mn, 1.2% Mg, 0.4% Ca, 0.3% Cr, 1.4% Si, 6.8% Cl (Caravaca et al. 1994).

2.4.3 FOUNDRY SAND

A metal casting foundry produces a metal casting shape by pouring liquid metal into a mold and allowing it to solidify and take the form of the mold. This process comprises five steps: pattern making, core making, molding, melting and pouring, and cleaning and reclaiming. The types of molding and core-making process include the sand-casting process, permanent mold process, die-casting process, centrifugal process, shell mold process, investment process, lost-foam process, plaster process, and graphite process.

The molding operation is performed by shaping a suitable refractory material to form the cavity, which is filled with the molten metal. After the metal obtains the desired shape, the casting is separated from the mold by means of the shakeout operation. The material employed for the molding operation depends on the type of metal being cast and the desired shape of the final product. High-quality size-specific silica sands are used to form molds that host the molten metal (Lee et al. 2004). Sand is used as it can absorb and transmit heat and allows gases evolved during binder breakdown to pass between the grains. In addition, the grains can hold together and give strength to the structure, and can also withstand high heat with moderate breaking down or fusing (Ekey 1958).

After the casting process is completed, resin-bound sands can be thermally reclaimed to make new molds and cores, while green sands require the addition of new bentonite clay and carbonaceous materials (Carey 2002). The sand is reused numerous times within the foundry process. However, over time, sand grains begin to fracture due to mechanical abrasion, sand reclamation, and exposure to high temperature. This eventually changes the shape of the sand grains, rendering them unsuitable for continued use in the foundry. Further new sand is added to the process in order to maintain proper tolerances and prevent casting defects. About 30%–60% of cast foundry solid wastes is made up of core and mold sand (EPA 1981). This removed sand, known as discarded, spent, or waste foundry sand (WFS), is commonly disposed of at foundry landfills or at off-site municipal landfills (Alves et al. 2014). Due to this practice, foundries generate approximately 9–12 million metric tons of WFS around the world (Abichou et al. 2004, Guney et al. 2006). Table 2.3 presents the chemical composition of a typical sample of WFS.

TABLE 2.3
Chemical Composition of Waste Foundry Sand

Constituent	Value (%)
SiO_2	87.91
Al_2O_3	4.70
Fe_2O_3	0.94
CaO	0.14
MgO	0.30
SO_3	0.09
Na_2O	0.19
K_2O	0.25
TiO_2	0.15
Mn_2O_3	0.02
SrO	0.03

Source: Reprinted from *Resour., Conserv. Recycl.*, 54(12), Siddiquea, R., Kaur, G., and Rajor, A., Waste foundry sand and its leachate characteristics, 1027–1036, doi:10.1016/j.resconrec.2010.04.006, Copyright 2010, with permission from Elsevier.

2.4.4   ELECTRIC ARC FURNACE DUST

Electric arc furnace (EAF) dust is a waste product which is generated when steel scrap is melted in an EAF. EAF generates more amount of dust during operation. This dust contains many metal oxides and chlorides along with toxic substances. It includes about 30% zinc as zincite and franklinite; 21.30% iron as magnetite, franklinite, and hematite; 7.02% various chlorides; and 3.17% lead as lead oxide and lead chlorides, along with dioxins. It also contains small amounts of Na, K, Mn, Mg, Cr, Cu, and Cd (Chen et al. 2011). Various chlorides present include NaCl, KCl, and PbOHCl (lead hydroxyl-halide) (Li et al. 1993, Leclerc et al. 2002). During steel production in EAF, around 15–20 kg of dust per ton of produced steel is generated as a by-product (Huber et al. 1999, Oustadakis et al. 2010).

In the steel production process, volatile compounds are fumed off during the melting of scrap and are collected along with particulate matter in an off-gas cleaning system (Barrett et al. 1992, Martins et al. 2008). The temperature of EAF can rise up to 1600°C, or even higher, during the metal fusing process, resulting in the volatilization of iron, zinc, and lead in the vapor phase. A large quantity of dust is generated when the vapor is cooled and collected.

When galvanized scrap is used in EAF, most of its zinc content ends up in the dust and fume due to its very low solubility in molten steel and slag, and also because zinc vapor pressure is higher at the steelmaking temperature as compared with the iron vapor pressure. This vapor zinc combines with other gaseous or particulate compounds generated during steel-making reactions, producing compounds such as ZnO and $ZnFe_2O_4$. The largest metallic portion of the EAF dust is zinc, which varies between 7% and 40%, depending on the ratio of galvanized scrap utilized (Orhan 2005, Pereira et al. 2007, Salihoglu and Pinarli 2008).

2.5   SOLAR PHOTOELECTRICITY WASTES

2.5.1   SOLAR PANELS

Solar energy is a promising technology that is significant in replacing renewable sources of energy because of its nonpolluting operation, extremely long operation time, minimum maintenance, robust technique, and aesthetic aspects (Klugmann-Radziemska 2012). Unprecedently, its use is hampered by its low energy density, which requires a significant area to concentrate it. Thus, solar cells were generated to convert solar rays directly to electricity (Grandell and Höök 2015). They have a long lifetime of about 25–30 years of power generation. Hence, their uses and applications are increasing tremendously. On the other side, when these panels will reach their end of life, substances used in the panels may cause environmental and health hazards (Berger et al. 2010).

Moreover, solar panels rely on the use of some rare metals, which are essential parts of certain solar energy technologies, such as indium, gallium, and ruthenium, most of which are unavailable as primary ores and mostly constitute as by-products related with primary ores. Indium is a by-product of zinc production (Fthenakis 2009), whereas tellurium is obtained as a by-product of copper, lead, gold, and bismuth ore processing (Fthenakis et al. 2009, Fthenakis and Anctil 2013, Rocchetti and Beolchini 2015).

Silicon-based solar cells are a first-generation technology that mainly comprise silicon and use silver as an electrode material. Thin-film cells are a second-generation technology, whereas the third-generation technology includes dye-sensitized solar cells (DSSCs), organic solar cells, etc.

CdTe thin films use a critical metal, tellurium. CdTe solar cells consist of four layers: a transparent conducting oxide layer consisting of SnO_2—a front contact on a glass substrate, a cadmium sulfide (CdS)—an n-type layer, CdTe film—an absorber layer and a back contact consisting of a copper layer (Klugmann-Radziemska 2012). CIGS thin film cells can be categorized as copper/indium/selenide (CIS) and copper/gallium/selenide (CGS) cells (Grandell and Höök 2015). CIS thin-film solar cells comprise glass (84%), an aluminum frame (12%), a polymer encapsulant (3%, e.g., ethylene vinyl acetate), along with valuable metals (Mo, Cu, In, Ga, Se, Cd, Zn, S).

The semiconductor layer of CIGS cells is made of a thin film of Cu(In, Ga) Se_2. This film is manufactured by a coevaporation and selenization process (Goetzberger et al. 2003). Evaporation sources are heated in vacuum and the evaporated atoms are deposited on the substrate in the coevaporation process, whereas in the selenization process, a film of copper, indium, and gallium is deposited by sputtering and then selenized with hydrogen selenide (Patniak 2003). However, in both processes, nearly 34% of the starting material is lost in different waste streams (Fthenakis 2009).

The cumulative amount of photovoltaic (PV) waste in 2017 is estimated to be 870 tons, and the total amount of solar panels waste is estimated to increase by 1,957,099 tons by 2038 (Paiano 2015). Despite silicon being the most common solar cell material, CdTe and CIGS solar cells are used in advanced technologies as they require much less semiconductor material and hence are cost-effective (Green 2007, Mehta 2009). However, the presence of expensive, rare, and toxic elements makes their recycling an essential way to lower production costs (Miles et al. 2007). Additionally, some solar modules contain hazardous materials such as cadmium, tellurium, lead, bismuth, and selenium. Cadmium compounds are, for example, currently regulated in many countries because of their toxicity to fish and wildlife and because they can pass into humans through the food chain (McDonald and Pearce 2010, Kang et al. 2012, Marwede et al. 2013, Jung et al. 2016).

2.6   POWER PLANT AND INCINERATION PLANT WASTES

2.6.1   THERMAL POWER PLANT BOTTOM AND FLY ASH

Thermal power plants that burn fossil fuel produce huge quantities of ash as waste, which is disposed of in ash ponds or landfills, thus causing environmental pollution. Figure 2.2 shows the formation of fly ash. Fly ash contains various heavy metals such as Al, Ga, Ge, Ca, Cd, Fe, Hg, Mg, Na, Ni, Pb, Ra, Th, V, Zn, Si, Cr, Mn, B, As, Cu, Zn, Mo, etc. (Adriano et al. 1980, Wong and Wong 1990, Rai et al. 2004, Meawad et al. 2010).

The composition of thermal power plant fly ash varies depending on the type of coal burned, boiler type, and collector setup. Fly ash is mainly composed of silica, alumina, iron oxide, and calcium, with varying amounts of carbon. The elements

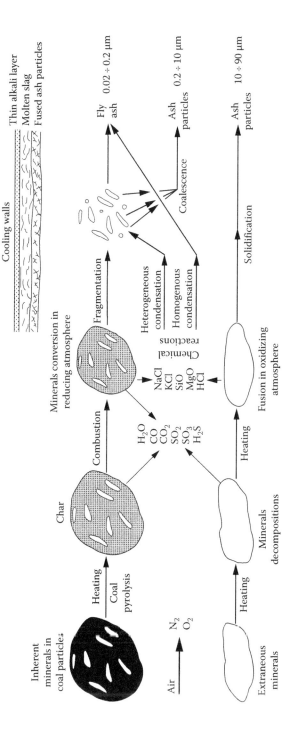

FIGURE 2.2 Formation of fly ash from pulverized fuel combustion. (Reprinted from *Fuel*, 81(10), Tomeczek, J. and Palugniok, H., Kinetics of mineral matter transformation during coal combustion, 1251–1258, doi:10.1016/S0016-2361(02)00027-3, Copyright 2002, with permission from Elsevier.)

that become volatile and then partly condense in the flue gases of the combustion system are mostly present in fly ash and include As, B, Bi, Cd, Ge, Hg, Mo, Pb, S, Se, Sb, Sn, and Ti (Querol et al. 1995). The concentration of trace elements in fly ash increases with a decrease in particle size and these elements can be enriched up to 30 times due to combustion depending on the coal feed (Gieré et al. 2003, Goodarzi and Sanei 2009).

Bottom ash is continuously discarded from the plant and contains large agglomerated ash particles that are not carried by the flue gas and fall through open grates into an ash hopper at the bottom of the furnace. Bottom ash can be wet or dry depending on the type and design of the boiler. Wet bottom ash is produced when molten coal residue from a wet bottom boiler is released into a water-filled hopper below the boiler, whereas dry bottom ash is produced in the dry bottom boiler (Halstead 1986, Lohtia and Joshi 1995). Bottom ash is mainly composed of silica, alumina, and iron, with a smaller content of calcium, magnesium, sulfates, and other compounds. Ba, Be, Co, Mn, Cs, Cu, Ni, Sr, Ta, V, W, Eu, Hf, and Zr are concentrated in bottom ash due to density aggregation effects.

Many thermal power plants utilize crude oil as the source of fuel. The incomplete combustion of this fuel oil in power plant furnaces results in the production of high amounts of ash, which contains a relatively high amount of valuable metals such as vanadium and nickel (about 2%–4%) as compared with natural ores (Navarro et al. 2007, Rasoulnia and Mousavi 2016). Thus, this ash can be used as a secondary source for metals such as V and Ni.

2.6.2 MUNICIPAL SOLID WASTE

Growing population, uncontrolled urbanization, and rapid industrialization have resulted in a rapid increase in global solid waste generation in recent years. It has been projected that by 2025, about 19 billion tons of solid waste will be generated every year (Yoshizawa et al. 2004).

Incineration is the most common method used for treating large amounts of wastes (such as municipal solid waste, MSW), thereby generating a new waste in the form of bottom and fly ash as well as waste gases, wherein the volume of waste is reduced by about 90% and weight of waste is reduced by 75%, and energy is recovered in the process in the form of heat and electricity (Hong et al. 2000). Such wastes are usually classified as a hazardous material due to their high content of toxic metals and soluble components (Wiles 1996).

The grate incinerator is the most commonly used incinerator for municipal waste, where the waste is conveyed from the incineration zone to the discharge zone by means of moving grate elements. The metals complexed with plastics and organic matter are released as the organic matter gets burnt in the incineration zone. The noncombustible residue remaining thereafter is a mixture of metal and minerals and termed as ash, and the ash which remains in the furnace is called as bottom ash. The air required for combustion is blown from below through passages in the grate elements and carries out fine ash particles, which are termed as fly ash. Later, after recovery of heat in the boiler, the fly ash and other pollutants are removed from the flue gas by a series of filters/washers (Bunge 2015).

Bottom ash is the noncombustible material remaining at the bottom of the incinerator furnace and constitutes about 10% volume of the waste. Fly ash comprises dust particles that are retained in the flue gas cleaning system and is recognized as more hazardous than bottom ash, since it easily becomes airborne and contains a high concentration of volatile elements, including hazardous heavy metals (e.g., Zn, Cu, Pb, Mn, and Fe) (Eighmy et al. 1995).

The nature of the metal species in fly or bottom ash largely depends on the type of waste, which consequently depends on the locality or municipality, the incineration process, the type of incineration furnace, etc. During the incineration process, all the organic matter is oxidized and the less volatile inorganic waste remains in the bottom ash, whereas more volatile ones are collected as fly ash in the ESP (Buekens 2013). MSW fly ash consists of heavy metals such as Cd, Cu, Ni, Pb, Zn, and As, along with chlorides of sodium and potassium and poisonous organic materials such as dioxins, which are concentrated and accumulated in the ash (Nagib and Inoue 2000). Some metals such as Cu and Zn are present in a high concentration, which can be recovered economically (Bosshard et al. 1996). The use of calcium for removal of gases such as SO_x, NO_x, and hydrogen chloride from the effluent gases makes the residue alkaline.

Fly ash comprises dust particles that are carried away from the combustion chamber with the flue gas and get filtrated (e.g., dry/wet scrubbers, ESP, chemical bag filters) through the air pollution control (APC) system, before being released into the atmosphere. It contains large amounts of soluble salts (e.g., Cl, Na) and hazardous metals (e.g., As, Cd, Pb) even after dilution with unreacted additives (e.g., lime and/ or soda addition in bag filters) and their neutralization capacity (Astrup et al. 2006, Funari et al. 2015, 2016). According to the literature, every year, thousands of tons of MSW fly ash produced contain 10^3 kg of Cu, Sn, and Sb, 10^4 kg of Al and Zn (Funari et al. 2015); further, their landfilling results in a loss of metals such as Al, Cu, Sn, and Zn, and critical raw materials (e.g., Cr, Co, Ga, Mg, Nb, Sb, and lanthanides) (Morf et al. 2013, Funari et al. 2015).

2.6.3 Coal Cleaning or Beneficiation Waste

Raw unprocessed coal contains a variety of metals, sometimes enriched concentrations of valuable metals. Moreover, combustion of coal results in the enrichment of metal concentrations in coal wastes, which in some cases is similar to mineral ores. Raw coal contains noncoal minerals, and to reduce this quantity, coal is subjected to coal preparation, known as coal washing or beneficiation. Washing of coal is mainly based on the differences in the specific gravity of coal and its impurities, and the different unit processes depend on the washability characteristics of a particular coal. The characteristics of wastes from a coal washing plant depend on the raw coal utilized and the final product. Generally, water is used for the washing process and the effluent contains a high load of total dissolved solids, calcium carbonate, and heavy metals (Dhar et al. 1986).

Heavy metals present in coal include Mn, Ca, Mg, Zn, Pb, Fe, Ni, Cu, Co, Al, etc. Various operations performed on run-of-mine (ROM) coal to make it suitable for end-use application without disturbing its physical characteristics are termed as coal beneficiation and include coal preparation and washing (Sahu et al. 2011).

2.7 PETROCHEMICAL WASTES

2.7.1 SPENT PETROLEUM CATALYSTS

A significant amount of waste is generated by the petrochemical and petroleum refining industry in the form of spent catalysts, which are used in the conventional production of gasoline, diesel fuel, jet fuel, heavy hydrocarbons, petrochemicals, and plastics in the petroleum distillation and petrochemical industry, and correspond to about 4 %wt. of the overall refinery waste (Liu et al. 2005, Marafi and Stanislaus 2008a, Amiri et al. 2011a). Catalysts contain chemicals such as metals, metal oxides, and metal sulfides, which facilitate in hydrocarbon transformation (Marafi and Stanislaus 2003, Silvy 2004). Hydroprocessing or hydrotreating catalysts (e.g., hydrodesulfurization, hydrodenitrogenation, and hydrogenation) constitute more than 90% of the total fixed-bed catalysts used during petroleum refinement (Kim et al. 2009). The life of a catalyst varies between 3 months and 6 years depending on the feed (Park et al. 2006, Amiri et al. 2011b). Hydrocracking, dehydrogenation, reforming, and hydrogenation catalysts are used extensively to purify and upgrade various petroleum feeds and to enhance the process efficiency, which lose their activity due to deposition of metal sulfides and oxides and metal organic components on their surfaces (Furimsky 1996, Rapaport 2000, Silvy 2004). Deactivation of catalysts also occurs due to structural changes by thermal degradation, phase separation, or phase transformation, which restricts their reactivation. Although they are reactivated and resued for many cycles, catalysts lose their effectiveness and are discarded after a few cycles of activation–deactivation (Marafi and Stanislaus 2008a,b, Macaskie et al. 2010, Pradhan et al. 2010). The amount of spent catalyst that is discarded is always greater than that of the fresh catalyst due to the deposition of coke, sulfur, and metals during the refining process. The literature suggests that nearly 150,000–170,000 tons per year of spent petroleum catalysts are generated worldwide, and this amount will increase in the coming years due to the elevated processing of heavier feedstock and the inflated demand for clean fuels (Dufresne 2007, Marafi and Stanislaus 2008b).

Fluid catalytic cracking (FCC) is a major secondary conversion process in petroleum refining which generates about 4×10^8 kg of spent catalysts annually on a global scale (Furimsky 1996). FCC is used for the conversion of relatively high-temperature boiling hydrocarbons to lower-temperature boiling hydrocarbons in the heating oil and gasoline range. During this process, nickel- and vanadium-containing porphyrins and porphyrin-like complexes from the hydrocarbon feed decompose, leaving metal contaminants on the FCC catalyst surface. These metals increase the production of coke, hydrogen, and light gases at the expense of the highly desired gasoline (Cimbalo et al. 1972). After reaction in the fluidized catalytic cracking unit (FCCU), spent catalysts are removed from the unit and replaced by fresh catalysts in order to maintain the desired catalytic activity.

Hydroprocessing catalysts are usually known to consist of molybdenum (Mo) or tungsten (W) supported on an alumina or silica carrier, along with promoters such as cobalt (Co) or nickel (Ni), and they enhance the removal of undesirable impurities such as sulfur, nitrogen, and metals such as vanadium (V) in feedstocks by promoting hydrodesulfurization, hydrodenitrogenation, and hydrodemetallization (Marafi and

Stanislaus 2008a,b, Ferella et al. 2011). Heavy metals such as Co and Ni (in hydro-processing catalysts) may be present originally or get added during the process (such as Ni and V in FCC catalysts) (Aung and Ting 2005). Thus, petrochemical wastes comprise valuable metals such as Pt, Re, Fe, Ni, Co, Mo, W, V, Cr, Cu, Zn, Sb, and Al (Santhiya and Ting 2006, Mishra et al. 2007, 2008, Kasikov and Petrova 2009, Amiri et al. 2011b, Bharadwaj and Ting 2013, Srichandan et al. 2013).

2.8 SLUDGE

2.8.1 MUNICIPAL SEWAGE SLUDGE

Increased urbanization and industrialization have resulted in a pronounced eleva-tion in the volume of municipal wastewater produced worldwide. Sewage sludge is produced as a result of treatment of municipal wastewater. Sludge contains solids present in the wastewater, along with suspended solids which are produced due to the removal of dissolved solids from the wastewater (Vesilind et al. 1986). Of late, the quantity of this sludge is increasing markedly and is expected to rise in the com-ing years. Depending on the nature of wastewater and the treatment procedure, the solid content of waste varies from 0.25% to 12% by weight (Metcalf and Eddy 1995). Wastewater is treated in three stages—mechanical, biological, and, if required, an additional stage for nutrient removal. In the mechanical stage, untreated wastewater passes through racks and screens for removal of coarse solids and thereafter through grid chambers, where sand, gravel, cinders, and other heavy solid materials are sepa-rated. Further, settleable solids and floating materials are removed in sedimentation tanks. Nearly 50%–70% of solids is removed in the mechanical stage, producing primary sludge having about 3%–7% solids (Chobanoglous 1987, Mudrack and Kunst 1988). Further, this primary sludge is treated biologically using microbes, wherein nonsettleable solids are coagulated and removed and the organic matter is stabilized; this sludge is referred to as secondary sludge. In the third stage, nutrients such as N and P are removed. Nitrogen is removed by nitrification or denitrifica-tion (Mathias 1994), while phosphorus is eliminated through chemical precipitation using additives, followed by sedimentation of the sludge formed, or through bio-logical treatment (Solter et al. 1994, Werther and Ogada 1999). During this process, nearly 50%–80% of heavy metals present in sewage gets complexed in the sludge by physicochemical and biological interactions (Lester et al. 1983).

Sewage sludge is generally applied to soils as a fertilizer due to its content of N, P, and other micronutrients such as copper, zinc, molybdenum, boron, iron, magnesium, and calcium, along with organic matter, which improves the physical, chemical, and biological properties of the soil (Beck et al. 1996, Pathak et al. 2009). However, nearly 0.5%–2% of the heavy metal content on a dry weight basis in sewage sludge, which can even increase up to 6%, especially for metals such as Cu and Zn (Jain and Tyagi 1992, Karvelas et al. 2003), confines its use as a fertilizer (Lester 1983). The repeated application of a heavy metal–containing sludge to the soil elevates the heavy metal concentration in the soil due to the degradation of organic matter. The forms of heavy metals in sludge vary according to the type of sludge, the character-istics of the metal, and the treatment method employed.

Sources of heavy metals in MSW sludge include household dust, batteries, disposable household materials (e.g., bottle tops), plastics, paints and inks, body care products, and medicines and household pesticides (National Household Hazardous Waste Forum (NHHWF) 2000, Bardos 2004).

2.8.2 Industrial Sludge

Electroplating is the application of a metal coating to a metallic or another conducting surface by an electrochemical process, in order to amend its appearance, impart specific mechanical properties, augment its anticorrosion property, its smoothness, and the luster of surfaces (Rehman and Lee 2014). This procedure includes connecting the desired materials to the cathode and immersing the cathode in an acidic solution, such as H_2SO_4, HCl, and HNO_3, containing metal cations to coat one or more thin layers of the metal (Wang 2006). The most widely used method for electroplating is the precipitation of metal ions as hydroxides, which however generates large quantities of heavy metal–loaded sludge (Chernicharo and von Speling 1996, de Souza e Silva et al. 2006, Adamczyk-Szabela 2013, Huang et al. 2013). This industrial galvanic sludge is abundant in metals such as Ni, Cr, Cu, Zn, Fe, etc., all together in a complicated liquid–solid mixture (Świerk et al. 2007, Dhal et al. 2013, Sochacki et al. 2014, Su et al. 2016). The metals used in coating the material accumulate in the wastewater of pretreatment, coating, finishing, and purging baths in the respective process stage. Further, after the detoxification of cyanide-, nitrite-, or Cr(VI)-containing waters, the metals are precipitated as more or less insoluble hydroxides or alkaline salts (Hartinger 1991), which then sediment as electroplating sludge (Wazeck 2013). Electroplating sludge comprises 34%–92% water and a mixture of degreasers, brighteners, and neutralizers (Huang 2001). Su et al. (2016) reported that more than 100,000 tons of electroplating sludge are produced yearly in China alone. About 1.3 million wet tons of electroplating sludge was generated in a year in the United States, most of which was disposed of in permitted hazardous waste landfills (USEPA 1998).

Ni and Cr are immensely used as an electroplating material because of their well-known corrosion resistance property (Lin et al. 1992). Ni coatings are known to render ductility and good wear resistance (Lampke et al. 2006), and Cr coatings are used to protect carbonic steel and various alloys from chemical attack and to improve their wear resistance (Wang et al. 2007). The presence of these metals in the sludge can cause environmental hazards, but the sludge can be regarded as a secondary source of such metals.

2.9　LIGHTING WASTES

2.9.1　Light-Emitting Diodes

Light-emitting diodes (LEDs) are considered as environmentally friendly light sources because they are energy efficient and mercury free (Schubert et al. 2006, Kim and Schubert 2008, Matthews et al. 2009, Mottier 2009). Thus, their use has increased dramatically in the last few years by replacing incandescent light sources and compact fluorescent lamps (CFLs) (Lim et al. 2013). This rapid growth implies that, eventually, they will contribute to the waste stream and impact the environment and human health (Ogunseitan et al. 2009).

LEDs are made of semiconductor diodes which emit energy in the form of photons when crossed by an electric current, where the wavelength and color of the light emitted depend on the chemical composition of the semiconductor material (Nardelli et al. 2016).

Waste LEDs will contribute to more metals in the waste as they need more metal-containing components for their functioning (Lim et al. 2013). LED semiconductor chips belong to groups III–V semiconductors (Schubert 2006), which contain arsenic, gallium, indium, and/or antimony (Schubert 2006, Mottier 2009). Various drivers are required for generating a direct current output for LEDs, which supplies the power to the LED semiconductor emitting the light. (Fimiani 2009, Szolusha 2009). Furthermore, an LED chip is assembled into a usable pin-type device through the application of leads, wires, solders, glues, and adhesives, as well as heat sinks for thermal dissipation management (Schubert 2006, Yung et al. 2008, Mottier 2009, Ribarich 2009). LEDs also contain additional metals such as copper, gold, silver, iron, arsenic, nickel, and lead (Lim et al. 2011). All these valuable and toxic metals significantly pose a risk to the environment and human health (Ogunseitan et al. 2009).

2.9.2 MERCURY LIGHT BULB WASTES

The use of fluorescent lamps, both tubes and CFLs, is increasing in homes around the world to improve energy efficiency (Thaler et al. 1995, Hildenbrand et al. 2003). Mercury is an essential component in all types of fluorescent lamps, which is a highly toxic heavy metal and is a chief source of waste, especially cold cathode fluorescent lamps (CCFLs), ultraviolet (UV) lamps, and high-pressure mercury lamps (SHPs). Mercury in bulbs converts electrical energy to radiant energy in the UV range and then reradiates it in the visible spectrum (Skarup and Lubeck Christensen 2004). The concentration of mercury in these lamps varies depending upon the manufacturer, lamp type, and year of manufacturing (Chang and Yen 2006, Rey-Raap and Gallardo 2012). As per a U.S. Environmental Protection Agency survey, the mercury contained in one type of fluorescent lamp (produced by the Sylvania lamp manufacturer) probably consists of 0.2% (0.042 mg) elemental mercury (vapor phase) and 99.8% (20.958 mg) divalent mercury incorporated into the phosphorus powder (USEPA 1998, Aucott et al. 2003). When such mercury-containing lamps are discarded, the mercury is released into the environment (Raposo et al. 2003). When a CFL is broken, elemental mercury (including vapor) and inorganic mercury compounds are released.

The amount of mercury released as mercury vapor or associated with the phosphor powder will depend on the age of the lamp and quantity of mercury vapor in the lamp. CFLs contain several species of mercury depending on the species of the mercury added by the manufacturer and the age of the lamp (United Nations Environment Programme (UNEP) 2005). After prolonged use, elemental mercury in lamp gets oxidized to form inorganic mercury compounds, mainly HgO

(Aucott et al. 2003). New lamps will release more mercury vapor, whereas in older or spent (used) lamps the mercury will have been oxidized and or have partitioned to lamp components (Jang et al. 2005).

A fluorescent lamp consists of a glass tube, a cathode, an anode, phosphor, electronic modules, aluminum end-caps, and other constituents, for example, alumina (barrier layer), inert gas, and mercury vapor (Harris 2001). When current is applied to the electrodes, mercury vapor gets excited and the emitted UV radiation is converted to visible light by the phosphor coating. During the operation of the lamp, mercury vapor impinges onto the glass, the phosphor powder, and the metal components, and hence requires an appropriate amount of mercury vapor to attain long lamp lifetimes. Milligrams to tens of milligrams of mercury are added to lamps according to their types and their manufacturers. Fluorescent lamps typically last for over 6000 h, but are ultimately burned out due to the exhaustion of mercury (Dunmire et al. 2003, Cain et al. 2007). In spent fluorescent lamps, mercury is primarily bound to the glass, the phosphor powder, and the metal end-parts, with only a small fraction remaining in vapor form at room temperature. Albeit each fluorescent lamp only contains a small amount of mercury, if millions are dumped, the cumulative mass of mercury can be significant. The small amount of mercury sealed within the glass tubes of fluorescent lamps poses no risk as long as the lamps remain intact. However, spent and broken lamps release the contained mercury into the environment. Reports state that 1.2%–6.8% of the total mercury in fluorescent lamps could be released into the air after breakage of lamps (USEPA 1998). To reduce mercury consumption, targets of 5 mg per lamp for CFLs up to 25 W, and 6 mg for CFLs with higher wattages, were set by the National Electric Manufacturers Association (NEMA) in 2007 (Eckelman et al. 2008). Most CFLs contain 3–5 mg per bulb, with bulbs containing as little as 1 mg being labeled "eco-friendly."

2.10   AUTOMOBILE WASTES

Automobile production generates huge quantities of waste, which includes metal, solvents, batteries, plastic, and glass. Nearly 75% of the total vehicle weight is recycled, while the remaining 25%, known as auto shredder residue (ASR), is landfilled (Nawrocky et al. 2010, House et al. 2011, Cheng et al. 2012, PRNewswire 2015). ASR is the final product of end-of-life vehicles (EVLs) recycling and recovery processes (Joung et al. 2007). ASR consists of foam and fluff (40%–52%), plastics (20%–27%), rubbers (18%–22%), and metals (4%–15%) (Sivakumar et al. 2014, Sharma et al. 2016). More than 70% of vehicle compounds consists of metals, steel, and aluminum being predominant in structural applications (Sullivan et al. 1998). Steel is mostly used in the frame, chassis, and body parts, while aluminum is used in the body, parts of engine and chassis, interior parts, airbags, etc. Other metals are found in auxiliary parts of the car, for example, copper in cables, connectors, radiators, etc. (Lundqvist et al. 2004); lead in vehicle battery and in alloying of steel and aluminum for machining properties; and zinc for steel coating in order to avoid corrosion (Sander et al. 2000).

End-of-life vehicles are treated at various stages such as pretreatment, dismantling, shredding, and shredder residue treatment (Lampman 1991, Year 2011). In the pretreatment step, toxic and dangerous compounds such as oils, fuels, battery, oil filters, and mercury-containing components are removed and recycled (Lundqvist et al. 2004). In the next step of dismantling, the individual parts are disassembled and separated into different material and component fractions to be recycled, such as tyres, catalytic converters, etc. (Staudinger et al. 2001). Furthermore, the remaining waste is shredded to reduce the waste volume and separate more homogenous fractions. The parts of vehicles are shredded into small pieces and subjected to separation, by mechanical and physical processes such as magnetic separation, eddy current belt, and sink-floating methods, depending on their type and properties. Here, the materials are categorized as ferrous metals (iron, steel), nonferrous metals (aluminum, copper), and shredder residues, of which ferrous and nonferrous metals are recycled as scrap metals. Shredder residue is a mixture of different components with different properties and constitutes the remaining 25% of the waste volume (Poulikidou 2010).

ASR is further classified as low-weight ASR, weighty ASR, and fines (soil, sand, etc.) (Simic and Dimitrijevic 2012). The literature suggests that ASR with particle size <6.0 mm is predominantly polluted with heavy metals and others hazardous materials (mineral oils and hydrocarbons). Residual metal pieces, solder, plasticizers, and paint are key sources of heavy metals in ASR (Kurose et al. 2006, Gonzalezfernandez et al. 2008, Lopes et al. 2009, Santini et al. 2011, Singh and Lee 2016). According to Singh et al. (2016b), ASR is rich in metals such as Zn, Cu, Mn, Fe, Ni, Pb, Cd, and Cr, and their concentrations in ASR are as follows: Zn at 5571.3 mg/kg, Cu at 1957.1 mg/g, Mn at 1068.8 mg/kg, Fe at 151.8 g/kg, Ni at 95.9 mg/kg, Pb at 1202.8 mg/kg, and Cr at 98.3 mg/kg.

2.11   METAL SPECIATION

Metal speciation is an important factor that influences the efficiency of metal bioleaching. It has been noted that the determination of the total heavy metal content is not enough to indicate heavy metals' bioavailability, mobility, and toxicity (Hsu and Lo 2001, Amir et al. 2005), which are often dependent on their chemical forms and their binding state (precipitated with primary or secondary minerals, complexed by organic ligands, etc.) (Nomeda et al. 2008, He et al. 2009a,b, Dong et al. 2013).

Speciation refers to aspects of the chemical and physical forms of an element. Oxidation state, stoichiometry, coordination (including the number and type of ligands), and physical state or association with other phases, all contribute in defining speciation. These properties govern the chemical behavior of elements, whether in environmental settings or in human organs, and play a crucial role in determining toxicity.

Metal ions form complexes with naturally occurring complexing agents or ligands released from industrial activity. These metal complexes are thereby mobilized and transported within environmental and biological systems. The impact of such metal complexes depends on the metal complex species that are kinetically and thermodynamically stable in these homogeneous and heterogeneous systems. Also, the organic fraction in the industrial waste undergoes biological, oxidation–reduction, or photochemical reactions and gets converted into metal-binding ligands.

Heavy metals from the waste react with these ligands and perturb biogeochemical cycles and eventually are exposed to humans.

It is not surprising that organic ligands introduced into the environment that have appropriate functional groups can mobilize and transport certain elements by complex formation. The term "complex formation" is used here in its broadest sense, that is, the successive formation of metal–ligand complexes, such as ML, ML2, ML3, etc. (where M and L, with charges omitted, represent the metal ion and the ligand, respectively), the formation of mixed ligand complexes (e.g., MLX, MLY, etc., where X and Y represent auxiliary ligands) in homogeneous systems, and the formation of adsorption complexes with particulate matter or with cell surfaces in heterogeneous systems. The presence of these complexes in the environment is governed by their thermodynamic stability, that is, their formation constants, and by their rates of formation and dissociation at concentrations that are found in the environment, and not under experimental conditions that are convenient for laboratory studies.

It is important to determine experimentally, or to deduce from the body of published data, the nature of these heavy metal complexes. The bioavailability and toxicity of these heavy metal complexes are critically dependent on their thermodynamic and kinetic stability. Identifying these species (i.e., speciation) and obtaining reasonable estimates of their thermodynamic and kinetic stabilities are the problems that continue to plague environmentalists and toxicologists (Fernando 1995).

The high concentrations of heavy metals in waste get adsorbed and accumulated in the soil. The distribution, mobility, and bioavailability of heavy metals in the environment depend both on their total concentration and on the association form in the solid phase to which they are bound (Filgueiras et al. 2002).

Depending on their nature, metals are associated in a variable manner with different phases in the waste and sequential leaching or extraction can be used to differentiate them. A precise knowledge of heavy metal speciation can be used to know their effect on plants and other species (García-Delgado et al. 2007). The procedure includes multiple extractions using any number of steps and extraction agents, where the aggressiveness of the extraction agent increases with each stage and is thus the best method to obtain information on the origin, the manner of occurrence, bioavailability, mobilization, and transport of heavy metals (Ma and Rao 1997). Tessier is considered to be the founder of speciation analysis, and the procedure for the speciation extraction procedure test (SEP test) is given in Table 2.4. (Tessier et al. 1979, Batley 1989, Tessier and Turner 1995, Xie and Zhu 2013). According to Tessier et al. (1979), there are four forms: (1) exchangeable ions (susceptible to changes in the ionic composition of water), (2) forms bound to carbonates (susceptible to pH changes), (3) forms bound to Mn and Fe oxides (unstable in reducing conditions), and (4) forms bound to organic matter (decomposed under oxidizing conditions and result in a release of the metals into the soil solution) (Ciba et al. 2003), and a residual fraction form: metals are stably bound in the crystal lattice of the mineral components of the compost. As metals from the residual fraction cannot enter the food chain, they contribute very little in heavy metal bioavailability.

To discriminate between the forms of pollutants bound to Mn oxides, hydrated Fe oxides (amorphous), and Fe oxides (crystalline), a speciation analysis procedure was developed step by step. Exchangeable ions are bound to the material surface either

TABLE 2.4
Procedure for Sequential Extraction Procedure (SEP) Test

Fraction	Reagent per 2 g Sample	Operation
Exchangeable	16 mL 1 M NaOAc, pH 8.2	Continuous agitation at room temperature for 1 h
Carbonate	16 mL 1 M NaOAc, pH 5.0	Continuous agitation at room temperature for 1 h
Fe–Mn oxides	40 mL 0.04 M $NH_2OH \cdot HCl$ in 25% (v/v) HOAc	Occasional agitation at $96°C \pm 3°C$ for 5 h
Organic combined	6 mL 0.02 M HNO_3 + 10 mL 30% H_2O_2, pH 2	Occasional agitation at $85°C \pm 2°C$ for 2 h
	6 mL 30% H_2O_2, pH 2	Intermittent agitation at $85°C \pm 2°C$ for 3 h
	10 mL 3.2 M NH_4OAc in 20% HNO_3	Continuous agitation at $85°C \pm 2°C$ for 20 min
Residual	HNO_3/$HClO_4$/HF digestion	

Source: With kind permission from Springer Science + Business Media: *J. Mater. Cycles Waste Manage.*, Leaching toxicity and heavy metal bioavailability of medical waste incineration fly ash, 15, 440–448, Xie, Y. and Zhu, J.

by electrostatic forces or weak adsorption, along with ions bound in soluble minerals and hence are easily leached by water or weak electrolyte solutions (Spear et al. 1998). However, the carbonate-bound elements have rather strong bonds and need a slightly acidic solution for their release, whereas the oxidic phase is evaluated as part of the residual fraction. Secondary oxides are formed on the surface of Mn and Fe oxides, due to their great binding capacity, by coprecipitation, adsorption, ion exchange, penetration into the grid, or formation of surface complexes. Bond formation includes several processes and its type depends on the reaction conditions. This coating of exchangeable ions, carbonates, Mn and Fe oxides and organic matter on the oxide surfaces can be separated by applying various reducing agents.

Also, toxic substances can be directly bonded to the surface of organic substances or form organometallic compounds, which can be released by oxidizing agents. Other toxic phases that are bound in the form of oxides, carbides, silicates, or clay materials can be released after decomposition by strong mineral acids or fusion.

Tessier's procedure of speciation analysis has been modified by many authors in terms of the number of steps and agents used (Campanella et al. 1995). This procedure has since been applied for investigating various forms of pollutants (heavy metals) in soils, sediments (Sposito et al. 1982, Shuman and Hargrove 1985, Zeihen and Brummer 1989, Calvet et al. 1990, Cauwenberg and Maes 1997, Száková et al. 1997, Borovec 2000, Venditti et al. 2000), geological materials (Hall et al. 1996, Tossavainen and Forssberg 2000), coal and its ash (Belevi et al. 1992, Buchholz and Landsberger 1995, Raclavská et al. 1995, Querol et al. 1996, Bartoňová et al. 2001), and fly ash from waste incinerators (Theis and Padgett 1983, Kirby and Rimstidt 1993, Tan et al. 1997). It has also been compared with leaching procedures used by the USEPA or with leaching by acid solutions (Buchholz and Landsberger 1995). Kulveitová (1999) and Polyák et al. (1995) applied this test for analysis of metallurgical waste. The procedure developed by Polyák et al. (1995) distinguishes seven forms of pollutants: (1) exchangeable ions, (2) forms bound to carbonates, (3) forms bound to Mn oxides, (4) forms bound to organic matter, (5) forms bound to amorphous Fe oxides, (6) forms bound to crystalline Fe oxides, and (7) forms bound in the residual fraction.

Thus, sequential extraction methods to investigate metal speciation of environmental samples provide pertinent information about possible toxicity when these samples are discharged into the environment.

REFERENCES

Abichou, T., C. H. Benson, T. B. Edil, and K. Tawfiq. 2004. Hydraulic conductivity of foundry sands and their use as hydraulic barriers. In *Recycled Materials in Geotechnics*, A. H. Aydilek, and J. Wartman (Eds.), Geotechnical Special Publication 127. Baltimore, MD: ASCE.

Adamczyk-Szabela, D. 2013. Assessing heavy metal content in soils surrounding electroplating plants. *Polish Journal of Environmental Studies* 22: 1247.

Adriano, D. C., A. L. Page, A. A. Elseewi, A. C. Chang, and I. Straughan. 1980. Utilization and disposal of fly ash and other coal residues in terrestrial ecosystems: A review. *Journal of Environmental Quality* 9: 333–344. doi:10.2134/jeq1980.00472425000900030001x.

Aktas, S. 2010. Silver recovery from spent silver oxide button cells. *Hydrometallurgy* 104(1): 106–111. doi:10.1016/j.hydromet.2010.05.004.

Al-Thyabat, S., T. Nakamura, E. Shibata, and A. Iizuka. 2013. Adaptation of minerals processing operations for lithium-ion (LiBs) and nickel metal hydride (NiMH) batteries recycling: Critical review. *Minerals Engineering* 45: 4–17. doi:10.1016/j.mineng.2012.12.005.

Alp, İ., H. Deveci, and H. Süngün. 2008. Utilization of flotation wastes of copper slag as raw material in cement production. *Journal of Hazardous Materials* 159(2): 390–395. doi:10.1016/j.jhazmat.2008.02.056.

Alves, B. S. Q., R. S. Dungan, R. L. P. Carnin, R. Galvez, and C. R. S. De Carvalho Pinto. 2014. Metals in waste foundry sands and an evaluation of their leaching and transport to groundwater. *Water, Air, and Soil Pollution* 225(5): 1963–1974. doi:10.1007/s11270-014-1963-4.

Amir, S., M. Hafidi, G. Merlina, and J.-C. Revel. 2005. Sequential extraction of heavy metals during composting of sewage sludge. *Chemosphere* 59(6): 801–810. doi:10.1016/j. chemosphere.2004.11.016.

Amiri, F., S. M. Mousavi, and S. Yaghmaei. 2011a. Enhancement of bioleaching of a spent Ni/Mo hydroprocessing catalyst by *Penicillium simplicissimum*. *Separation and Purification Technology* 80(3): 566–576. doi:10.1016/j.seppur.2011.06.012.

Amiri, F., S. Yaghmaei, and S. M. Mousavi. 2011b. Bioleaching of tungsten-rich spent hydrocracking catalyst using *Penicillium simplicissimum*. *Bioresource Technology* 102(2): 1567–1573. doi:10.1016/j.biortech.2010.08.087.

Asadi Zeydabadi, B., D. Mowla, M. H. Shariat, and J. Fathi Kalajahi. 1997. Zinc recovery from blast furnace flue dust. *Hydrometallurgy* 47(1): 113–125. doi:10.1016/S0304-386X(97)00039-X.

Astrup, T., J. J. Dijkstra, R. N. J. Comans, H. A. van der Sloot, and T. H. Christensen. 2006. Geochemical modeling of leaching from MSWI air-pollution-control residues. *Environmental Science & Technology* 40(11): 3551–3557. doi:10.1021/es052250r.

Aucott, M., M. McLinden, and M. Winka. 2003. Release of mercury from broken fluorescent bulbs. *Journal of the Air and Waste Management Association* 53(2): 143–151.

Aung, K. M. M. and Y. P. Ting. 2005. Bioleaching of spent fluid catalytic cracking catalyst using *Aspergillus niger*. *Journal of Biotechnology* 116(2): 159–170. doi:10.1016/j.jbiotec.2004.10.008.

Baddour-Hadjean, R., J. P. Pereira-Ramos, M. Latroche, and A. Percheron-Guégan. 2003. In situ x-ray diffraction study on MmNi3.85Mn0.27Al0.37Co0.38 as negative electrode in alkaline secondary batteries. *Electrochimica Acta* 48(19): 2813–2821. doi:10.1016/S0013-4686(03)00416-X.

Baghalha, M., V. G. Papangelakis, and W. Curlook. 2007. Factors affecting the leachability of Ni/Co/Cu slags at high temperature. *Hydrometallurgy* 85(1): 42–52. doi:10.1016/j.hydromet.2006.07.007.

Bakhtiari, F., M. Zivdar, H. Atashi, and S. A. Seyed Bagheri. 2008. Bioleaching of copper from smelter dust in a series of airlift bioreactors. *Hydrometallurgy* 90(1): 40–45. doi:10.1016/j.hydromet.2007.09.010.

Bardos, P. 2004. *Composting of Mechanically Segregated Fractions of Municipal Solid Waste—A Review*. Falfield, U.K.: Sita Environmental Trust.

Barrett, E. C., E. H. Nenniger, and J. Dziewinski. 1992. A hydrometallurgical process to treat carbon steel electric arc furnace dust. *Hydrometallurgy* 30(1): 59–68. doi:10.1016/0304-386X(92)90077-D.

Bartoňová, L., Z. Klika, J. Seidlerová, and P. Danihelka. 2001. *Sborník vědeckých prací Vysoké školy báňské—Technické univerzity Ostrava/Transactions of the VŠB—Technical University of Ostrava*, 1st edn. Ostrava, Czech Republic: VŠB Technical University of Ostrava (in Czech).

Batley, G. E. 1989. *Trace Element Speciation: Analytical Methods and Problems*. Boca Raton, FL: CRC Press.

Baumann, W. and A. Muth. 1997. *Batterien—Daten und Fakten zum Umweltschutz*, p. 593, Berlin, Germany: Springer.

Bayat, O., E. Sever, B. Bayat, V. Arslan, and C. Poole. 2009. Bioleaching of zinc and iron from steel plant waste using *Acidithiobacillus ferrooxidans*. *Applied Biochemistry and Biotechnology* 152(1): 117–126. doi:10.1007/s12010-008-8257-5.

Beck, A. J., D. L. Johnson, and K. C. Jones. 1996. The form and bioavailability of non-ionic organic chemicals in sewage sludge-amended agricultural soils. *Science of the Total Environment* 185(1–3): 125–149. doi:10.1016/0048-9697(96)05047-4.

Belevi, H., D. Stampfli, and P. Baccini. 1992. Chemical behaviour of municipal solid waste incinerator bottom ash in monofills. *Waste Management & Research* 10(2): 153–167. doi:10.1016/0734-242X(92)90069-W.

Berger, W., F. G. Simon, K. Weimann, and E. A. Alsema. 2010. A novel approach for the recycling of thin film photovoltaic modules. *Resources, Conservation and Recycling* 54(10): 711–718. doi:10.1016/j.resconrec.2009.12.001.

Bernardes, A. M., D. C. R. Espinosa, and J. A. S. Tenório. 2004. Recycling of batteries: A review of current processes and technologies. *Journal of Power Sources* 130(1–2): 291–298. doi:10.1016/j.jpowsour.2003.12.026.

Bharadwaj, A. and Y. P. Ting. 2013. Bioleaching of spent hydrotreating catalyst by acidophilic thermophile *Acidianus brierleyi*: Leaching mechanism and effect of decoking. *Bioresource Technology* 130: 673–680. doi:10.1016/j.biortech.2012.12.047.

Borovec, Z. 2000. Speciace prvků v kontaminovaných půdách, kalech, říčních a jezerních sedimentech. *Vodní Hospodářství* 1: 1–5 (in Czech).

Bosshard, P. P., R. Bachofen, and H. Brandl. 1996. Metal leaching of fly ash from municipal waste incineration by *Aspergillus niger*. *Environmental Science & Technology* 30(10): 3066–3070. doi:10.1021/es960151v.

Brandl, H. and M. A. Faramarzi. 2006. Microbe-metal-interactions for the biotechnological treatment of metal-containing solid waste. *China Particuology* 4(2): 93–97. doi:10.1016/S1672-2515(07)60244-9.

Buchholz, B. A. and S. Landsberger. 1995. Leaching dynamics studies of municipal solid waste incinerator ash. *Journal of the Air and Waste Management Association* 45(8): 579–590. doi:10.1080/10473289.1995.10467388.

Buchmann, I. 2013. Global battery markets information—Battery University. http://batteryuniversity.com/learn/article/global_battery_markets (accessed April 25, 2013).

Buekens, A. 2013. *Incineration Technologies*. New York: Springer.

Bunge, R. 2015. Recovery of metals from incinerator bottom ash. https://www.umtec.ch/fileadmin/user_upload/umtec.hsr.ch/Dokumente/News/1504_Metals_from_MWIBA__R._Bunge.pdf (accessed March 8, 2016).

Cain, A., S. Disch, C. Twaroski, J. Reindl, and C. R. Case. 2007. Substance flow analysis of mercury intentionally used in products in the United States. *Journal of Industrial Ecology* 11: 61–75.

Calvet, R., S. Bourgeois, and J. J. Msaky. 1990. Some experiments on extraction of heavy metals present in soil. *International Journal of Environmental Analytical Chemistry* 39(1): 31–45. doi:10.1080/03067319008027680.

Campanella, L., D. D'Orazio, B. M. Petronio, and E. Pietrantonio. 1995. Proposal for a metal speciation study in sediments. *Analytica Chimica Acta* 309(1–3): 387–393. doi:10.1016/0003-2670(95)00025-U.

Caravaca, C., A. Cobo, and F. J. Alguacil. 1994. Considerations about the recycling of EAF flue dusts as source for the recovery of valuable metals by hydrometallurgical processes. *Resources, Conservation and Recycling* 10(1–2): 35–41. doi:10.1016/0921-3449(94)90036-1.

Carey, P. R. 2002. Sand/binders/sand preparation & coremaking. *Foundry Management & Technology* 130(1): 39–52.

Cauwenberg, P. and A. Maes. 1997. Influence of oxidation on sequential chemical extraction of dredged river sludge. *International Journal of Environmental Analytical Chemistry* 68(1): 47–57. doi:10.1080/03067319708030479.

Chang, T. C. and J. H. Yen. 2006. On-site mercury-contaminated soils remediation by using thermal desorption technology. *Journal of Hazardous Materials* 128(2): 208–217. doi:10.1016/j.jhazmat.2005.07.053.

Chatterjee, A. and J. Abraham. 2017. Efficient management of e-wastes. *International Journal of Environmental Science and Technology* 14(1): 211–222. doi:10.1007/s13762-016-1072-6.

Chauhan, R. and K. Upadhyay. 2015. Removal of heavy metal from e-waste: A review. *International Journal on Chemical Studies* 3: 15–21.

Chen, W.-S., Y.-H. Shen, M.-S. Tsai, and F.-C. Chang. 2011. Removal of chloride from electric arc furnace dust. *Journal of Hazardous Materials* 190(1): 639–644. doi:10.1016/j.jhazmat.2011.03.096.

Cheng, Y. W., J. H. Cheng, C. L. Wu, and C. H. Lin. 2012. Operational characteristics and performance evaluation of the ELV recycling industry in Taiwan. *Resources, Conservation and Recycling* 65: 29–35. doi:10.1016/j.resconrec.2012.05.001.

Chernicharo, C. A. L. and M. von Speling. 1996. A new approach for ambiental control industry. In *Proceedings of International Seminary of Tendencies in Simplified Treatment of Domestic and Industrial, Waste Water Brazil*, pp. 158–166.

Chobanoglous, G. 1987. *Wastewater Engineering, Treatment, Disposal and Reuse*. New Delhi, India: Tata McGraw-Hill.

Ciba, J., M. Zołotajkin, J. Kluczka, K. Loska, and J. Cebula. 2003. Comparison of methods for leaching heavy metals from composts. *Waste Management* 23(10): 897–905. doi:10.1016/S0956-053X(03)00128-4.

Cimbalo, R. N., R. L. Foster, and S. J. Wachtel. 1972. Deposited metals poison FCC catalyst. *Oil and Gas Journal* 70: 112–122.

Cui, J. and E. Forssberg. 2003. Mechanical recycling of waste electric and electronic equipment: A review. *Journal of Hazardous Materials* 99(3): 243–263. doi:10.1016/S0304-3894(03)00061-X.

Das, B., S. Prakash, P. S. R. Reddy, and V. N. Misra. 2007. An overview of utilization of slag and sludge from steel industries. *Resources, Conservation and Recycling* 50(1): 40–57. doi:10.1016/j.resconrec.2006.05.008.

Davenport, W. G. L., M. King, M. Schlesinger, and A. K. Biswas. 2002. *Extractive Metallurgy of Copper*, 4th edn. Oxford, U.K.: Elsevier Science Ltd./Pergamon Press.

Davis, G. and S. Herat. 2008. Electronic waste: The local government perspective in Queensland, Australia. *Resources, Conservation and Recycling* 52(8): 1031–1039. doi:10.1016/j.resconrec.2008.04.001.

de Souza, M., C. C. Bueno, D. C. de Oliveira, and J. A. S. Tenório. 2001. Characterization of used alkaline batteries powder and analysis of zinc recovery by acid leaching. *Journal of Power Sources* 103(1): 120–126. doi:10.1016/S0378-7753(01)00850-3.

de Souza e Silva, P. T., N. T. de Mello, M. M. M. Duarte, C. B. S. M. Montenegro, A. N. Araújo, B. de Barros Neto, and V. L. da Silva. 2006. Extraction and recovery of chromium from electroplating sludge. *Journal of Hazardous Materials* 128(1): 39–43. doi:10.1016/j.jhazmat.2005.07.026.

Deng, T. and Y. Ling. 2004. Processing of copper converter slag for metals reclamation. Part II: Mineralogical study. *Waste Management & Research* 22: 376–382.

Dhal, B., H. N. Thatoi, N. N. Das, and B. D. Pandey. 2013. Chemical and microbial remediation of hexavalent chromium from contaminated soil and mining/metallurgical solid waste: A review. *Journal of Hazardous Materials* 250: 272–291. doi:10.1016/j.jhazmat.2013.01.048.

Dhar, B. B., S. Ratan, and A. Jamal. 1986. Impact of opencast coal mining on water environment: A case study. *Journal of Mines, Metals and Fuels* 34: 596–601.

Dong, B., X. Liu, L. Dai, and X. Dai. 2013. Changes of heavy metal speciation during high-solid anaerobic digestion of sewage sludge. *Bioresource Technology* 131: 152–158. doi:10.1016/j.biortech.2012.12.112.

Dufresne, P. 2007. Hydroprocessing catalysts regeneration and recycling. *Applied Catalysis A: General* 322: 67–75. doi:10.1016/j.apcata.2007.01.013.

Dunmire, C., C. Calwell, A. Jacob, M. Ton, T. Reeder, and V. Fulbright. 2003. Mercury in fluorescent lamps: Environmental consequences and policy implications for NRDC, Final report prepared for Natural Resources Defense Council. Durango, CO: Ecos Consulting.

Eckelman, M. J., P. T. Anastas, and J. B. Zimmerman. 2008. Spatial assessment of net mercury emissions from the use of fluorescent bulbs. *Environmental Science & Technology* 42(22): 8564–8570. doi:10.1021/es800117h.

Eighmy, T. T., J. D. Eusden, J. E. Krzanowski, D. S. Domingo, D. Staempfli, J. R. Martin, and P. M. Erickson. 1995. Comprehensive approach toward understanding element speciation and leaching behavior in municipal solid waste incineration electrostatic precipitator ash. *Environmental Science & Technology* 29(3): 629–646.

Ekey, D. C. 1958. *Introduction to Foundry Technology*. New York: McGraw-Hill.

Environmental Protection Agency (EPA). 1981. Summary of factors affecting compliance by ferrous foundries, Vol. 1, EPA-340/1-80-020, pp. 34/49.

E-Waste Fact Sheet. 2009. Clean up Australia. https://www.cleanup.org.au/PDF/au/clean-up-australia---e-waste-factsheet-final.pdf (accessed August 3, 2016).

Federal Highway Administration (FHWA). 1997. User guidelines for waste and byproduct materials in pavement construction: FHWA-RD-97-148, US Departement of Transportation, McLean, VA. http://www.fhwa.dot.gov/publications/research/infrastructure/structures/97148/ (accessed February 2012).

Ferella, F., A. Ognyanova, I. De Michelis, G. Taglieri, and F. Vegliò. 2011. Extraction of metals from spent hydrotreating catalysts: Physico-mechanical pre-treatments and leaching stage. *Journal of Hazardous Materials* 192(1): 176–185. doi:10.1016/j.jhazmat.2011.05.005.

Fernando, Q. 1995. Metal speciation in environmental and biological systems. *Environmental Health Perspectives* 103: 13–16.

Ferreira, D. A., L. M. Z. Prados, D. Majuste, and M. B. Mansur. 2009. Hydrometallurgical separation of aluminium, cobalt, copper and lithium from spent Li-ion batteries. *Journal of Power Sources* 187(1): 238–246. doi:10.1016/j.jpowsour.2008.10.077.

Filgueiras, A. V., I. Lavilla, and C. Bendicho. 2002. Chemical sequential extraction for metal partitioning in environmental solid samples. *Journal of Environmental Monitoring* 4: 823–857.

Fimiani, S. 2009. Solving the LED-driver challenge for light-bulb replacement. *EDN* 54(7): 33–35.

Frenay, J. and S. Feron. 1990. Domestic battery recycling in western Europe. In *Proceedings of the Second International Symposium on Recycling of Metals and Engineered Materials*, Vol. 2, pp. 639–647. The Minerals, Metals and Materials Society.

Fthenakis, V. 2009. Sustainability of photovoltaics: The case for thin-film solar cells. *Renewable and Sustainable Energy Reviews* 13(9): 2746–2750. doi:10.1016/j.rser.2009.05.001.

Fthenakis, V. and A. Anctil. 2013. Direct Te mining: Resource availability andimpact on cumulative energy demand of CdTe PV life cycles. *IEEE Journal of Photovoltaics* 3(1): 433–438.

Fthenakis, V., W. Wang, and H. C. Kim. 2009. Life cycle inventory analysis of the production of metals used in photovoltaics. *Renewable and Sustainable Energy Reviews* 13(3): 493–517. doi:10.1016/j.rser.2007.11.012.

Funari, V., R. Braga, S. N. H. Bokhari, E. Dinelli, and T. Meisel. 2015. Solid residues from italian municipal solid waste incinerators: A source for 'critical' raw materials. *Waste Management* 45: 206–216. doi:10.1016/j.wasman.2014.11.005.

Funari, V., T. Meisel, and R. Braga. 2016. The potential impact of municipal solid waste incinerators ashes on the anthropogenic osmium budget. *Science of the Total Environment* 541: 1549–1555. doi:10.1016/j.scitotenv.2015.10.014.

Furimsky, E. 1996. Spent refinery catalysts: Environment, safety and utilization. *Catalysis Today* 30(4): 223–286. doi:10.1016/0920-5861(96)00094-6.

Gallardo, A. and N. Rey-Raap. 2012. Determination of mercury distribution inside spent compact fluorescent lamps by atomic absorption spectrometry. *Waste Management* 32: 944–948.

García-Delgado, M., M. S. Rodríguez-Cruz, L. F. Lorenzo, M. Arienzo, and M. J. Sánchez-Martín. 2007. Seasonal and time variability of heavy metal content and of its chemical forms in sewage sludges from different wastewater treatment plants. *Science of the Total Environment* 382(1): 82–92. doi:10.1016/j.scitotenv.2007.04.009.

Gieré, R., L. E. Carleton, and G. R. Lumpkin. 2003. Micro and nanochemistry of fly ash from a coal-fired power plant. *American Mineralogist* 88: 1853–1865.

Goetzberger, A., C. Hebling, and H.-W. Schock. 2003. Photovoltaic materials, history, status and outlook. *Materials Science & Engineering R: Reports* 40(1): 1–46. doi:10.1016/S0927-796X(02)00092-X.

Goldrig, D. C. and L. M. Jukes. 1997. Petrology and stability of steel slag. *Iron Making and Steel Making* 24(6): 447–456.

Gonzalez-fernandez, O., M. Hidalgo, E. Margui, M. L. Carvalho, and I. Queralt. 2008. Heavy metals' content of automotive shredder residues (ASR): Evaluation of environmental risk *Environmental Pollution* 153(January 1980): 476–482. doi:10.1016/j.envpol.2007.08.002.

Goodarzi, F. and H. Sanei. 2009. Plerosphere and its role in reduction of emitted fine fly ash particles from pulverized coal-fired power plants. *Fuel* 88(2): 382–386. doi:10.1016/j.fuel.2008.08.015.

Gorai, B., R. K. Jana, and Premchand. 2003. Characteristics and utilisation of copper slag—A review. *Resources, Conservation and Recycling* 39(4): 299–313. doi:10.1016/S0921-3449(02)00171-4.

Grandell, L. and M. Höök. 2015. Assessing rare metal availability challenges for solar energy technologies. *Sustainability (Switzerland)* 7(9): 11818–11837. doi:10.3390/su70911818.

Gratz, E., Q. Sa, D. Apelian, and Y. Wang. 2014. A closed loop process for recycling spent lithium ion batteries. *Journal of Power Sources* 262: 255–262. doi:10.1016/j.jpowsour.2014.03.126.

Green, M. A. 2007. Thin-film solar cells: Review of materials, technologies and commercial status. *Journal of Materials Science: Materials in Electronics* 18(Suppl. 1): S15–S19.

Gschneidner, K. A., B. J. Beaudry, and J. Capellen. 1990. *Properties and Selection: Nonferrous Alloys and Special-Purpose Materials*, Vol. 2, ASM Handbook. ASM International ISBN 978-0-87170-378-1.

Guidelines for Environmentally Sound Management of E-Waste. 2008. As approved vide MoEF letter No. 23-23/2007-HSMD dt. March 12, 2008. http://www.moef.nic.in/divisions/hsmd/guidelines-e-waste.pdf.

Guney, Y., A. H. Aydilek, and M. Melih Demirkan. 2006. Geoenvironmental behavior of foundry sand amended mixtures for highway subbases. *Waste Management* 26(9): 932–945. doi:10.1016/j.wasman.2005.06.007.

Guo, J., J. Guo, and Z. Xu. 2009. Recycling of non-metallic fractions from waste printed circuit boards: A review. *Journal of Hazardous Materials* 168(2): 567–590. doi:10.1016/j.jhazmat.2009.02.104.

Guo, Z. H., Y. Cheng, L. Y. Chai, and J. Song. 2007. Mineralogical characteristics and environmental availability of non-ferrous slag. *Journal of Central South University (Science and Technology)* 38(6): 1100–1105 (in Chinese).

Hagelüken, C. 2006. Improving metal returns and eco-efficiency in electronics recycling. In *Proceedings of the 2006 IEEE International Symposium on Electronics and the Environment*, May 8–11, 2006, Scottsdale, AZ, pp. 218–223.

Hall, G. E. M., G. Gauthier, J. C. Pelchat, P. Pelchat, and J. E. Vaive. 1996. Application of a sequential extraction scheme to 10 geological certified reference materials for the determination of 20 elements. *Journal of Analytical Atomic Spectrometry* 11: 787–796. doi: 10.1039/ja9961100787.

Hall, W. J. and P. T. Williams. 2007. Separation and recovery of materials from scrap printed circuit boards. *Resources, Conservation and Recycling* 51(3): 691–709. doi:10.1016/j.resconrec.2006.11.010.

Halstead, W. J. 1986. Use of fly ash in concrete. National Highway Research Program, Synthesis of Highway Practice #127. Washington, DC: Transportation Research Board.

Harris, T. 2001. How fluorescent lamps work. HowStuffWorks.com. http://home.howstuffworks.com/fluorescent-lamp.htm (accessed on January 9, 2015).

Hartinger, L. 1991. Handbuch der Abwasser und Recyclingtechnik.—1–144, München, Wien (Carl Hanser).

He, M. M., W. H. Li, X. Q. Liang, D. L. Wu, and G. M. Tian. 2009a. Effect of composting process on phytotoxicity and speciation of copper, zinc andlead in sewage sludge and swine manure. *Waste Management* 29: 590–597.

He, M.-M., G.-M. Tian, and X.-Q. Liang. 2009b. Phytotoxicity and speciation of copper, zinc and lead during the aerobic composting of sewage sludge. *Journal of Hazardous Materials* 163(2): 671–677. doi:10.1016/j.jhazmat.2008.07.013.

Hildenbrand, V. D., C. J. M. Denissen, A. J. H. P. van der Pol et al. 2003. Reduction of mercury loss in fuorescent lamps coated with thin metal-oxide films. *Journal of the Electrochemical Society* 150(7): H147–H155.

Hong, K. J., S. Tokunaga, Y. Ishigami, and T. Kajiuchi. 2000. Extraction of heavy metals from MSW incinerator fly ash using saponins. *Chemosphere* 41(3): 345–352. doi:10.1016/S0045-6535(99)00489-0.

House, B. W., D. W. Capson, and D. C. Schuurman. 2011. Towards real-time sorting of recyclable goods using support vector machines. In *Sustainable Systems and Technology, IEEE International Symposium*, Chicago, IL, pp. 1–6.

Hsu, J.-H. and S.-L. Lo. 2001. Effect of composting on characterization and leaching of copper, manganese, and zinc from swine manure. *Environmental Pollution* 114(1): 119–127. doi:10.1016/S0269-7491(00)00198-6.

Huang, H.-L. 2001. Thermal treatment of Cr-Zn containing plating sludge. Taichung, Taiwan: Department of Environmental Science and Engineering, Tunghai University.

Huang, R., K.-L. Huang, Z.-Y. Lin, J.-W. Wang, C. Lin, and Y.-M. Kuo. 2013. Recovery of valuable metals from electroplating sludge with reducing additives via vitrification. *Journal of Environmental Management* 129: 586–592. doi:10.1016/j.jenvman.2013.08.019.

Huang, R. and Y. Ikuhara. 2012. STEM characterization for lithium-ion battery cathode materials. *Current Opinion in Solid State and Materials Science* 16(1): 31–38. doi:10.1016/j.cossms.2011.08.002.

Huber, J.-C., Patisson, F., Rocabois, P., Birat, J.-C., and Ablitzer, D. 1999. Some means to reduce emissions and improve the recovery of electric arc furnace dust by controlling the formation mechanisms. In *Proceedings of the Rewas'99—Global Symposium on Recycling, Waste Treatment and Clean Technology*, Vol. II, pp. 1483–1492. San Sebastian, Spain.

Ijadi Bajestani, M., S. M. Mousavi, and S. A. Shojaosadati. 2014. Bioleaching of heavy metals from spent household batteries using *Acidithiobacillus ferrooxidans*: Statistical evaluation and optimization. *Separation and Purification Technology* 132: 309–316. doi:10.1016/j.seppur.2014.05.023.

Jain, D. K. and R. D. Tyagi. 1992. Leaching of heavy metals from anaerobic sewage sludge by sulfur-oxidizing bacteria. *Enzyme and Microbial Technology* 14(5): 376–383. doi:10.1016/0141-0229(92)90006-A.

Jang, M., S. M. Hong, and J. K. Park. 2005. Characterization and recovery of mercury from spent fluorescent lamps. *Waste Management* 25(1): 5–14. doi:10.1016/j.wasman.2004.09.008.

Jena, P. K., C. S. K. Mishra, D. K. Behera, S. Mishra, and L. B. Sukla. 2012. Does participatory forest management change household attitudes towards forest conservation and management? *African Journal of Environmental Science and Technology* 6(5): 208–213. doi:10.5897/AJEST11.362.

Jha, M. K., V. Kumar, and R. J. Singh. 2001. Review of hydrometallurgical recovery of zinc from industrial wastes. *Resources, Conservation and Recycling* 33(1): 1–22. doi:10.1016/S0921-3449(00)00095-1.

Jossen, A., J. Garche, and D. U. Sauer. 2004. Operation conditions of batteries in PV applications. *Solar Energy* 76(6): 759–769. doi:10.1016/j.solener.2003.12.013.

Joung, H.-T., Y.-C. Seo, and K. -H. Kim. 2007. Distribution of dioxins, furans, and dioxin-like PCBs in solid products generated by pyrolysis and melting of automobile shredder residues. *Chemosphere* 68(9): 1636–1641. doi:10.1016/j.chemosphere.2007.04.003.

Jung, B., J. Park, D. Seo, and N. Park. 2016. Sustainable system for raw-metal recovery from crystalline silicon solar panels: From noble-metal extraction to lead removal. *ACS Sustainable Chemistry & Engineering* 4(8): 4079–4083. doi:10.1021/acssuschemeng.6b00894.

Kahhat, R., J. Kim, M. Xu, B. Allenby, E. Williams, and P. Zhang. 2008. Exploring e-waste management systems in the United States. *Resources, Conservation and Recycling* 52(7): 955–964. doi:10.1016/j.resconrec.2008.03.002.

Kang, S., S. Yoo, J. Lee, B. Boo, and H. Ryu. 2012. Experimental investigations for recycling of silicon and glass from waste photovoltaic modules. *Renewable Energy* 47: 152–159. doi:10.1016/j.renene.2012.04.030.

Karvelas, M., A. Katsoyiannis, and C. Samara. 2003. Occurrence and fate of heavy metals in the wastewater treatment process. *Chemosphere* 53(10): 1201–1210. doi:10.1016/S0045-6535(03)00591-5.

Kasikov, L. G. and A. M. Petrova. 2009. Processing of deactivated platinum-rhenium catalysts. *Theoretical Foundations of Chemical Engineering* 43(4): 544–552.

Kasper, A. C., G. D. T. Derselli, D. D. Freitas, J. A. S. Tenório, A. M. Bernardes, and H. M. Veit. 2011. Printed wiring boards for mobile phones: Characterization and recycling of copper. *Waste Management* 31(12): 2536–2545. doi:10.1016/j.wasman.2011.08.013.

Kierkegaard, S. 2007. Charging up the batteries: Squeezing more capacity and power into the new EU battery directive. *Computer Law & Security Review* 23(4): 357–364. doi:10.1016/j.clsr.2007.05.001.

Kim, H.-I., K.-H. Park, and D. Mishra. 2009. Sulfuric acid baking and leaching of spent Co-Mo/Al_2O_3 catalyst. *Journal of Hazardous Materials* 166. doi:10.1016/j.jhazmat.2008.11.051.

Kim, J. K. and E. F. Schubert. 2008. Transcending the replacement paradigm of a solid-state lighting. *Optics Express* 16(26): 21835.

Kirby, C. S. and J. D. Rimstidt. 1993. Mineralogy and surface properties of municipal solid waste ash. *Environmental Science & Technology* 27(4): 652–660. doi:10.1021/es00041a008, American Chemical Society.

Kirchner, G. 1991. Aluminium recycling in Germany. *Aluminium Today* 21.

Klugmann-Radziemska, E. 2012. Current trends in recycling of photovoltaic solar cells and modules waste. *Chemistry Didactics Ecology Metrology* 17(1–2): 89–95. doi:10.2478/cdem-2013-0008.

Kulveitová, H. 1999. *Chemická speciace zinku, kadmia a olova a jejich vyluhování z tuhých metalurgických emisí.* Habilitation. Ostrava, Czech Republic: VŠB-Technical University of Ostrava (in Czech).

Kurose, K., T. Okuda, W. Nishijima, and M. Okada. 2006. Heavy metals removal from automobile shredder residues (ASR). *Journal of Hazardous Materials* 137(3): 1618–1623. doi:10.1016/j.jhazmat.2006.04.049.

LaDou, J. 2006. Printed circuit board industry. *International Journal of Hygiene and Environmental Health* 209(3): 211–219. doi:10.1016/j.ijheh.2006.02.001.

Laforest, G. and J. Duchesne. 2006. Characterization and leachability of electric arc furnace dust made from remelting of stainless steel. *Journal of Hazardous Materials* 135(1): 156–164. doi:10.1016/j.jhazmat.2005.11.037.

Lampke, T., B. Wielage, D. Dietrich, and A. Leopold. 2006. Details of crystalline growth in co-deposited electroplated nickel films with hard (nano)particles. *Applied Surface Science* 253(5): 2399–2408. doi:10.1016/j.apsusc.2006.04.060.

Lampman, S. 1991. *Tuning Up the Metals in Auto Engines*, p. 17. Advanced Materials & Processes.

Leclerc, N., E. Meux, and J.-M. Lecuire. 2002. Hydrometallurgical recovery of zinc and lead from electric arc furnace dust using mononitrilotriacetate anion and hexahydrated ferric chloride. *Journal of Hazardous Materials* 91(1): 257–270. doi:10.1016/S0304-3894(01)00394-6.

Leclerc, N., E. Meux, and J.-M. Lecuire. 2003. Hydrometallurgical extraction of zinc from zinc ferrites. *Hydrometallurgy* 70(1): 175–183. doi:10.1016/S0304-386X(03)00079-3.

Lee, C. K. and K.-I. Rhee. 2002. Preparation of $LiCoO_2$ from spent lithium-ion batteries. *Journal of Power Sources* 109(1): 17–21. doi:10.1016/S0378-7753(02)00037-X.

Lee, T., J.-W. Park, and J.-H. Lee. 2004. Waste green sands as reactive media for the removal of zinc from water. *Chemosphere* 56(6): 571–581. doi:10.1016/j.chemosphere.2004.04.037.

Lepawsky, J. and C. Mcnabb. 2010. Mapping international flows of electronic waste. *The Canadian Geographer/Le Géographe Canadien* 54(2): 177–195. doi:10.1111/j.1541-0064.2009.00279.x.

Lester, J. N. 1983. Significance and behaviour of heavy metals in waste water treatment processes I. Sewage treatment and effluent discharge. *Science of the Total Environment* 30(September): 1–44. doi:10.1016/0048-9697(83)90002-5.

Lester, J. N., R. M. Sterritt, and P. W. W. Kirk. 1983. Significance and behaviour of heavy metals in waste water treatment processes II. Sludge treatment and disposal. *Science of the Total Environment* 30(September): 45–83. doi:10.1016/0048-9697(83)90003-7.

Leung, A. O. W., W. J. Luksemburg, A. S. Wong, and M. H. Wong. 2007. Spatial distribution of polybrominated diphenyl ethers and polychlorinated dibenzo-P-dioxins and dibenzofurans in soil and combusted residue at guiyu, an electronic waste recycling site in Southeast China. *Environmental Science & Technology* 41(8): 2730–2737. doi:10.1021/es0625935.

Li, C.-L. and M.-S. Tsai. 1993a. Mechanism of spinel ferrite dust formation in electric arc furnace steelmaking. *ISIJ International* 33(2): 284–290. doi:10.2355/isijinternational.33.284.

Li, C.-L. and M.-S. Tsai. 1993b. A crystal phase study of zinc hydroxide chloride in electric arc-furnace dust. *Journal of Materials Science* 28: 4562–4570.

Li, J., H. Lu, J. Guo, Z. Xu, and Y. Zhou. 2007. Recycle technology for recovering resources and products from waste printed circuit boards. *Environmental Science & Technology* 41(6): 1995–2000. doi:10.1021/es0618245.

Li, J., P. Shrivastava, Z. Gao, and H. C. Zhang. 2004. Printed circuit board recycling: A stateof-the-art survey. *IEEE Transactions on Electronics Packaging Manufacturing* 27: 1.

Li, L., J. B. Dunn, X. X. Zhang, L. Gaines, R. J. Chen, F. Wu, and K. Amine. 2013. Recovery of metals from spent lithium-ion batteries with organic acids as leaching reagents and environmental assessment. *Journal of Power Sources* 233: 180–189. doi:10.1016/j.jpowsour.2012.12.089.

Lim, S. R., D. Kang, O. A. Ogunseitan, and J. M. Schoenung. 2011. Potential environmental impacts of light-emitting diodes (LEDs): Metallic resources, toxicity, and hazardous waste classification. *Environmental Science and Technology* 45(1): 320–327. doi:10.1021/es101052q.

Lim, S. R., D. Kang, O. A. Ogunseitan, and J. M. Schoenung. 2013. Potential environmental impacts from the metals in incandescent, compact fluorescent lamp (CFL), and light-emitting diode (LED) bulbs. *Environmental Science and Technology* 47(2): 1040–1047. doi:10.1021/es302886m.

Lin, K. L., C. J. Hsu, I. M. Hsu, and J. T. Chang. 1992. Electroplating of Ni-Cr on steel with pulse plating. *Journal of Materials Engineering and Performance* 1: 359–361.

Liu, C., Y. Yu, and H. Zhao. 2005. Hydrodenitrogenation of quinoline over Ni-Mo/Al$_2$O$_3$ catalyst modified with fluorine and phosphorus. *Fuel Processing Technology* 86(4): 449–460. http://cat.inist.fr/?aModele=afficheN&cpsidt=16429628#.WJ27NNCIWCU. mendeley (accessed February 10).

Lohtia, R. P. and R. C. Joshi. 1995. Mineral admixtures. In *Concrete Admixtures Handbook: Properties Science and Technology*, V. S. Ramachandran (Ed.), pp. 657–739. New York: William Andrew.

Lopes, M. H., M. Freire, M. Galhetas, I. Gulyurtlu, and I. Cabrita. 2009. Leachability of automotive shredder residues burned in a fluidized bed system. *Waste Management* 29(5): 1760–1765. doi:10.1016/j.wasman.2008.11.005.

López, F. A., E. Sáinz, A. Formoso, and I. Alfaro. 1994. The recovery of alumina from salt slags in aluminium remelting. *Canadian Metallurgical Quarterly* 33(1): 29–33. doi:10.1179/cmq.1994.33.1.29.

Ludwig, C., S. Hellweg, and S. Stucki. 2003. *Municipal Solid Waste Management: Strategies and Technologies for Sustainable Solutions*, pp. 320–322. Berlin, Germany: Springer.

Lundqvist, U., B. Andersson, M. Axsäter et al. 2004. *Design for Recycling in the Transportation Sector—Future Scenarios and Challenges*, p. 7. Göteborg, Sweden: Department of Physical Resource Theory, Chalmers University of Technology, Göteborg University.

Ma, H. and J. Suhling. 2009. A review of mechanical properties of lead-free solders for electronic packaging. *Journal of Materials Science* 44: 1141–1158.

Ma, L. Q. and G. N. Rao. 1997. Chemical fractionation of cadmium, copper, nickel, and zinc in contaminated soils. *Journal of Environmental Quality (USA)* 26(1): 259–264.

Macaskie, L. E., I. P. Mikheenko, P. Yong et al. 2010. Hydrometallurgy today's wastes, tomorrow's materials for environmental protection. *Hydrometallurgy* 104(3–4): 483–487. doi:10.1016/j.hydromet.2010.01.018.

Mantuano, D. P., G. Dorella, R. C. A. Elias, and M. B. Mansur. 2006. Analysis of a hydrometallurgical route to recover base metals from spent rechargeable batteries by liquid–liquid extraction with Cyanex 272. *Journal of Power Sources* 159(2): 1510–1518. doi:10.1016/j.jpowsour.2005.12.056.

Marafi, M. and A. Stanislaus. 2003. Options and processes for spent catalyst handling and utilization. *Journal of Hazardous Materials* 101(2): 123–132. doi:10.1016/S0304-3894(03)00145-6.

Marafi, M. and A. Stanislaus. 2008a. Spent catalyst waste management: A review: Part I—Developments in hydroprocessing catalyst waste reduction and use. *Resources, Conservation and Recycling* 52(6): 859–873. doi:10.1016/j.resconrec.2008.02.004.

Marafi, M. and A. Stanislaus. 2008b. Spent hydroprocessing catalyst management: A review: Part II. Advances in metal recovery and safe disposal methods. *Resources, Conservation and Recycling* 53(1): 1–26. doi:10.1016/j.resconrec.2008.08.005.

Martins, F. M., J. M. dos Reis Neto, and C. J. da Cunha. 2008. Mineral phases of weathered and recent electric arc furnace dust. *Journal of Hazardous Materials* 154(1): 417–425. doi:10.1016/j.jhazmat.2007.10.041.

Marwede, M., W. Berger, M. Schlummer, A. Mäurer, and A. Reller. 2013. Recycling paths for thin-film chalcogenide photovoltaic waste—Current feasible processes. *Renewable Energy* 55: 220–229. doi:10.1016/j.renene.2012.12.038.

Mathias, B. 1994. *Basiswissen Umwelttechnik [Basic Knowledge of Environmental Technology]*, 2nd edn. Würzburg, Germany: Vogel Buchverlag.

Matthews, D. H., H. S. Matthews, P. Jaramillo, and C. L. Weber. 2009. Energy consumption in the production of high-brightness light emitting diodes. *IEEE International Symposium on Sustainable Systems and Technology (ISSST)*, Tempe, AZ, Vol. 6.

McDonald, N. C. and J. M. Pearce. 2010. Producer responsibility and recycling solar photovoltaic modules. *Energy Policy* 38(11): 7041–7047. doi:10.1016/j.enpol.2010.07.023.

Meawad, A. S., D. Y. Bojinova, and Y. G. Pelovski. 2010. An overview of metals recovery from thermal power plant solid wastes. *Waste Management* 30(12): 2548–2559. doi:10.1016/j.wasman.2010.07.010.

Mehta, S. 2009. *Global PV Cell and Module Production Analysis*. GTM Research, Greentech Media Inc, Massachusetts.

Metcalf and Eddy. 1995. *Wastewater Engineering: Treatment, Disposal and Reuse*, 3rd edn. New York: McGraw-Hill Publishing Company Ltd.

Miles, R. W., G. Zoppi, and I. Forbes. 2007. Inorganic photovoltaic cells. *Materials Today* 10(11): 20–27. doi:10.1016/S1369-7021(07)70275-4.

Mishra, D., D. J. Kim, D. E. Ralph, J. G. Ahn, and Y. H. Rhee. 2007. Bioleaching of vanadium rich spent refinery catalysts using sulfur oxidizing lithotrophs. *Hydrometallurgy* 88(1–4): 202–209. doi:10.1016/j.hydromet.2007.05.007.

Mishra, D., D. J. Kim, D. E. Ralph, J. G. Ahn, and Y. H. Rhee. 2008. Bioleaching of spent hydro-processing catalyst using acidophilic bacteria and its kinetics aspect. *Journal of Hazardous Materials* 152(3): 1082–1091. doi:10.1016/j.jhazmat.2007.07.083.

Morf, L. S., R. Gloor, O. Haag, M. Haupt, S. Skutan, F. Di Lorenzo, and D. Böni. 2013. Precious metals and rare earth elements in municipal solid waste—Sources and fate in a swiss incineration plant. *Waste Management* 33(3): 634–644. doi:10.1016/j.wasman.2012.09.010.

Mottier, P. 2009. *LEDs for Lighting Applications*. London, U.K.: ISTE Ltd./John Wiley & Sons, Inc.

Mudrack, K. and S. Kunst 1988. *Biologie der Abwasserreinigung (Biology of Wastewater Treatment)*. Gustav-Fischer.

Mukherjee, R., R. Krishnan, T.-M. Lu, and N. Koratkar. 2012. Nanostructured electrodes for high-power lithium ion batteries. *Nano Energy* 1(4): 518–533. doi:10.1016/j.nanoen.2012.04.001.

Nagib, S. and K. Inoue. 2000. Recovery of lead and zinc from fly ash generated from municipal incineration plants by means of acid and/or alkaline leaching. *Hydrometallurgy* 56(3): 269–292. doi:10.1016/S0304-386X(00)00073-6.

Nardelli, A., E. Deuschle, L. D. de Azevedo, J. L. N. Pessoa, and E. Ghisi. 2016. Assessment of light emitting diodes technology for general lighting: A critical review. *Renewable and Sustainable Energy Reviews*. doi:10.1016/j.rser.2016.11.002.

National Household Hazardous Waste Forum (NHHWF). 2000. *Issues Surrounding the Collection and Identification of Post-consumer Batteries*. Leeds, England: NHHWF.

Navarro, R., J. Guzman, I. Saucedo, J. Revilla, and E. Guibal. 2007. Vanadium recovery from oil fly ash by leaching, precipitation and solvent extraction processes. *Waste Management* 27(3): 425–438. doi:10.1016/j.wasman.2006.02.002.

Nawrocky, M., D. C. Schuurman, and J. Fortuna. 2010. Visual sorting of recyclable goods using a support vector machine. *Electrical and Computer Engineering, Canada: IEEE* 23(10): 1–4.

Nishi, Y. 2001. Lithium ion secondary batteries; past 10 years and the future. *Journal of Power Sources* 100(1): 101–106. doi:10.1016/S0378-7753(01)00887-4.

Niu, Z., Y. Zou, B. Xin, S. Chen, C. Liu, and Y. Li. 2014. Process controls for improving bioleaching performance of both Li and Co from spent lithium ion batteries at high pulp density and its thermodynamics and kinetics exploration. *Chemosphere* 109: 92–98. doi:10.1016/j.chemosphere.2014.02.059.

Nogueira, C. A. and F. Margarido. 2007. Chemical and physical characterization of electrode materials of spent sealed Ni–Cd batteries. *Waste Management* 27(11): 1570–1579. doi:10.1016/j.wasman.2006.10.007.

Nomeda, S., P. Valdas, S.-Y. Chen, and J.-G. Lin. 2008. Variations of metal distribution in sewage sludge composting. *Waste Management* 28(9): 1637–1644. doi:10.1016/j.wasman.2007.06.022.

Norgate, T. and N. Haque. 2012. Using life cycle assessment to evaluate some environmental impacts of gold production. *Journal of Cleaner Production* 29: 53–63. doi:10.1016/j.jclepro.2012.01.042.

Ogunseitan, O. A. 2013. The basel convention and e-waste: Translation of scientific uncertainty to protective policy. *The Lancet* 1(6): e313–e314. doi:10.1016/S2214-109X(13)70110-4.

Ogunseitan, O. A., J. M. Schoenung, J. D. M. Saphores, and A. A. Shapiro. 2009. The electronics revolution: From e-wonderland to ewasteland. *Science* 326(5953): 670–671.

Orhan, G. 2005. Leaching and cementation of heavy metals from electric arc furnace dust in alkaline medium. *Hydrometallurgy* 78(3): 236–245. doi:10.1016/j.hydromet.2005.03.002.

Oustadakis, P., P. E. Tsakiridis, A. Katsiapi, and S. Agatzini-Leonardou. 2010. Hydrometallurgical process for zinc recovery from electric arc furnace dust (EAFD): Part I: Characterization and leaching by diluted sulphuric acid. *Journal of Hazardous Materials* 179(1): 1–7. doi:10.1016/j.jhazmat.2010.01.059.

Paiano, A. 2015. Photovoltaic waste assessment in Italy. *Renewable and Sustainable Energy Reviews* 41: 99–112. doi:10.1016/j.rser.2014.07.208.

Pan, J., J. Wang, and D. Shaddock. 2005. Lead-free solder joint reliability-state of the art and perspectives. *Journal of Microelectronics and Electronic Packaging* 2: 72–83.

Panda, S., S. Mishra, D. S. Rao, N. Pradhan, U. Mohapatra, S. Angadi, and B. K. Mishra. 2015. Extraction of copper from copper slag: Mineralogical insights, physical beneficiation and bioleaching studies. *Korean Journal of Chemical Engineering* 32(4): 667–676. doi:10.1007/s11814-014-0298-6.

Pant, D., D. Joshi, M. K. Upreti, and R. K. Kotnala. 2012. Chemical and biological extraction of metals present in e waste: A hybrid technology. *Waste Management* 32(5): 979–990. doi:10.1016/j.wasman.2011.12.002, Elsevier Ltd.

Park, K. H., B. Ramachandra Reddy, D. Mohapatra, and C.-W. Nam. 2006. Hydrometallurgical processing and recovery of molybdenum trioxide from spent catalyst. *International Journal of Mineral Processing* 80. doi:10.1016/j.minpro.2006.05.002.

Park, Y. J. and D. J. Fray. 2009. Recovery of high purity precious metals from printed circuit boards. *Journal of Hazardous Materials* 164(2): 1152–1158. doi:10.1016/j.jhazmat.2008.09.043.

Pathak, A., M. G. Dastidar, and T. R. Sreekrishnan. 2009. Bioleaching of heavy metals from sewage sludge: A review. *Journal of Environmental Management* 90(8): 2343–2353. doi:10.1016/j.jenvman.2008.11.005.

Patniak, P. 2003. *Handbook of Inorganic Chemistry*. New York: MacGraw-Hill.

Peng, B. and J. Chen. 2009. Functional materials with high-efficiency energy storage and conversion for batteries and fuel cells. *Coordination Chemistry Reviews* 253(23): 2805–2813. doi:10.1016/j.ccr.2009.04.008.

Pereira, C.F., Y. L. Galiano, M. A. Rodríguez-Piñero, and J. Vale Parapar. 2007. Long and short-term performance of a stabilized/solidified electric arc furnace dust. *Journal of Hazardous Materials* 148(3): 701–707. doi:10.1016/j.jhazmat.2007.03.034.

Perrine, C., R. Jérôme, D. Jérémie, and B. Jean-Yves. 2006. Speciation of Cr and V within BOF steel slag reused in road constructions. *Journal of Geochemical Exploration* 88: 10–14.

Petter, P., H. M. Veit, and A. M. Bernardes. 2014. Evaluation of gold and silver leaching from printed circuit board of cellphones. *Waste Management* 34(2): 475–482. doi:10.1016/j.wasman.2013.10.032.

Piatak, N. M., M. B. Parsons, and R. R. Seal. 2015. Characteristics and environmental aspects of slag: A review. *Applied Geochemistry* 57: 236–266. doi:10.1016/j.apgeochem.2014.04.009.

Polák, M. and L. Drápalová. 2012. Estimation of end of life mobile phones generation: The case study of the Czech Republic. *Waste Management* 32(8): 1583–1591. doi:10.1016/j.wasman.2012.03.028.

Polyák, K., I. Bodog, and J. Hlavay. 1995. Speciation of metalions in solid samples. 2. Investigation on fly ashes. *Magyar Kémiai Folyóirat* 101: 24–30.

Poulikidou, S. 2010. Identification of the main environmental challenges in a sustainability perspective for the automobile industry. Thesis in the Master Programme, Industrial Ecology, Department of Energy and Environment Division of Physical Resource Theory, Chalmers University of Technology, Göteborg, Sweden.

Pradhan, D., D.-J. Kim, J.-G. Ahn, G. R. Chaudhury, and S.-W. Lee. 2010. Kinetics and statistical behavior of metals dissolution from spent petroleum catalyst using acidophilic iron oxidizing bacteria. *Journal of Industrial and Engineering Chemistry* 16(5): 866–871. doi:10.1016/j.jiec.2010.03.006.

Proctor, D. M., K. A. Fehling, E. C. Shay et al. 2000. Physical and components of incineration slag. *Aufbereitungs-Technik/Mineral Processing* 37(4): 149–157.

PRNewswire. 2015. Waste management in the automotive industry 2015–2019. http://www.prnewswire.com/news-releases/waste-management-in-the-automotive-industry-2015-2019-300173788.html.

Putois, F. 1995. Market for nickel-cadmium batteries. *Journal of Power Soures* 57: 67–70.

Querol, X., J. Fernández-Turiel, and A. López-Soler. 1995. Trace elements in coal and their behaviour during combustion in a large power station. *Fuel* 74(3): 331–343. doi:10.1016/0016-2361(95)93464-O.

Querol, X., R. Juan, A. Lopez-Soler, J. L. Fernandez-Turiel, and C. R. Ruiz. 1996. Mobility of trace elements from coal and combustion wastes. *Fuel* 75(7): 821–838. doi:10.1016/0016-2361(96)00027-0.

Raclavská, H., D. Matýsek, and V. Zamarský. 1995. *Nové trendy v úpravnictví*, 1st edn. Ostrava, Czech Republic: VŠB Technical University of Ostrava (in Czech).

Rai, U. N., K. Pandey, S. Sinha, A. Singh, R. Saxena, and D. K. Gupta. 2004. Revegetating fly ash landfills with *Prosopis juliflora* L.: Impact of different amendments and rhizobium inoculation. *Environment International* 30(3): 293–300. doi:10.1016/S0160-4120(03)00179-X.

Rapaport, D. 2000. Spent hydroprocessing catalyst listed as hazardous wastes. *Hydrocarbon Processing* 79: 11–22.

Raposo, C., C. C. Windmöller, and W. A. D. Júnior. 2003. Mercury speciation in fluorescent lamps by thermal release analysis. *Waste Management* 23(10): 879–886. doi:10.1016/S0956-053X(03)00089-8.

Rasoulnia, P. and S. M. Mousavi. 2016. RSC advances V and Ni recovery from a vanadium-rich power plant residual ash using acid producing fungi: *Aspergillus niger* and *Penicillium simplicissimum. RSC Advances* 6: 9139–9151. doi:10.1039/C5RA24870A, Royal Society of Chemistry.

Rayovac Corp. 1999. Material safety data sheet. http://rayovac.com.

Ribarich, T. 2009. How compact fluorescent lamps work-and how to dim them? *Electronics Engineering Times* 1564: 39.

Rehman, A. and S. Lee. 2014. Review of the potential of the Ni/Cu plating technique for crystalline silicon solar cells. *Materials* 7(2): 1318–1341. doi:10.3390/ma7021318.

Rey-Raap, N. and A. Gallardo. 2012. Determination of mercury distribution inside spent compact fluorescent lamps by atomic absorption spectrometry. *Waste Management* 32(5): 944–948. doi:10.1016/j.wasman.2011.12.001.

Rocchetti, L. and F. Beolchini. 2015. Recovery of valuable materials from end-of-life thin-film photovoltaic panels: Environmental impact assessment of different management options, *Journal of Cleaner Production* 89: 59–64. doi:10.1016/j.jclepro.2014.11.009.

Ruetschi, P., F. Meli, and J. Desilvestro. 1995. Nickel-metal hydride batteries. The preferred batteries of the future? *Journal of Power Sources* 57(1–2): 85–91. doi:10.1016/0378-7753(95)02248-1.

Rydh, C. J. and B. Svärd. 2003. Impact on global metal flows arising from the use of portable rechargeable batteries. *Science of the Total Environment* 302(1): 167–184. doi:10.1016/S0048-9697(02)00293-0.

Sahu, H. B., S. Dash, and A. K. Swar. 2011. Environmental impact of coal beneficiation and its mitigation measures. *Indian Journal of Environmental Protection* 31(8): 691–698.

Salihoglu, G. and V. Pinarli. 2008. Steel foundry electric arc furnace dust management: Stabilization by using lime and portland cement. *Journal of Hazardous Materials* 153(3): 1110–1116. doi:10.1016/j.jhazmat.2007.09.066.

Salgado, A. L., A. M. O. Veloso, D. D. Pereira, G. S. Gontijo, A. Salum, and M. B. Mansur. Recovery of zinc and manganese from spent alkaline batteries by liquid-liquid extraction with Cyanex 272. 2003. *Journal of Power Sources* 115: 367–373.

Sander, K., J. Lohse, and U. Pirntke. 2000. *Heavy Metals in Vehicles*. Hamburg, Germany: Ökopol, Institute for Environmental Startegies.

Santhiya, D. and Y. P. Ting. 2006. Use of adapted *Aspergillus niger* in the bioleaching of spent refinery processing catalyst. *Journal of Biotechnology* 121(1): 62–74. doi:10.1016/j.jbiotec.2005.07.002.

Santini, A., L. Morselli, F. Passarini, I. Vassura, S. Di Carlo, and F. Bonino. 2011. End-of-life vehicles management: Italian material and energy recovery efficiency. *Waste Management* 31(3): 489–494. doi:10.1016/j.wasman.2010.09.015.

Sathaiyan, N. 2014. Reclamation of mercury from used silver oxide watch batteries. *Advances in Chemical Engineering and Science* 4(1): 1–6. doi:10.4236/aces.2014.41001.

Sathaiyan, N., V. Nandakumar, and P. Ramachandran. 2006. Hydrometallurgical recovery of silver from waste silver oxide button cells. *Journal of Power Sources* 161. doi:10.1016/j.jpowsour.2006.06.011.

Sayilgan, E., T. Kukrer, G. Civelekoglu, F. Ferella, A. Akcil, F. Veglio, and M. Kitis. 2009. A review of technologies for the recovery of metals from spent alkaline and zinc–carbon batteries. *Hydrometallurgy* 97(3): 158–166. doi:10.1016/j.hydromet.2009.02.008.

Scarr, R. F., J. C. Hunter, and P. J. Slezak. 2001. Alkaline-manganese dioxide batteries. In *Handbook of Batteries*, D. Linden (Ed.), McGraw-Hill, Chapter 10, pp. 253–284.

Schluep, M., C. Hagelüken, R. Magalini et al. 2009. *Recycling—From E-Waste to Resources*. Berlin, Germany: United Nations Environment Programme (UNEP).

Schubert, E. F. 2006. *Light Emitting Diodes*, 2nd edn. New York: Cambridge University Press.

Schubert, E. F., J. K. Kim, H. Luo, and J.-Q. Xi. 2006. Solid-state lighting—A benevolent technology. *Reports on Progress in Physics* 69(12): 3069–3099. doi:10.1088/0034-4885/69/12/R01.

Schumm, J. B. 2013. Battery electronics—Britannica online encyclopedia. http://www.britannica.com/EBchecked/topic/56126/battery/45858/Development-of-batteries (accessed December 10, 2013).

Secretariat, R. S. 2011. *E-Waste in India*. New Delhi, India: India Research Unit (Larrdis), Rajya Sabha Secretariat. http://rajyasabha.nic.in/rsnew/publication_electronic/E-Waste_in_india.pdf.

Shapek, R. A. 1995. Local government household battery collection programs: Costs and benefits. *Resources, Conservation and Recycling* 15(1): 1–19. doi:10.1016/0921-3449(95)00025-E.

Sharma, P., A. Sharma, A. Sharma, and P. Srivastava, 2016. Automobile waste and its management. *Research Journal of Chemical and Environmental Sciences* 4(2): 1–7.

Shen, H., E. Forssberg, and U. Nordström. 2004. Physicochemical and mineralogical properties of stainless steel slags oriented to metal recovery. *Resources, Conservation and Recycling* 40(3): 245–271. doi:10.1016/S0921-3449(03)00072-7.

Shin, S.-M., J.-G. Kang, D.-H. Yang, T.-H. Kim, and J.-S. Sohn. 2007. Comparison of acid and alkaline leaching for recovery of valuable metals from spent zinc-carbon battery. *Geosystem Engineering* 10(2): 21–26. doi:10.1080/12269328.2007.10541267.

Shin, S.-M., N. H. Kim, J. S. Sohn, D. H. Yang, and Y. H. Kim. 2005. Development of a metal recovery process from Li-ion battery wastes. *Hydrometallurgy* 79: 172–181.

Shin, S. M., G. Senanayake, J.-s. Sohn, J.-g. Kang, D.-h. Yang, and T.-h. Kim. 2009. Separation of zinc from spent zinc-carbon batteries by selective leaching with sodium hydroxide. *Hydrometallurgy* 96. doi:10.1016/j.hydromet.2008.12.010.

Shuman, L. M. and W. L. Hargrove. 1985. Effect of tillage on the distribution of manganese, copper, iron, and zinc in soil fractions. *Soil Scientific Society of America Journal* 49: 1117–1121. doi:10.2136/sssaj1985.03615995004900050009x.

Siddiquea, R., G. Kaur, and A. Rajor. 2010. Waste foundry sand and its leachate characteristics. *Resources, Conservation and Recycling* 54(12): 1027–1036. doi:10.1016/j.resconrec.2010.04.006.

Silvy, R. P. 2004. Future trends in the refining catalyst market. *Applied Catalysis A: General* 261(2): 247–252. doi:10.1016/j.apcata.2003.11.019.

Simic, V. and B. Dimitrijevic. 2012. Production planning for vehicle recycling factories in the EU legislative and global business environments. *Resources, Conservation and Recycling* 60: 78–88. doi:10.1016/j.resconrec.2011.11.012.

Singh, J. and B.-K. Lee. 2016. Kinetics and extraction of heavy metals resources from automobile shredder residue. *Process Safety and Environmental Protection* 99: 69–79. doi:10.1016/j.psep.2015.10.010.

Singh, J., J.-k. Yang, and Y.-y. Chang. 2016a. Synthesis of nano zero-valent metals from the leaching liquor of automobile shredder residue: A mechanism and potential applications for phenol degradation in water. *Process Safety and Environmental Protection* 102: 204–213. doi:10.1016/j.psep.2016.03.013.

Singh, J., J.-K. Yang, and Y.-Y. Chang. 2016b. Quantitative analysis and reduction of the eco-toxicity risk of heavy metals for the fine fraction of automobile shredder residue (ASR) using H_2O_2. *Waste Management* 48: 374–382. doi:10.1016/j.wasman.2015.09.030.

Sivakumar, G. D., S. Godwin Barnabas, and S. Anatharam. 2014. Indian automobile material recycling management. *International Journal of Innovative Research in Science Engineering and Technology* 3(3): 2754–2758.

Skarup, S. and C. Lubeck Christensen. 2004. Mass flow analyses of mercury 2001, Environmental Project Nr. 917 2004, Danish EPA.

Sochacki, A., J. Surmacz-Górska, O. Faure, and B. Guy. 2014. Polishing of synthetic electroplating wastewater in microcosm upflow constructed wetlands: Metals removal mechanisms. *Chemical Engineering Journal* 242(April): 43–52. doi:10.1016/j. cej.2013.12.075.

Solter, K., N. Peschen, K. Rohricht, and R. Gehrke. 1994. Simultanfallung mit Kalk im großtechnischen Betrieb-Ergebnisse, Kosten, Vorteile [Simultaneous precipitation using lime in largescale operation—Results, costs and advantages]. *AwtAbwassertechnik* 1: 19–23.

Spear, T. M., W. Svee, J. H. Vincent, and N. Stanisch. 1998. Chemical speciation of lead dust associated with primary lead smelting. *Environmental Health Perspectives* 106: 565–571.

Sposito, G., L. J. Lund, and A. C. Chang. 1982. Trace metal chemistry in arid-zone field soils amended with sewage sludge: I. Fractionation of Ni, Cu, Zn, Cd, and Pb in solid phases. *Soil Scientific Society of America Journal* 46: 260–264. doi:10.2136/sssaj1982.036159 95004600020009x.

Srichandan, H., D. J. Kim, C. S. Gahan, S. Singh, and S. W. Lee. 2013. Bench-scale batch bioleaching of spentpetroleum catalyst using mesophilic iron and sulfur oxidizing acidophiles. *Korean Journal of Chemical Engineering* 30(5): 1076–1082.

Staudinger, J., G. A. Keoleian, and M. S. Flynn. 2001. Management of end-of-life vehicles (ELVs) in the US. Centre for sustainable systems, A report of the center for sustainable systems report no. CSS01-01. Ann Arbor, MI: University of Michigan.

Su, R., B. Liang, and J. Guan. 2016. Leaching effects of metal from electroplating sludge under phosphate participation in hydrochloric acid medium. *Procedia Environmental Sciences* 31: 361–365. doi:10.1016/j.proenv.2016.02.048.

Subramanian, V. R. 1995. Galvanisers ash—A raw material for zinc sulphate industry. *ILZIC Quarterly* 3(3): 59–61.

Sullivan, J., R. Williams, S. Yester, et al. 1998. Life cycle inventory of a generic U.S. family Sedan, overview of results USCAR AMP Project. SAE Technical Paper No. 982160, Society of Automotive Engineers (SAE) Inc. doi:10.4271/982160.

Sum, E. Y. L. 2005. The recovery of metals from electronic scrap. *The Journal of the Minerals, Metals & Materials Society* 43: 53–61. doi:10.1007/BF03220549.

Świerk, K., A. Bie licka, I. Bojanowska, and Z. Maćkiewicz. 2007. Investigation of heavy metals leaching from industrial wastewater sludge. *Polish Journal of Environmental Studies* 16(3): 447–451.

Száková, J., P. Tlustoš, D. Pavlíková, and J. Balík. 1997. Použitelnost různých extrakčních činidel pro stanovení podílu půdního arsenu využitelného rostlinami. *Chemicke Listy* 91: 580–584, (in Czech).

Szekely, J. 1996. Steelmaking and industrial ecology—Is steel a green material (review). *ISIJ International* 36(1): 21–32. doi:10.2355/isijinternational.36.121.

Szolusha, K. 2009. High power-factor LED driver converts AC input to power halogen replacement. *Electronic Design* 57(20): 57–58.

Tan, L. C., V. Choa, and J. H. Tay. 1997. The influence of pH on mobility of heavy metals from municipal solid waste incinerator fly ash. *Environmental Monitoring and Assessment* 44: 275–284.

Tessier, A., P. G. C. Campbell, and M. Bisson. 1979. Sequential extraction procedure for the speciation of particulate trace metals. *Analytical Chemistry* 51(7): 844–851. doi:10.1021/ac50043a017.

Tessier, A. and D. Turner. 1995. *Metal Speciation and Bioavailability in Aquatic Systems.* Chichester, U.K.: Wiley.

Thakur, P. K. 2000. Utilization of steel melting slag to generate wealth from waste. In *Proceedings of Conference on Environmental Management in Metallurgical Industries*, pp. 187–193, Dec 14–16, BHU, Varanasi, India.

Thaler, E. G., R. H. Wilson, D. A. Doughty, and W. W. Beers. 1995. Measurement of mercury bound in the glass envelope during operation of fluorescent lamps. *Journal of the Electrochemical Society* 142(6): 1968–1970.

Theis, T. L. and L. E. Padgett. 1983. Factors affecting the release of trace metals from municipal sludge ashes. *Journal of the Water Pollution Control Federation* 55: 1271–1279.

Tomeczek, J. and H. Palugniok. 2002. Kinetics of mineral matter transformation during coal combustion. *Fuel* 81(10): 1251–1258. doi:10.1016/S0016-2361(02)00027-3.

Tossavainen, M., F. Engstrom, Q. Yang, N. Menad, M. Lidstrom Larsson, and B. Bjorkman. 2007. Characteristics of steel slag under different cooling conditions. *Waste Management* 27(10): 1335–1344. doi:10.1016/j.wasman.2006.08.002.

Tossavainen, M. and E. Forssberg. 2000. Leaching behavior of rock material and slag used in road construction-A mineralogical interpretation. *Steel Research* 71: 442–448.

Townsend, T. G. 2011. Environmental issues and management strategies for waste electronic and electrical equipment. *Journal of the Air & Waste Management Association* 61(6): 587–610. doi:10.3155/1047-3289.61.6.587.

Tuncuk, A., V. Stazi, A. Akcil, E. Y. Yazici, and H. Deveci. 2012. Aqueous metal recovery techniques from e-scrap: Hydrometallurgy in recycling. *Minerals Engineering* 25(1): 28–37. doi:10.1016/j.mineng.2011.09.019.

Turan, M., H. Deniz, S. Altundoğan, and F. Tümen. 2004. Recovery of zinc and lead from zinc plant residue. *Hydrometallurgy* 75(1–4): 169–176. doi:10.1016/j.hydromet.2004.07.008.

Turbini, L., C. Munie, D. Bernier, J. Gamalski, and D. Bergman. 2004. Examining the environmental impact of lead-free. *IEEE Transactions on Electronics Packaging Manufacturing* 24: 4–8.

United Nations Environment Programme (UNEP). 2005. Toolkit for identification and quantification of mercury releases: Pilot Draft Inter-Organization. Programme for the Sound Management of Chemicals, United Nations Environmental Programme

United States Environmental Protection Agency (USEPA). 1998. Mercury emissions from the disposal of fluorescent lamps, revised model, Final report. Washington, DC: USEPA.

USEPA. 1998. Common sense initiative, metal finishing sector, Workgroup Report: F006 Benchmarking Study.

Vassura, I., L. Morselli, E. Bernardi, and F. Passarini. 2009. Chemical characterisation of spent rechargeable batteries. *Waste Management* 29(8): 2332–2335. doi:10.1016/j.wasman.2009.03.033.

Venditti, D., J. Durécu, and J. Berthelin. 2000. Multidisciplinary approach to assess history, environmental risks, and remediation feasibility of soils contaminated by metallurgical activities. Part A: Chemical and physical properties of metals and leaching ability. *Archives of Environmental Contamination and Toxicology* 38: 411–420. doi:10.1007/s002449910055.

Vesilind, P. A., G. C. Hartman, and M. E. T. Skene. 1986. *Sludge Management and Disposal.* Chelsea, UK: Lewis Publishing.

Wang, D.-L. 2006. *Practical Electroplating.* Hsu Foundation.

Wang, L., K. S. Nam, and S. C. Kwon. 2007. Effect of plasma nitriding of electroplated chromium coatings on the corrosion protection C45 mild steel. *Surface and Coatings Technology* 202(2): 203–207. doi:10.1016/j.surfcoat.2007.05.027.

Wang, R. C., Y. C. Lin, and S. H. Wu. 2009. A novel recovery process of metal values from the cathode active materials of the lithium-ion secondary batteries. *Hydrometallurgy* 99(3–4): 194–201. doi:10.1016/j.hydromet.2009.08.005.

Wazeck, J. 2013. Heavy metal extraction from electroplating sludge using *Bacillus subtilis* and *Saccharomyces cerevisiae. Geologica Saxonica* 59: 251–258. www.senckenberg.de/geologica-saxonica.

Werther, J. and T. Ogada. 1999. Sewage sludge combustion. *Progress in Energy and Combustion Science* 25(1): 55–116. doi:10.1016/S0360-1285(98)00020-3.

Whittingham, M. S. 2004. Lithium batteries and cathode materials. *Chemical Reviews* 104(10): 4271–4302. doi:10.1021/cr020731c.

Wiles, C. C. 1996. Municipal solid waste combustion ash: State-of-the-knowledge. *Journal of Hazardous Materials* 47(1): 325–344. doi:10.1016/0304-3894(95)00120-4.

Wong, J. W. C. and M. H. Wong. 1990. Effects of fly ash on yields and elemental composition of two vegetables, *Brassica parachinensis* and *B. chinensis. Agriculture, Ecosystems & Environment* 30(3): 251–264. doi:10.1016/0167-8809(90)90109-Q.

Worrell, E. and M. A. Reuter (Eds.). 2014. *Handbook of Recycling.* Amsterdam, the Netherlands: Elsevier. Epba (2010) European Portable Battery Association. Epba sustainability Report.

Xie, Y. and J. Zhu. 2013. Leaching toxicity and heavy metal bioavailability of medical waste incineration fly ash. *Journal of Material Cycles and Waste Management* 15: 440–448.

Xin, B., W. Jiang, H. Aslam, K. Zhang, C. Liu, R. Wang, and Y. Wang. 2012. Bioleaching of zinc and manganese from spent Zn-Mn batteries and mechanism exploration. *Bioresource Technology* 106: 147–153. doi:10.1016/j.biortech.2011.12.013.

Xu, J., H. R. Thomas, R. W. Francis, K. R. Lum, J. Wang, and B. Liang. 2008. A review of processes and technologies for the recycling of lithium-ion secondary batteries. *Journal of Power Sources* 177(2): 512–527. doi:10.1016/j.jpowsour.2007.11.074.

Yadav, U. S., B. K. Das, and A. Kumar. 2001. Recovery of mineral values from integrated steel plant waste. In *Proceedings of Sixth Southern Hemisphere Meeting on Mineral Technology*, pp. 719–725, Rio de Janeiro, Brazil.

Yamane, L. H., V. T. de Moraes, D. C. R. Espinosa, and J. A. S. Tenório. 2011. Recycling of WEEE: Characterization of spent printed circuit boards from mobile phones and computers. *Waste Management* 31(12): 2553–2558. doi:10.1016/j.wasman.2011.07.006.

Yang, T., Z. Xu, J. Wen, and L. Yang. 2009. Factors influencing bioleaching copper from waste printed circuit boards by *Acidithiobacillus ferrooxidans. Hydrometallurgy* 97(1): 29–32. doi:10.1016/j.hydromet.2008.12.011.

Year, G. (Ed.) 2011. Good year webpage. In *History: The Strange Story of Rubber.*

Yoo, K., J.-c. Lee, K.-s. Lee, B.-S. Kim, M.-s. Kim, S.-k. Kim, and B. D. Pandey. 2012. Recovery of Sn, Ag and Cu from waste Pb-free solder using nitric acid leaching. *Materials Transactions* 53(12): 2175–2180. doi:10.2320/matertrans.M2012268.

Yoshizawa, S., M. Tanaka, and A. Shekdar. 2004. *Global Trends in Waste Generation. Recycling, Waste Treatment and Clean Technology.* Spain: TMS Mineral, Metals and Materials Publishers.

Yung, K. C., J. Wang, and T. M. Yue. 2008. Thermal management for boron nitride filled metal core printed circuit board. *Journal of Composite Materials.* http://journals.sagepub.com/doi/abs/10.1177/0021998308096326#.WJ3GEpQjVRQ.mendeley.

Zeihen, H. and G. W. Brummer 1989. Chemische Extraktionen zur Bestimmung von Schwermetallbindungsformen in Boden. *Mitteilungen der Deutschen Bodenkundlichen Gesellschaft* 59: 505–510, (in German).

Zeng, X., L. Zheng, H. Xie, B. Lu, K. Xia, K. Chao, W. Li, J. Yang, S. Lin, and J. Li. 2012. Current status and future perspective of waste printed circuit boards recycling. *Procedia Environmental Sciences* 16: 590–597. doi:10.1016/j.proenv.2012.10.081.

Zhang, P., T. Yokoyama, O. Itabashi, Y. Wakui, T. M. Suzuki, and K. Inoue. 1998. Hydrometallurgical process for recovery of metal values from spent nickel-metal hydride secondary batteries. *Hydrometallurgy* 50(1): 61–75. doi:10.1016/S0304-386X(98)00046-2.

Zhang, T., D. Li, Z. Tao, and J. Chen. 2013. Understanding electrode materials of rechargeable lithium batteries via DFT calculations. *Progress in Natural Science: Materials International* 23(3): 256–272. doi:10.1016/j.pnsc.2013.04.005.

Zunkel, D. 1996. What to do with your EAF dust steel? *Times International* 20: 46–50.

3 Conventional Metal Recycling Techniques

3.1 INTRODUCTION

Excessive growth in population and developing industrialization are the prime reasons for the exploitation of natural resources and generation of huge amounts of metal-laden waste. Due to extensive and uncontrolled use, nonrenewable resources are on the verge of exhaustion and, at the same time, several environmental problems are increasing at an alarming rate, which will have a hazardous global impact. Hence, it is a need of the time to reduce the dependence on nonrenewable resources and demand for primary resources, and to develop a cleaner technique based on renewable natural resources. Besides the rising cost of energy and stringent environmental regulations, there is an extensive loss of valuable metals due to the dumping of wastes (Shen et al. 2007, Xin et al. 2011). New resources of metals for tomorrow's need are themselves present in today's waste, which needs to be exploited and recovered with the aid of innovative technologies. Therefore, it has become obligatory to increase the recovery of metals from secondary raw materials, which as a matter of fact may contain metals of much higher grade than primary ores, or at least the raw material is more readily acquirable. Despite being seen as a promising metal resource, there are many practical difficulties in utilizing secondary raw materials, concerning mainly economics, logistics, technology, and complexity of raw materials (Reuter et al. 2013).

Various conventional methods are being applied since many years for the retrieval of these metals from the waste. Some methods are advantageous over other, but sometimes, combinatorial methods appear to be the best option from the perspective of economics as well as the environment and human health.

The chapter aims to encompass the different aspects of metal recovery such as different methods employed, different mechanisms that regulate the potential of metal removal, the plethora of organisms involved, the recent trends, and future prospects. The recovery of metals from wastes has a massive future potential and will play a remarkable role in altering the situation of ever-decreasing grades of ores. Also, the major areas that need to be explored in the future are highlighted, along with different available tools for facilitating the decision-making processes in view of large-scale applications.

This chapter presents an overview of the various methodologies and their mechanisms, employed for the recovery of metals from secondary sources.

3.2 METHODS FOR METAL RECOVERY

3.2.1 Physical/Mechanical Methods

Prior to proceeding to any method of metal recovery, the obtained waste is dismantled and separated according to the reuse of components, hazardous components, etc. and then conveyed for further processing. The physical/mechanical processes include screening, shape separation, magnetic separation, electric conductivity–based separation, and density-based separation.

3.2.1.1 Screening

The main aim of this method is to prepare a uniformly sized feed for further mechanical processing and upgrade of metal contents and is necessary to separate metals from plastics and ceramics, as they differ in both particle size and shape. Various screening methods such as rotating screen, trommel, and vibratory screening are used in the processing of municipal as well as industrial wastes (Wilson et al. 1994).

3.2.1.2 Shape Separation

Shape separation controls the properties of particles and is categorized into four groups depending on the velocity of particles on the tilted solid wall, the time required by the particles to pass through a mesh aperture, their cohesive force to a solid wall, and their settling velocity in liquid (Furuuchi and Gotoh 1992). Of all these methods, the use of tilted plates and sieves has been the most basic method employed in the recycling industry (Koyanaka et al. 1997, Gungor and Gupta 1998).

3.2.1.3 Magnetic Separation

Magnetic separation is employed to recover ferromagnetic metals from nonferromagnetic metals and other nonmagnetic wastes. An intense field magnetic separation is achieved using high-intensity separators (Cui and Forssberg 2003).

3.2.1.4 Electric Conductivity–Based Separation

Electric conductivity–based separation separates materials that vary in their electric conductivity (or resistivity). This method can be divided into three types: (1) eddy current separation, (2) corona electrostatic separation, and (3) triboelectric separation (Van Der Valk et al. 1986, Schubert and Warlitz 1994, Stahl and Beier 1997, Higashiyama and Asano 1998, Meier-Staude and Koehnlechner 2000).

3.2.1.4.1 Eddy Current Separation

Eddy current separators operate using rare earth permanent magnets and recover nonferrous metals from scrap and other waste materials and handle relatively coarse-sized feeds (Norrgran and Wernham 1991, Wilson et al. 1994, Dalmijn and van Houwelingen 1995, Gesing et al. 1998). They are widely used in the reclamation of municipal and industrial wastes, along with other applications such as foundry casting sand, polyester polyethylene terephthalate (PET), electronic scrap, glass cullet, shredder fluff, and spent potliner (Mathieu et al. 1990, Hoberg 1993, Wernham et al. 1993, Schubert 1994, Meyer 1995, Dalmijn and van Houwelingen 1996).

3.2.1.4.2 Corona Electrostatic Separation

Corona electrostatic separation involves a rotor-type separator used to segregate raw materials into conductive and nonconductive fractions. The separator separates metal and nonmetal wastes depending on the difference in their electric conductivity or specific electric resistance. This method has been successfully implemented in the recovery of copper, aluminum, and other precious metals (Iuga et al. 1989, 1998, Schubert and Warlitz 1994, Dascalescu et al. 1994a,b, Higashiyama and Asano 1998, Zhang and Forssberg 1998, Meier-Staude and Koehnlechner 2000).

3.2.1.4.3 Triboelectric Separation

Triboelectric separation separates plastics according to the difference in their electric properties and is independent of particle shape, low energy consumption, and high throughput (Stahl and Beier 1997).

3.2.1.5 Density-Based Separation

Density-based separation separates on the basis of differential density. Gravity concentration separates materials of different specific gravity by their relative movement in response to the gravitational force and one or more other forces, the latter often being the resistance to motion offered by a fluid, such as water or air (Wills 1988). The motion of a particle in a fluid is dependent on the particle's density, size, and shape, where large particles are affected more than smaller ones.

3.2.2 PYROMETALLURGICAL METHOD

Pyrometallurgical processing is a traditional method used for the recovery of metals from ores and wastes. This process uses heat to separate metals from other materials, employing the differences between oxidation potentials, melting points, vapor pressures, densities, and/or miscibility of the ore components when melted (Roto 1998). This method comprises incineration, smelting in a plasma arc furnace or blast furnace, drossing, sintering, melting, etc. for the treatment of wastes (Sum 1991, Hoffmann 1992, Lee et al. 2007). The heating is carried out in a blast furnace at temperatures above 1500°C to convert waste to a refined form. Carbon in the form of coke or coal is added as a reducing agent during heating of the oxide waste, where the oxygen of the metal combines with the carbon and is removed in carbon dioxide gas and metals are extracted by converting sulfides into oxides and then reducing these oxides to metals. In the case of nonmetallic parts of waste, that is, gangue, it is removed by means of flux, which when heated combines with gangue to form a molten mass called slag. Being lighter than the metal, the slag floats on the top and can be skimmed or drawn off (Singh and Cameotra 2015).

Although pyrometallurgical treatment is more economical and efficient, and recovers maximum amounts of metals, it does have some limitations, which include the generation of hazardous emissions such as dioxins and furans due to the incineration of wastes containing plastic and other organic matter, which are known to be carcinogenic and increase the risk of contracting respiratory diseases, besides loss of plastic which can be recycled otherwise. Metals such as Fe and Al end up in the waste slag as oxides. Along with this, incinerators also produce a high amount of greenhouse effect gases such as SO_2, Cl_2, HCl (hydrochloric acid), and NO_x and ash with a high concentration of heavy metals such as lead, arsenic, and cadmium, which are known to cause birth defects, cancer, respiratory ailments, and reproductive dysfunction. Metals are recovered partially and contain many impurities; thus, they need to be processed further to attain the desired purity by employing other treatments. Further, incomplete combustion releases carbon monoxide and other volatile compounds such as formaldehyde and acetaldehyde, whose treatment requires large capital investment in advanced technologies and equipment (Sum 1991, Shuey and Taylor 2005, Antrekowitsch et al. 2006, Hagelüken 2006, Chiang et al. 2007, Dalrymple et al. 2007, Cui and Zhang 2008, Khaliq et al. 2014).

After all the waste is burnt, a residue ash is left behind. It is estimated that for every three tons of waste that is incinerated, one ton of ash is generated. This ash is very toxic and contains high concentrations of heavy metals and dioxins, which when landfilled or buried eventually leach out into the soil, polluting the environment.

Many environmental concerns are related to smelting activities, as they spew extremely harmful pollutants into the atmosphere. However, smelting with pollution control equipment is extremely expensive, which contributes directly to the high cost.

3.2.3   HYDROMETALLURGICAL METHOD

In the hydrometallurgical method, desired metals are separated from other metals depending on the differences between constituent solubilities and/or electrochemical properties while in aqueous solutions, employing acid or caustic leaching at much lower temperatures (Roto 1998). Figure 3.1 illustrates various types of hydrometallurgical techniques. The pregnant solution is separated and metal is purified by removing impurities as gangue materials. The isolation of the metal of interest is conducted through precipitation, solvent extraction, adsorption, and ion exchange enrichment processes. Finally, metals are recovered from solution through electrorefining

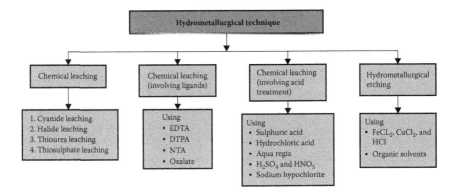

FIGURE 3.1 Types of hydrometallurgical techniques. (Reprinted from *Waste Manage.*, 32, Pant, D., Joshi, D., Upreti, M.K., and Kotnala, R.K., Chemical and biological extraction of metals present in E-waste: A hybrid technology, 979–990, Copyright (2012), with permission from Elsevier.)

(electrometallurgy) or chemical reduction processes or crystallization (Tavlarides et al. 1985, Shamsuddin 1986, Sum 1991, Yang 1994, Ritcey 2006, Sadegh Safarzadeh et al. 2007). Chemical treatments help minimize the impurities and toxic compounds may be replaced for less harmful constituents. As compared with the pyrometallurgical process, this process is exact, predictable, and easily controlled (Paretsky et al. 2004). Solvents employed for metal leaching include cyanides, halides, thiourea, and thiosulfates (Kołodzziej and Adamski 1984, Chmielewski et al. 1997, Quinet et al. 2005, Sheng and Etsell 2007). Further recovery of metals is attained by cementation, solvent extraction, adsorption on activated carbon, and ion exchange methods (Khaliq et al. 2014).

Chemical leaching techniques for precious metals (Zhong et al. 2006, Cui and Zhang 2008, Wu et al. 2009, Ha et al. 2010) use various leaching agents such as cyanide, halide, thiourea, and thiosulfate as follows:

Cyanide leaching

$$4Au + 8CN^- \rightarrow 4Au(CN)_2^- + 4e^- \tag{3.1}$$

$$O_2 + 2H_2O + 4e^- \rightarrow 4OH^- \tag{3.2}$$

Halide leaching

$$2HNO_3 + 6HCl \rightarrow 2NO + 4H_2O + 3Cl_2 \tag{3.3}$$

$$2Au + 11HCl + 3HNO_3 \rightarrow 2HAuCl_4 + 3NOCl + 6H_2O \tag{3.4}$$

Thiourea leaching

$$Au + 2CS(NH_2)_2 \rightarrow Au(CS(NH_2)_2)^{2+} + e^- \tag{3.5}$$

Thiosulfate leaching

$$Au + 5S_2O_3^{2-} + Cu(NH_3)_4^{2+} \rightarrow Au(S_2O_3)_2^{3-} + 4NH_3 + Cu(S_2O_3)_3^{5-} \tag{3.6}$$

$$2Cu(S_2O_3)_3^{5-} + 8NH_3 + \frac{1}{2}O_2 + H_2O \rightarrow 2Cu(NH_3)_4^{2+} + 2OH^- + 6S_2O_3^{2-} \tag{3.7}$$

Cyanide is being used predominantly as a leaching agent for the recovery of metals (Au, Ag, etc.) from mines and secondary sources because of its high efficiency and low cost. Dilute solutions of sodium cyanide (100–500 ppm) are used in tank and heap leaching processes. Cyanide ions (CN^-) play a major role in the dissolution of metals present in complexes with other materials. This dissolution is an electrochemical process and can be given by general Equations 3.1 and 3.2 (Dorin and Woods 1991, Marsden and House 1992, Cui and Zhang 2008, Syed 2016). However, a series of environmental accidents at various gold mines that resulted in severe environmental contamination has spread concern over the use of cyanide as a leach-ing agent (Hilson and Monhemius 2006). Halide leaching (Equations 3.3 and 3.4) employs various halides such as fluorine, chlorine, bromine, iodine, and astatine in metal dissolution (La Brooy et al. 1994). In thiourea leaching, thiourea ($(NH_2)_2CS$) dissolves metals in acidic conditions by forming a cationic complex (Equation 3.5) (Yannopoulos 1991). However, successful leaching using thiourea depends on the optimization and control of pH, redox potential, thiourea concentration, and leach-ing time. Alkaline thiosulfate dissolves metals slowly (Equations 3.6 and 3.7) and the rate of dissolution is affected by thiosulfate concentration, dissolved oxygen, and process temperature, and can be enhanced by adding copper ions (Yannopoulos 1991, Kuzugudenli and Kantar 1999). The process can be enhanced in the presence of ammonia using copper(II) as an oxidant (Abbruzzese et al. 1995, Aylmore 2001). In spite of the potential environmental benefits of thiosulfate, the slow process and high reagent consumption render its use uneconomical (Feng and Van Deventer 2002).

Chemical leaching can also be achieved by complexing ligands with metals using various chelating agents such as EDTA, DTPA (diethylene triamine pentaacetate), and NTA (nitrilotriacetic acid), along with organic acids like oxalate, citrate, and tartrate for the extraction of different metals such as Cr, Cu, Zn, and Pb (Elliott and Shastri 1999, Peters 1999, Hong et al. 2000, Cheikh et al. 2010). Various inorganic acids such as sulfuric acid (H_2SO_4) (Bayat and Sari 2010, Yang et al. 2011), HCl (Baba et al. 2010, Lu et al. 2011), aqua regia (Sheng and Etsell 2007), and a solution of H_2SO_4 and HNO_3 (Nnorom and Osibanjo 2009), along with sodium hypochlorite (in combination with an acid or alkali) (Li et al. 2011), are also used for the recovery of metals. In hydrometallurgical etching, chemicals such as $FeCl_3$, $CuCl_2$, and HCl are used to extract precious metals (Barbieri et al. 2009).

Once the metals have been separated from each other, pure metal can be obtained by precipitation or cementation, or electrolytically and may be further refined by electrochemical methods, if necessary. As the concentrations of metals in secondary raw materials are often low, and also highly variable, hydrometallurgical recovery methods are extremely suitable for the treatment of such materials, since they are much more adjustable than pyrometallurgical methods.

Precipitation is based on the solubility and/or insolubility of the product. It is enabled by hydroxide (CaO or $Ca(OH)_2$, $Mg(OH)_2$, NaOH, NH_4OH) (Blais et al. 2008), sulfide (FeS, CaS, Na_2S, NaHS, NH_4S, H_2S, $Na_2S_2O_3$) (Lewis 2010), or car-bonate (Hetherington 2006). Precipitation is pH dependent and hence the process is controlled by monitoring and adjusting pH (Hetherington 2006, Lewis 2010). Cementation is an electrochemical process in which a metal in an aqueous solution is reduced to its elemental form by some more electropositive metal in solid form. In solvent extraction, metals are extracted with a suitable reagent to a water-insoluble organic phase, which is typical, for example, a kerosene-based hydrocarbon solvent. While mixing, one of the phases is dispersed within the other, creating large surface area, over which the extractable species moves to the other phase. In ion exchange, an ion with a fixed charge in the ion exchange material is replaced by another ion. The process is stoichiometric. And because the ion exchange process often involves sorption of electrolytes, etc., and vice versa, sorption processes may be accompanied by ion exchange (Helfferich 1995).

Nevertheless, hydrometallurgical treatment of wastes generates some hazard-ous gases such as chlorine, hydrogen cyanide, and other noxious gases, along with huge amounts of wastewater, which otherwise is not generated in furnace processes. However, as compared with pyrometallurgy, this treatment is more environmentally friendly. Gases formed during the process are not allowed to escape and solvents are fully trapped at room temperature, where they are not in a position to produce dioxins or other greenhouse effects. Also, sulfur is presented as either a stable sulphate or elemental sulfur rather than sulfur dioxide emissions. Almost all waste components formed during or after treatment are segregated and recovered for further recycling or reuse. Each component refining stage could be accomplished in one process, with-out the need for diversion to another process. Leaching processes produce residues, while effluent treatment results in sludge, which can be sent for metal recovery.

One of the limitations of hydrometallurgy is the slow and time-consuming pro-cess, which may affect the economy of the process when compared with pyrometal-lurgy. Some leaching agents employed for the process, for example, cyanide, are dangerous, and thus the process needs high safety standards. The use of halides as leachants is difficult due to strong corrosive acids and oxidizing conditions. Also, there is the risk of losing the valuable metals during dissolution and recovery steps (La Brooy et al. 1994, Hilson and Monhemius 2006, Cui and Zhang 2008, Khaliq et al. 2014).

3.2.4 Microbiological and Biochemical Recycling

Industrial wastes contain high concentrations of heavy metals, which are not only a threat to the environment but also a huge waste of metal deposits. These toxic and nonbiodegradable metals pollute the soil and water. Hence, there is a pressing need to extract these metals in an eco-friendly manner. The answer lies in nature itself, that is, bioleaching. Natural bioleaching has existed as long as the history of the earth, but only a few decades ago did humans realize the importance of this event and that it is responsible for acid production in some mining wastes and that the involved bacte-rial activity liberates the metals. From then onward, humans have been benefited from what microbes do naturally. The mobilization of metal cations from insoluble ores by biological oxidation and complexation processes is referred to as bioleaching (Rohwerder et al. 2003).

In view of the present scenario of high-grade ore exhaustion and enforcement of strict antipollution laws, biohydrometallurgy, that is, the use of diverse microor-ganisms for leaching, is the most promising, superior, and green approach toward replacing conventional methods and a revolutionary solution to the treatment of wastes, overburdens, and low-grade ores. Different factors such as microbiological, physicochemical, and process parameters synergistically influence the leaching of metals by microbes.

Various advantages of the bioleaching process include

1. Safe and environmentally friendly process
2. Indigenous availability of microorganisms
3. No secondary pollution by emission of dust or gases

4. Low capital investment
5. Reduced energy consumption
6. Reduced operating cost
7. Reduced technical sophistication
8. Applicability to low/complex ores or wastes
9. Waste recycling
10. Site-specific process

Moreover, the process can be applied for selective metal leaching and can handle complex materials and wastes of low grade as well as high concentrates.

Furthermore, the use of microorganisms has many advantages, as follows:

1. They have the ability to grow and inhabit extreme environmental conditions like extreme pH, extreme temperature, high metal concentrations, etc.
2. They can grow at a faster rate under controlled conditions as well as under stressful conditions and generate higher biomass yield than plants and macroalgae.
3. Biomass handling is easy with unicellular microorganisms.
4. They require comparatively smaller and less expensive infrastructure facilities regarding equipment and land space, as they can be cultured in laboratory incubators.
5. They can be modified genetically.

Unlike chemical leaching, the main advantage of bioleaching is the biological production of the agent required for metal mobilization, and in terms of large-scale applications, there is no need of a continuous delivery to the plant, becoming an obvious advantage both for the process economics and for the environment in terms of carbon emissions.

3.2.4.1 Microbial/Bioleaching Process

3.2.4.1.1 Bacteria

Primarily chemolithotrophic bacteria are predominantly involved in bioleaching. They are extremely acidophilic (growing at a pH below 3) and utilize either S_0 or reduced inorganic sulfur compounds (RISCs) or Fe(II) iron (some use both) as an energy source (Quatrini et al. 2006, 2009, Bonnefoy and Holmes 2012, Liu et al. 2012) and grow autotrophically by fixing CO_2 from the atmosphere. They are further divided into obligate and facultative chemolithoautotrophs (mixotrophs) (utilize both organic and inorganic carbon sources) and are either obligate aerobes or facultative anaerobes. In nature, these bacteria are found exhibiting synergistic interactions to intensify growth and the kinetics of oxidation reactions.

The classic example of bioleaching bacteria belongs to the genus *Acidithiobacillus* (formerly called *Thiobacillus*). They are mesophilic (growing at 20°C–40°C), Gram-negative proteobacteria comprising the iron- and sulfur-oxidizing *Acidithiobacillus ferrooxidans*, the sulfur-oxidizing *At. thiooxidans* and *At. caldus*, and the iron-oxidizing *Leptospirillum ferrooxidans* and *L. ferriphilum* (Hallberg and Johnson 2003, Mangold et al. 2011). *At. ferrooxidans* has been widely used as a model microorganism to investigate metal solubilization mechanisms mediated by leaching bacteria, as they can live aerobically using either Fe(II) or reduced inorganic sulfur compounds as electron donors and anaerobically through the oxidation of hydrogen (or sulfur) coupled with Fe(III) reduction (Ohmura et al. 2002, Hallberg et al. 2011, Johnson 2012, Osorio et al. 2013). *At. thiooxidans* derives energy solely through the oxidation of reduced sulfur compounds, as it cannot oxidize iron or pyrite. *At. caldus* grows on reduced sulfur substrates and can also use molecular hydrogen as an electron donor. *Leptospirillum* grows on iron substrates, but cannot oxidize sulfur or thiosulfate.

Moderately thermophilic Gram-positive bacteria (growing at 40°C–60°C) belong to the genera *Acidimicrobium* (*Acidimicrobium ferrooxidans*), *Ferromicrobium*, and *Sulfobacillus* (*Sulfobacillus thermosulfidooxidans*) (Johnson 1998). And the thermophiles (growing at 60°C–80°C) belong to the genera *Sulfolobus metallicus*, *Sulfobacillus* sp., and *Metallosphaera sedula* (Erüst et al. 2013).

Archaebacteria from genera such as *Sulfolobus*, *Acidianus*, *Metallosphaera*, and *Sulfurisphaera* are also involved in bioleaching, which are extremely thermophilic in nature and also sulfur and iron oxidizers (Norris et al. 2000). Additionally, there are some reports of bioleaching by mesophilic and acidophilic iron-oxidizing archaebacteria from *Thermoplasmales*, such as *Ferroplasma acidiphilum* and *F. acidarmanus* (Edwards et al. 2000).

Various salient features that make these organisms suitable for bioleaching include their autotrophic nature, their utilization of ferrous iron or reduced inorganic sulfur compounds (some use both) as an energy source, their ability to grow at extremely acidic conditions (pH 1.4–1.6), and their tolerance to widely ranging concentrations of metal contaminants (Fonti et al. 2016).

3.2.4.1.2 Fungi

Although the literature is burgeoning with the use of bacteria for leaching of metals, fungi are also being applied since a long time. To date, several fungi such as *Penicillium simplicissimum*, *Penicillium chrysogenum*, *Aspergillus niger*, etc. have been reported for the recovery of heavy metals from different wastes such as municipal solid waste incinerator fly ash (Wu and Ting 2006, Yang et al. 2008, 2009, Xu and Ting 2009), spent catalyst (Wu and Ting 2006, Yang et al. 2008, Qu et al. 2013), electronic scrap (Brandl et al. 1999), and red mud (Qu and Lian 2013, Qu et al. 2013).

Bacterial leaching is preferred over fungal, as fungi require a constant supply of nutrients for growth, handling of fungi in turnover is difficult, and the method requires a long processing time. However, fungal leaching has several advantages over bacterial bioleaching. Fungi are able to grow at high pH, making them more effective in bioleaching of alkaline materials (Burgstaller and Schinner 1993). Moreover, metal leaching is rapid, with a short lag phase, and secreted fungal metabolites such as organic acids cause chelation with metal ions (Wu and Ting 2006, Gadd 2007, 2010, Kim et al. 2016). These organic acids form soluble metal complexes and chelates or also replace metal ions from the waste with hydrogen ions (Ren et al. 2009). Thus, these fungal metabolites act as lixiviants by forming metal complexes or precipitants and thus reduce the toxicity of heavy metals in fungal biomass (Santhiya and Ting 2006, Wu and Ting 2006).

3.2.4.2 Mechanisms Underlying the Bioleaching Process

Microorganisms mobilize metals via the following processes:

1. Oxidation/solubilization of sulfides and some oxide/carbonate minerals
2. Organic compounds production by organotrophic microorganisms that can solubilize minerals either by oxidation or reduction
3. Adsorption of the dissolved metal species on the bacterial surface or precipitation on the cell wall, intracellular accumulation (bioaccumulation) (Hoque and Philip 2011)

3.2.4.2.1 Mechanism of Bacterial Oxidation

A general biological oxidation reaction of mineral sulfides during leaching can be expressed as follows (Sand et al. 2001, Preston and Natarajan 2004, Rohwerder and Sand 2007, Schippers 2007, Giaveno et al. 2011):

$$MS + 2O_2 \rightarrow MSO_4 \tag{3.8}$$

where M is a bivalent metal.

The two major mechanisms of bacterial leaching are the direct and indirect mechanisms (Sand et al. 2001, Giaveno et al. 2011). In the direct mechanism, bacteria adhere to the insoluble metal sulfide and oxidize it by an enzyme system to sulfate and metal cations (as shown in Equations 3.9 and 3.10). Initially, the adherence occurs by van der Waals–type interaction forces. This interaction intensifies the formation of extracellular exopolysaccharides (EPS). The adhered bacteria start reproducing and form microcolonies to cover the material like a multilayered film, known as a biofilm (Giaveno et al. 2011). The sulfur moiety of the mineral is biologically oxidized to sulfate without the production of any detectable intermediate.

The indirect mechanism involves the oxidative dissolution of a metal sulfide by Fe(III) ions. During this reaction, Fe(II) ions and elemental sulfur are generated, which are further biologically oxidized to Fe(III) ions and sulfate (as shown in Equations 3.11 through 3.13). This mechanism does not require the attachment of cells to the sulfide mineral (Sand et al. 2001).

a. *Direct mechanism*:

$$FeS_2 + 3.5O_2 + H_2O \rightarrow Fe^{2+} + 2H^+ + 2SO_4^{2-} \tag{3.9}$$

$$2Fe^{2+} + 0.5O_2 + 2H^+ \rightarrow 2Fe^{3+} + H_2O \tag{3.10}$$

b. *Indirect mechanism*:

$$FeS_2 + 14Fe^{3+} + 8H_2O \rightarrow 15Fe^{2+} + 16H^+ + 2SO_4^{2-} \tag{3.11}$$

$$MS + 2Fe^{3+} \rightarrow M^{2+} + S^0 + 2Fe^{2+} \tag{3.12}$$

$$S^0 + 1.5O_2 + H_2O \rightarrow 2H^+ + SO_4^{2-} \tag{3.13}$$

Acidithiobacillus ferrooxidans produces ferric sulfide (Equation 3.14), which acts as a strong oxidizing agent, dissolving the wide variety of metal sulfide minerals.

$$4FeSO_4 + O_2 + 2H_2SO_4 \rightarrow 2Fe_2(SO_4)_3 + 2H_2O \tag{3.14}$$

This leaching is termed as indirect leaching because it proceeds in the absence of both oxygen and viable bacteria. Several minerals can be leached by this method (Equations 3.15 through 3.17).

$$CuFeS_2(chalcopyrite) + 2Fe_2(SO_4)_3 \rightarrow CuSO_4 + 5FeSO_4 + 2S^0 \tag{3.15}$$

$$FeS_2(pyrite) + Fe_2(SO_4)_3 \rightarrow 3FeSO_4 + 2S^0 \tag{3.16}$$

$$UO_2 + Fe_2(SO_4)_3 + 2H_2SO_4 \rightarrow UO_2(SO_4)_3 + 2FeSO_4 + 4H^+ \tag{3.17}$$

Further, elemental sulfur generated during indirect leaching is converted to H_2SO_4 by *At. ferrooxidans* as follows:

$$2S^0 + 3O_2 + 2H_2O \rightarrow 2H_2SO_4 \tag{3.18}$$

The formed H_2SO_4 helps maintain a favorable pH for bacterial growth and also aids in the effective bioleaching of oxide minerals, as shown in Equations 3.19 and 3.20:

$$CuO\ (tenorite) + 2H_2SO_4 \rightarrow CuSO_4 + H_2O \tag{3.19}$$

$$UO_3 + 3H_2SO_4 \rightarrow UO_2(SO_4)_3 + H_2O + 4H^+ \tag{3.20}$$

The traditional hypothesis of the direct and indirect mechanisms is a complex description. Also, mere attachment of bacteria on mineral surfaces may not be indicative of the direct mechanism. Hence, the term "contact" leaching was proposed by Tributsch (2001) instead of "direct" leaching to describe the bacterial association with the mineral surface rather than by means of attack. According to Rohwerder et al. (2003), along with the contact and noncontact mechanisms, the cooperative mechanism also facilitates mineral bio-oxidation and leaching, where planktonic iron and sulfur oxidizers oxidize colloidal sulfur, other sulfur intermediates, and ferrous iron in the leaching solution, releasing protons and ferric ion, which is further used in noncontact leaching (Rohwerder et al. 2003).

Many authors support the indirect mechanism as the sole mechanism of metal sulfide dissolution (Fowler and Crundwell 1998, Schippers und Sand 1999, Sand et al. 2001, Tributsch 2001, Donati and Sand 2007). Previously, Sand et al. (1995, 1997, 2001) suggested a new model for bioleaching by investigating the degradation products formed during the course of metal sulfide dissolution and the analysis of extracellular polymeric substances, EPS, produced by the microorganism for cell attachment and biofilm formation, in addition to considering the previous knowledge of sulfur chemistry, mineralogy, and solid-state physics. The author propose the hypothesis that iron(III) ions and/or protons are the only (chemical) agents dissolving a metal sulfide, suggesting the mechanism being an indirect one. The main role of bacteria is to regenerate Fe(III) ions and/or protons and to concentrate them at the mineral/water or mineral/bacterial cell interface in order to magnify the degradation/attack. This reaction takes place in the exopolymer layer, the glycocalyx, surrounding the bacterial cell. EPS, along with Fe(III) ion complexes with glucuronic acid residues, play a vital role in the primary attachment of cells. The concentration of the degrading agents at the interface accelerates metal dissolution by bacteria, thus explaining the hypothesis. Moreover, thiosulfate is formed as an intermediate product during the oxidation of sulfur compounds. Also, sulfur or polythionate granules are formed in the periplasmic space or in the cell envelope. On the basis of these key intermediates formed during the process, this indirect mechanism is categorized as the thiosulfate and polysulfide mechanisms (Figure 3.2) (Schippers et al. 1996, 1999, Schippers and Sand 1999, Sand et al. 2001, Rohwerder et al. 2003).

Different reduced inorganic sulfur compounds get accumulated depending on the electronic configuration of metal sulfides (i.e., acid soluble vs. acid insoluble) and the geochemical conditions of the environment (e.g., pH, different oxidants' availability) (Schippers and Sand 1999, Sand et al. 2001, Schippers 2004, Giaveno et al. 2011).

Microorganisms play a major role in the oxidation of these intermediate sulfur compounds formed during the dissolution of metal sulfides. They oxidize Fe(II) to Fe(III) ions during bioleaching, under oxic or acidic conditions, which act as oxidants for both metal sulfides and intermediate sulfur compounds. Further, these intermediate sulfur compounds are oxidized to H_2SO_4 by microbes.

Metal sulfides are conductors, semiconductors, or insulators, and their metal and sulfur atoms are bound in the crystal lattice (Vaughan and Craig 1978, Xu and Schoonen 2000). According to the molecular orbital and valence band theory, the orbitals of single atoms or molecules form electron bands with different energy levels. The metal sulfides FeS_2 (pyrite), MoS_2 (molybdenite), and WS_2 (tungstenite) consist of pairs of sulfur atoms (Vaughan and Craig 1978) which form nonbonding orbitals. Consequently, the valence bands of these metal sulfides are only derived from orbitals of metal atoms, whereas the valence bands of all other metal sulfides are derived from both metal and sulfur orbitals (Borg and Dienes 1992). Thus, the valence bands of FeS_2, MoS_2, and WS_2 do not contribute to the bonding between the metal and the sulfur moiety of the metal sulfide. This is the reason behind the resistance of these metal sulfides against proton attack. These bonds are broken by multistep electron transfers with an oxidant such as the Fe(III) ion.

For the other metal sulfides, electrons from the valence band are removed by the action of both Fe(III) ions and protons, resulting in the cleavage of the bonds between the metal and the sulfur moiety of the metal sulfide. Hence, these metal sulfides are readily soluble in acid, whereas FeS_2, MoS_2, and WS_2 are insoluble (Singer and Stumm 1970, Tributsch and Bennett 1981a,b, Crundwell 1988, Rossi 1993, Sand et al. 2001).

The thiosulfate and polysulfide mechanisms are able to explain the occurrence of all inorganic sulfur compounds, which have been documented for bioleaching environments.

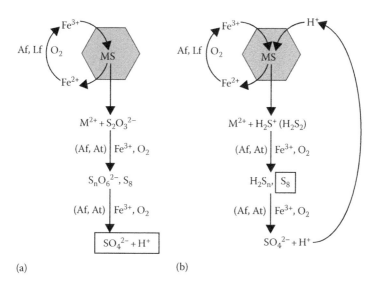

FIGURE 3.2 Schematic comparison of the thiosulfate (a) and polysulfide (b) mechanisms in (bio)leaching of metal sulfides. Iron(III) ions attack metal sulfides (MS) by electron extraction and are thereby reduced to the iron(II) ion form. As a result, the metal sulfide crystal releases metal cations (M^{2+}) and water-soluble intermediary sulfur compounds. Iron(II)-oxidizing bacteria such as *At. ferrooxidans* (Af) and *L. ferrooxidans* (Lf) catalyze the recycling of iron(III) ions in acidic solutions. In the case of acid-soluble metal sulfides (b), an additional attack is performed by protons, which can bind the valence band electrons of these metal sulfides. The liberated sulfur compounds are oxidized abiotically and by sulfur compound–oxidizing bacteria such as *At. ferrooxidans* and *At. thiooxidans* (At). In the case of mainly abiotic reactions, the contribution of sulfur compound oxidizers is indicated within brackets. The main electron acceptors of oxidation reactions other than the initial iron(III) ion attack on the metal sulfide are given to the right of the arrows. The main reaction products that accumulate in the absence of sulfur compound oxidizers are boxed, that is, sulfuric acid in (a) and elemental sulfur in (b). The equations given are not stoichiometric. For details, see text and Pronk et al. (1992), Schippers and Sand (1999), Rohwerder and Sand (2003), and Schippers (2004). (With kind permission from Springer Science + Business Media: *Appl. Environ. Microbiol.*, Bacterial leaching of metal sulfides proceeds by two indirect mechanisms via thiosulfate or via polysulfides and sulfur, 65(1), 1999, 319–321, Schippers, A. and Sand, W.; *Appl. Microbiol. Biotechnol.*, Progress in bioleaching: Fundamentals and mechanisms of bacterial metal sulfide oxidation—Part A, 97, 2013, 7529–7541, Vera, M., Schippers, A., and Sand, W.)

3.2.4.2.1.1 Thiosulfate Pathway The metal sulfides which are insensitive to proton attack directly transfer an electron from their metal to the Fe(III) ion, and after a series of oxidation reaction, by breaking the chemical bond between the metal and the sulfur and losing electrons, produce thiosulfate. This solubilization by a combinatorial proton and oxidative attack can be shown by the following reactions (Equations 3.21 through 3.26):

$$FeS_2 + 6Fe^{3+} + 3H_2O \rightarrow S_2O_3^{2-} + 7Fe^{2+} + 6H^+ \tag{3.21}$$

$$S_2O_3^{2-} + 8Fe^{3+} + 5H_2O \rightarrow 2SO_4^{2-} + 8Fe^{2+} + 10H^+ \tag{3.22}$$

As seen in Equations 3.21 and 3.22, sulfate is the major product (90%); other products (1%–2%) such as sulfite, sulfate trithionate, tetrathionate, and other polythionates can be detected. The formation and regeneration of thiosulfate and/or Fe(III) occur in these reactions. In particular, tetrathionate forms from thiosulfate oxidation (Equation 3.23) across a wide range of pH due to different oxidants (Schippers et al. 1996, Schippers and Jørgensen 2002). Further, thiosulfate decomposes to sulfite and elemental sulfur, which is a very slow reaction at pH 2, and hence, tetrathionate is the main product of thiosulfate reactions. This tetrathionate decomposes to sulfate (Equation 3.24) and trithionate (Equation 3.25):

$$2S_2O_3^{2-} + 2Fe^{3+} + 5H_2O \rightarrow S_4O_6^{2-} + 2Fe^{2+} \tag{3.23}$$

$$S_4O_6^{2-} + H_2O \rightarrow HS_3O_3^- + SO_4^{2-} + H^+ \tag{3.24}$$

$$S_3O_3^{2-} + 1.5O_2 \rightarrow S_3O_6^{2-} \tag{3.25}$$

Sometimes, pentathionate, elemental sulfur, and sulfite are also formed as side products. As for tetrathionate (Equation 3.24), trithionate is hydrolyzed to thiosulfate and sulfate (Equation 3.26) (Wentzien et al. 1994):

$$S_3O_6^{2-} + H_2O \rightarrow S_2O_3^{2-} + SO_4^{2-} + H^+ \tag{3.26}$$

Although Fe(III) ions are absent in the solution phase, thiosulfate can be formed in the presence of O_2 and MnO_2 at circumneutral pH. Schippers (2004) suggests that O_2 can promote the chemical dissolution of pyrite as a Fe(II)/Fe(III) shuttle if Fe(III) complexing organic substances reduce the pyrite oxidation rate, whereas under anoxic conditions, MnO_2 can promote the chemical dissolution of pyrite (Schippers 2004). Equation 3.25 can occur with Fe(III) as an alternative oxidant. These reactions thus comprise the cyclic degradation of thiosulfate to sulfate (Schippers et al. 1996, 1999).

The example of thiosulfate pathway can be given by FeS_2 and its oxidation, as a representative of the three metal sulfides—FeS_2, MoS_2, and WS_2, as it is studied in detail (Dutrizac and MacDonald 1974, Lowson 1982, Nordstrom 1982, Evangelou 1995, Rimstidt and Vaughan 2003, Schippers 2004, Druschel and Borda 2006). The sulfur moiety of pyrite is initially oxidized by oxidant Fe(III) ion to soluble sulfur intermediates. The details of these reactions are described by Moses et al. (1987) and Luther (1987), where thiosulfate is formed as the first soluble sulfur intermediate. As per this mechanism, the disulfide moiety of pyrite (FeS_2) is oxidized by hydrated Fe(III) ions to a sulfonic acid group by the removal of several electrons. During this transformation, the bonds between iron and the two sulfur atoms are broken to form hydrated Fe(II) ions and thiosulfate.

This thiosulfate is further completely oxidized to tetrathionate (Williamson and Rimstidt 1994, Schippers et al. 1996). Tetrathionate is then degraded to various sulfur compounds, that is, trithionate, pentathionate, elemental sulfur, and sulfite

(Schippers et al. 1996, Druschel 2002, Schippers 2004, Druschel and Borda 2006). All these sulfur compounds are finally oxidized to sulfate in chemical and/or biological reactions.

Moreover, Balci et al. (2007) confirmed the stoichiometry of the thiosulfate pathway during the bioleaching experiments with *At. ferrooxidans*, wherein the stable isotopes of oxygen and sulfur were determined in the pyrite oxidation reaction products.

3.2.4.2.1.2 Polysulfide Pathway   Overall, the polysulfide mechanism can be described by the following equations (Schippers and Sand 1999):

$$MS + Fe^{3+} + H^+ \rightarrow M^{2+} + 0.5H_2S_n + Fe^{2+} \quad (n \geq 2) \tag{3.27}$$

$$0.5H_2S_n + Fe^{3+} \rightarrow 0.125S_8 + Fe^{2+} + H^+ \tag{3.28}$$

$$0.125S_8 + 1.5O_2 + H_2O \rightarrow SO_4^{2-} + 2H^+ \tag{3.29}$$

Acid-soluble metal sulfides are readily solubilized by proton attack, for example, As_2S_3 (orpiment), As_4S_4 (realgar), $CuFeS_2$ (chalcopyrite), FeS (troilite), Fe_7S_8 (pyrrhotite), MnS_2 (hauerite), PbS (galena), and ZnS (sphalerite). This can be shown by the following reaction (3.30):

$$MS + H^+ \rightarrow M^{2+} + H_2S \tag{3.30}$$

In the presence of Fe(III) ions, hydrogen sulfide is oxidized concomitant to the proton attack:

$$H_2S + Fe^{3+} \rightarrow H_2S^{*+} + Fe^{2+} \tag{3.31}$$

The radical cation H_2S^{*+} dissociates into the radical HS^* and the formation of disulfide is favoured (Schippers and Sand 1999):

$$H_2S^{*+} + H_2O \rightarrow H_3O^+ + HS^* \tag{3.32}$$

$$2HS^* \rightarrow HS_2^- + H^+ \tag{3.33}$$

Fe(III) further oxidizes this disulfide to HS^*, which is then either dimerized to tetrasulfide or reacts with HS^* to form trisulfide, with reactions similar to Equations 3.31 through 3.33. Further chain elongation of polysulfides (H_2S_n) takes place by analogous reactions. Polysulfides decompose into rings of elemental sulfur as S_8 (Schippers 2004):

$$HS_9^- \rightarrow HS^- + S_8 \tag{3.34}$$

Elemental sulfur (>90%) is the main product formed, as the sulfur moiety of these metal sulfides is mainly oxidized to elemental sulfur, at low pH, along with thiosulfate (Dutrizac and MacDonald 1974, Zhang and Millero 1993, Schippers and Sand 1999, McGuire et al. 2001).

A series of reactions for acid-soluble metal sulfides inherently explain the formation of elemental sulfur via polysulfides (Schippers and Sand 1999), which have been detected during the dissolution of, for example, Fe_7S_8 (Thomas et al. 1998, 2001), PbS (Smart et al. 2000), and $CuFeS_2$ (Hackl et al. 1995). And hence, the oxidation mechanism for acid-soluble metal sulfides has been named the polysulfide mechanism (Schippers and Sand 1999). Although elemental sulfur is chemically inert in natural environments, it can be biologically oxidized to H_2SO_4.

The overall resulting reactions are as follows:

$$MS + Fe^{3+} + H^+ \rightarrow M^{2+} + 0.5H_2S_n + Fe^{2+} \quad n \geq 2 \tag{3.35}$$

$$0.5H_2S_n + nFe^{3+} \rightarrow 0.125S_8 + Fe^{2+} + H^+ \tag{3.36}$$

The stoichiometry of the polysulfide pathway was confirmed during the bioleaching experiments with *At. ferrooxidans*, wherein stable isotopes of oxygen and sulfur were determined in the products of chalcopyrite and sphalerite oxidation (Thurston et al. 2010, Balci et al. 2012).

In both pathways, the major bacterial role is the regeneration of the oxidant, Fe(III) ions (Figure 3.2). Thus, acidophilic sulfur oxidizers determine the Fe(III)/Fe(II) ratio and control the redox potential of the leaching environment, along with the transformation of intermediary sulfur compounds to H_2SO_4 (Schippers and Sand 1999, Schippers et al. 1999, Gonzalez-Toril et al. 2003, Rowe et al. 2007). However, the oxidation of elemental sulfur can only be carried out by microorganisms, as they are inert to abiotic oxidation in acidic environments (Figure 3.2b). Inhibition or the absence of S oxidizers may result in the accumulation of elemental sulfur. The generated protons are utilized in the first phases of the polysulfide pathway, and hence, sulfur oxidizers are required for their regeneration through the production of H_2SO_4 from reduced sulfur compounds and to stabilize the pH (Figure 3.2b) (Schippers 2004). In the absence of acidophilic sulfur oxidizers, S_0 accumulates in both the pathways, whereas various polythionates accumulate in the thiosulfate pathway. This elemental sulfur can form a layer on the metal sulfide surface or remain suspended as free aggregates and crystals (Mustin et al. 1993, Fowler and Crundwell 1999, Fowler et al. 1999). The formation of the sulfur layer results in a change in the electrochemical properties of the metal sulfide surface, forming a barrier, which results in a reduction in the diffusion rates of ions and oxygen, consequently hampering the leaching kinetics. This phenomenon was observed in the case of acid-soluble sphalerite at low redox potentials in the absence of sulfur oxidizers (Fowler et al. 1999) and also for chalcopyrite (Bevilaqua et al. 2002).

However, no inhibiting sulfur layer was observed at high redox potentials (about 750 mV vs. standard hydrogen electrode [SHE]), both with acid-soluble or acid-insoluble metal sulfide sphalerite (Fowler and Crundwell 1998, Fowler et al. 1999). Although elemental sulfur was formed in these cases, it was present as a free aggregate, which did not have any impact on the leaching rates under pH-controlled conditions (Vera et al. 2013).

Mostly, a consortium of Fe/S oxidizers is employed to enhance metal dissolution. For example, the sulfur-oxidizing autotroph *At. thiooxidans* is unable to dissolve pyrite alone but can improve its dissolution in a coculture with the Fe oxidizer *L. ferrooxidans* or *Ferrimicrobium acidophilum* (Kelly and Wood 2000, Okibe and Johnson 2004).

As the solubilization of metals is mostly dependent on the metabolic products of microbes, the microbes play a major role in establishing and maintaining the cycle of generation and regeneration of the leaching agents. The twofold role of microbes can be given as follows:

1. To catalyze the regeneration of the ferric ions consumed in the chemical oxidation of the intermediary hydrogen sulfide to elemental sulfur via the formation of polysulfides
2. To catalyze the generation of H_2SO_4 in order to maintain the supply of protons required in the first reaction step for the dissolution of the mineral

3.2.4.2.2 Mechanism of Fungal Leaching

Some microorganisms, for instance, fungi, are capable of producing and excreting metabolic by-products, such as polysaccharides, amino acids, and proteins, which help dissolve metals (Welch et al. 1999, Kisielowska and Kasinska-Pilut 2005). One such metabolite is soluble organic acids with a low molecular weight, which not only acidify the environment but also act as chelating agents, facilitating metal solubilization (Welch et al. 1999, Saidan et al. 2012, Calgoro et al. 2015). Fungi which exist in all ecological niches and are known to be active in many natural cycles produce large amounts of organic acids, such as citric acid, glycolic acid, oxalic acid, etc.

The glycolytic pathway converts glucose into a variety of products, including organic acids. The metal leaching process is mediated by a chemical attack by extracted organic acids on ores/wastes. The acids usually have the dual effect of increasing metal dissolution by lowering the pH and increasing the load of soluble metals by complexing/chelating into soluble organic metallic complexes (Ghorbani et al. 2007).

Leaching of metals by fungi is based on three mechanisms—acidolysis, complexolysis, and redoxolysis. In acidolysis, the oxygen atoms covering the surface of metallic compounds get protonated. The protons and oxygen are associated with water; thus, the metal is detached from the surface (Burgstaller and Schinner 1993), as shown in Equation 3.37, where nickel ions are produced as a result of the acidolysis reaction (Asghari et al. 2013):

$$NiO + 2H^+ \rightarrow Ni^{2+} + H_2O \qquad (3.37)$$

In complexolysis, metal ions are solubilized due to the formation of metal complexes and chelates (due to the complex capacity of a molecule) such as a complex of oxalic acid with aluminum, iron, and magnesium; a complex of citric acid with magnesium and calcium; and a complex of tartaric acid with iron, calcium, silicon, magnesium, and aluminum, as shown in Equation 3.38, where nickel citrate is produced due to the formation of a complex of Ni with citric acid (Asghari et al. 2013).

Moreover, metal ions solubilized into solution by acidolysis are stabilized in complexolysis (Burgstaller and Schinner 1993, Bosecker 1997:

$$Ni^{2+} + C_6H_8O_7 \rightarrow Ni(C_6H_5O_7)^- + 3H^+ \tag{3.38}$$

The oxidation–reduction processes aid fungal leaching by the redoxolysis mechanism, as shown in Equation 3.39, where manganese ion is produced as a result of the redoxolysis reaction (Asghari et al. 2013):

$$MnO_2 + 2e^- + 4H^+ \rightarrow Mn^{2+} + 2H_2O \tag{3.39}$$

Qu et al. (2013) described the related reactions between metal ions and different organic acids (where M^{n+} corresponds to the metal ions with certain valence), as follows:

$$\text{Gluconic acid: } C_6H_{12}O_7 \rightarrow C_6H_{11}O_7^- + H^+ \quad (pK_a = 3.86) \tag{3.40}$$

$$n[C_6H_{11}O_7^-] + M^{n+} \rightarrow M[C_6H_{11}O_7]_n \text{ (gluconic metallic complex)} \tag{3.41}$$

$$\text{Oxalic acid: } C_2H_2O_4 \rightarrow C_2HO_4^- + H^+ \quad (pK_{a1} = 1.25) \tag{3.42}$$

$$C_2HO_4^- \rightarrow C_2O_4^{2-} + H^+ \quad (pK_{a2} = 4.14) \tag{3.43}$$

$$n[C_2HO_4^-] + M^{n+} \rightarrow M[C_2HO_4]_n \text{ (oxalic metallic complex)} \tag{3.44}$$

$$n[C_2O_4^{2-}] + 2M^{n+} \rightarrow M_2[C_2O_4]_n \text{ (oxalic metallic complex)} \tag{3.45}$$

$$\text{Citric acid: } C_6H_8O_7 \rightarrow C_6H_7O_7^- + H^+ \quad (pK_{a1} = 3.09) \tag{3.46}$$

$$C_6H_7O_7^- \rightarrow C_6H_6O_7^{2-} + H^+ \quad (pK_{a2} = 4.75) \tag{3.47}$$

$$C_6H_6O_7^{2-} \rightarrow C_6H_5O_7^{3-} + H^+ \quad (pK_{a3} = 6.40) \tag{3.48}$$

$$n[C_6H_7O_7^-] + M^{n+} \rightarrow M[C_6H_7O_7]_n \text{ (citric metallic complex)} \tag{3.49}$$

$$n[C_6H_6O_7^{2-}] + 2M^{n+} \rightarrow M_2[C_6H_6O_7]_n \text{ (citric metallic complex)} \tag{3.50}$$

$$n[C_6H_5O_7^{3-}] + 3M^{n+} \rightarrow M_3[C_6H_5O_7]_n \text{ (citric metallic complex)} \tag{3.51}$$

These reactions demonstrate the prime role of fungal metabolic organic acids as leaching agents (Deng et al. 2013). The pH of and the metals in the culture medium significantly affect the type and concentration of organic acids produced, and hence, it is crucial to analyze organic acids in order to understand the mechanism of bioleaching (Yang et al. 2008, Horeh et al. 2016).

3.2.4.2.3 Bioaccumulation

Organisms such as bacteria, fungi, yeast, and algae sequester heavy metals from dilute wastewater solutions and accumulate them by bioaccumulation in live cells (Volesky 2003, 2007, Gonen and Aksu 2007, Aksu and Karabayur 2008, Naja et al. 2008).

Bioaccumulation has the advantage of reduced unit processes, including separate biomass cultivation, harvesting, drying, processing, and storage (Aksu and Dönmez 2005). However, the abiotic stress or toxicity of heavy metals being accumulated hampers the effectiveness of microoragnisms as toxic metal ions form complexes with cellular membrane, which causes the loss of organisms integrity and impairs their function (Yilmazer and Saracoglu 2009).

In this mechanism, microorganisms immobilize and/or remove metals through intracellular and extracellular sorption of metals. This involves

1. Extracellular accumulation of metals by binding or precipitation on the cell wall
2. Intracellular accumulation of metabolically essential metals such as K, Fe, Mg, and Mo, and traces of Cu, and Ni, wherein they are first bound to the cell wall and then transferred into the cells
3. Intracellular accumulation of nonessential metals such as Co, Ni, Cu, Cd, and Ag in large amounts primarily through the pathways involved in the uptake of essential metals

The process of metal uptake by bacteria proceeds in two steps. In the primary step, metal ions reversibly bind to highly reactive bacterial surfaces. At near-neutral pH, bacteria possess a net negative charge due to their cell wall components such as peptidoglycan (Gram-positive) and hydroxyl, carboxyl, and phosphate groups on the outer membrane (Gram-negative). These functional groups attract and bind metal ions of net positive charge by gaining or losing protons. These reversibly bound metals can be easily separated by a chelating agent or dilute acid. In the next step, these bound metals are transported into the cell cytoplasm. This process is energy dependent and relies on the organism's mode of respiration.

Bacteria have a greater initial metal-binding capacity to the cell surface than most yeast, whereas yeast can uptake and accumulate a high quantity of metals intracellularly. Hence, bacteria are preferred over yeast for the rapid removal of metals by binding (Hoque and Philip 2011).

Fungi can limit metal accumulation in their cytoplasm by a tolerance mechanism, as the cytoplasm contains the majority of enzymes and proteins that are susceptible to damage by heavy metals. Two mechanisms are involved in the process. First is the extracellular mechanism, which excludes the entry of metals into the cell by binding them to ligands (e.g., citrate, oxalate) (Bellion et al. 2006) which are excreted by the cell or components of the cell wall (chitin, glucan). Structural changes in transport proteins also alter the rate of metal entry into the cell. Second is the intracellular detoxification mechanism, wherein metals bind to intracellular compounds such as metallothionein (MT, an intracellular protein) and glutathione (GSH, a small sulfur-containing molecule) (Cánovas et al. 2004). These mechanisms can be used to enhance the ability of organisms to remove heavy metals from solutions.

3.2.4.2.4 Biosorption

Biosorption is another biological process which has been seen as a promising process for pollutant removal and/or metal recovery because of its simplicity and efficiency, and because of the availability of biomass and waste bioproducts

(Fomina and Gadd 2014). It is a passive physicochemical and metabolically independent process, performed either by dead biomass or fragments of cells and tissues, for example, the passive uptake of metals by microbial biomass (Volesky 1990, Malik 2004, Dobson and Burgess 2007, Gadd 2009). However, it can also be performed by live cells as passive uptake or metabolically independent adsorption of a sorbate via surface complexation onto cell walls and/or other outer layers being the first, fast and reversible adsorption step operating within a much slower and complex overall bioaccumulation mechanism (Volesky 1990, Malik 2004, Mao et al. 2009).

The mechanism is generally based on physicochemical interactions such as electrostatic interactions between metal ions and the functional groups present on the cell surface (e.g., carboxyl, carbonyl, amine, amide, hydroxyl, phosphate groups), ion exchange, complexation, coordination and chelation between metal ions and ligands, microprecipitation, and microbial reduction (Kratochvil and Volesky 1998, Kuyucak and Volesky 1988, Baldrian 2003, Mack et al. 2007), and differs according to the biomass type (live, or dead, or as a derived product).

Many bacteria, fungi, yeast, and algal strains (living or dead) and their components, seaweeds, plant materials, industrial and agricultural wastes, and natural residues are used as biosorbents (Fomina and Gadd 2014).

Cui and Zhang (2008) reviewed that acidic conditions favor the biosorption of precious metals on bacteria and chitosan derivatives, from solution. The adsorption capacities (Qmax) of precious metals on different types of biomass vary from 0.003 to 40 mmol/g (dry biomass), which suggest that more work input is required to select the perfect biomass. The biomass may be used in its natural state or also can be modified by cross-linking to ameliorate its biosorption efficiency.

3.2.4.2.5 Cyanide Leaching

A classic example of cyanogenic bacteria is *Chromobacterium violaceum*. It is a Gram-negative bacterium that can live in both aerobic and anaerobic conditions owing to its ability to utilize a wide range of energy sources (Creczynski-Pasa and Antônio 2004). Due to its cyanide-producing ability, it is best known for biodissolution of gold from different sources. It produces cyanide as a secondary metabolite. *C. violaceum* consists of an operon for HCN (hydrogen cyanide) synthase (hcnA, hcnB, and hcnC) in its genome, encoding a formate dehydrogenase and two amino acid oxidases, respectively, which are involved in cyanic acid synthesis. During cyanide synthesis, four electrons produced by HCN synthase are transferred to oxygen, throughout the respiratory chain. As these reactions occur at a low oxygen level, HCN is mainly generated in aerobic conditions (Creczynski-Pasa and Antônio 2004). The secondary metabolite HCN produced from glycine is converted to cyanide by the enzyme HCN synthase; however, the amount of this lixiviant produced is limited (20 mg of cyanide per liter of bacteria culture, approximately 1×10^{16} colony forming units). Moreover, the regulation of this operon by quorum control further hinders this bacterium's widespread industrial use (Motokawa and Kikuchi 1971, de Vasconcelos et al. 2003).

Cyanide is produced by these organisms in order to inhibit the competing microorganisms (Blumer and Haas 2000). Cyanide occurs in solution as free cyanide, which includes the cyanide anion (CN^-) and the nondissociated HCN. At physiological pH,

HCN has a pK_a of 9.3 at 26°C, and therefore, cyanide is largely present as volatile HCN. In the presence of salts, this value decreases to approximately 8.3 and the volatility is reduced (Faramarzi and Brandl 2006, Liu et al. 2016).

3.2.4.2.6 Mechanism of Cyanide Leaching

Leaching of gold by cyanogenic microorganisms is an indirect process where microbial metabolites play a crucial role. The gold from minerals is dissolved by metabolites through the formation of soluble metallic complexes (Faramarzi et al. 2004). Au dissolution in a cyanide solution consists of an anodic and a cathodic reaction, as summarized by Elsner's equation (Equations 3.1 and 3.2) (Hedley and Tabachnick 1958).

Biological cyanide leaching is very much similar to industrial Au cyanidation, wherein cyanogenic microorganisms produce the cyanide lixiviant, which then reacts with solid Au to complete the leaching process (Knowles and Bunch 1986, Rawlings 2002, Valenzuela et al. 2006, Cui and Zhang 2008).

3.3 CONCLUSION AND FUTURE PERSPECTIVE

Rapid industrialization and global development over the past few years have greatly elevated the rate of metal release into the environment. While some metals are integral to various life processes, others are nonessential and potentially toxic. Besides, unlike many organic pollutants, heavy metals are nonbiodegradable and accumulate in the environment.

Recent legislation in several countries has driven research into the treatment and recovery of metals from various waste sources, but these studies must be applied on a large scale. The recovery of metals from waste sources will help not only in reducing waste but also in protecting the environment.

As a traditional technology, pyrometallurgy has been used for the recovery of metals from waste materials as an economical and eco-efficient method. However, it has encountered some challenges from an environmental point of view. Hydrometallurgy requires a huge investment in leaching reagents and operations, and also creates secondary pollution. Considering this, biohydrometallurgical recovery of heavy metals from wastes appears to be an imperative solution. It is an effective and economical alternative to conventional methods for the recovery of metals from low-grade ores and wastes.

Although this technique has enormous potential, most of the research has remained confined to the laboratory scale; moreover, the use of this process in the real world, especially for metal recovery from secondary sources, is still limited. The process lags behind due to some disadvantages, such as slow processing time, resulting in less profit, difficulty in controlling the process, and greater applicability to acid-producing minerals and not acid consuming minerals.

The technology still needs to be refined in order to prosper at a commercial level. As compared with the vast microbial diversity, very few strains have been employed for the recovery of metals, and hence, there is a need to exploit different microbes, which might open new avenues in this field. Microbes inhabiting extreme environments have a great potential and, if implemented, can significantly contribute to the field and present new frontiers for biohydrometallurgy.

The scopes of improvement include the investigation of microbial community composition and diversity of metal-contaminated sites, bioheaps, reactors, etc. using advanced DNA and PCR techniques and further exploration of the interactions among the microbes and involved metabolic pathways.

The use of high-temperature-tolerant microorganisms can accelerate reaction rates and shorten the leaching time; additionally, mutation, genetic modification, and/or metabolic engineering of bioleaching microorganisms will help make the process more efficient.

Scale-up investigations with estimations of cost and environmental impacts are also obligatory. Also, biohydrometallurgy is an interdisciplinary work that requires collaboration between the microbiologist, metallurgical engineer, and miner in order to optimize operational aspects of the mining and resource recovery system. Above mentioned points represent the crucial aspects in the development of truly sustainable, effective, and ecologically compatible processes of bioleaching of wastes.

Every process has advantages and disadvantages, and there could be various technical, economic, and environmental reasons for choosing one process over the other. Hence, all aspects should be considered depending on the specific feed materials and final product. The development of a hybrid methodology by integrating bioleaching with other methods such as hydrometallurgy will provide a unique dimension to microbial metal recovery and also assist in overcoming the problems associated with either technique and help recover heavy metals in a faster, efficient, cost-effective, and, most importantly, eco-friendly manner.

REFERENCES

Abbruzzese, C., P. Fornari, R. Massidda, F. Vegliò, and S. Ubaldini. 1995. Thiosulphate leaching for gold hydrometallurgy. *Hydrometallurgy* 39(1–3): 265–276. doi:10.1016/0304-386X(95)00035-F.

Aksu, Z. and G. Dönmez. 2005. Combined effects of molasses sucrose and reactive dye on the growth and dye bioaccumulation properties of *Candida tropicalis*. *Process Biochemistry* 40(7): 2443–2454. doi:10.1016/j.procbio.2004.09.013.

Aksu, Z. and G. Karabayur. 2008. Comparison of biosorption properties of different kinds of fungi for the removal of Gryfalan Black RL metal-complex dye. *Bioresource Technology* 99(16): 7730–7741. doi:10.1016/j.biortech.2008.01.056.

Antrekowitsch, H., M. Potesser, W. Spruzina, and F. Prior. 2006. Metallurgical recycling of electronic scrap. In *Proceedings of the EPD Congress*, pp. 899–908, San Antonio, TX.

Asghari, I., S. M. Mousavi, F. Amiri, and S. Tavassoli. 2013. Bioleaching of spent refinery catalysts: A review. *Journal of Industrial and Engineering Chemistry* 19(4): 1069–1081. doi:10.1016/j.jiec.2012.12.005.

Aylmore, M. G. 2001. Treatment of a refractory gold-copper sulfide concentrate by copper ammoniacal thiosulfate leaching. *Minerals Engineering* 14(6): 615–637. doi:10.1016/S0892-6875(01)00057-7.

Baba, A., F. Adekola, and D. Ayodele. 2010. Study of metals dissolution from a brand of mobile phone waste. *Metalurgija—Journal of Metallurgy* 16(4): 269–277.

Balci, N., B. Mayer, W. C. Shanks, and K. W. Mandernack. 2012. Oxygen and sulfur isotope systematics of sulfate produced during abiotic and bacterial oxidation of sphalerite and elemental sulfur. *Geochimica et Cosmochimica Acta* 77: 335–351. doi:10.1016/j.gca.2011.10.022.

Balci, N., W. C. Shanks, B. Mayer, and K. W. Mandernack. 2007. Oxygen and sulfur isotope systematics of sulfate produced by bacterial and abiotic oxidation of pyrite. *Geochimica et Cosmochimica Acta* 71(15): 3796–3811. doi:10.1016/j.gca.2007.04.017.

Baldrian, P. 2003. Interactions of heavy metals with white-rot fungi. *Enzyme and Microbial Technology* 32(1): 78–91. doi:10.1016/S0141-0229(02)00245-4.

Barbieri, L., R. Goivanardi, I. Lancellotti, and M. Michelazzi. 2009. A new environmentally friendly process for the recovery of gold from electronic waste. *Environmental Chemistry Letters* 8(2): 171–178.

Bayat, B. and B. Sari. 2010. Comparative evaluation of microbial and chemical leaching processes for heavy metal removal from dewatered metal plating sludge. *Journal of Hazardous Materials* 174: 763–769. doi:10.1016/j.jhazmat.2009.09.117.

Bellion, M., M. Courbot, C. Jacob, D. Blaudez, and M. Chalot. 2006. Extracellular and cellular mechanisms sustaining metal tolerance in ectomycorrhizal fungi. *FEMS Microbiology Letters* 254(2): 173–181. doi:10.1111/j.1574-6968.2005.00044.x.

Bevilaqua, D., A. L. L. C. Leite, O. Garcia Jr., and O. H. Tuovinen. 2002. Oxidation of chalcopyrite by *Acidithiobacillus ferrooxidans* and *Acidithiobacillus thiooxidans* in shake flasks. *Process Biochemistry* 38: 587–592.

Blais, J. F., Z. Djedidi, C. Ben, R. D. Tyagi, and G. Mercier. 2008. Metals precipitation from effluents: Review. *Practice Periodical of Hazardous, Toxic, Waste Management* 12: 135–149.

Blumer, C. and D. Haas. 2000. Mechanism, regulation, and ecological role of bacterial cyanide biosynthesis. *Archives of Microbiology* 173(3): 170–177.

Bonnefoy, V. and D. S. Holmes. 2012. Genomic insights into microbial iron oxidation and iron uptake strategies in extremely acidic environments. *Environmental Microbiology* 14(7): 1597–1611. doi:10.1111/j.1462-2920.2011.02626.x.

Borg, R. J. and G. J. Dienes. 1992. *The Physical Chemistry of Solids*. Boston, MA: Academic Press.

Bosecker, K. 1997. Bioleaching: Metal solubilization by microorganisms. *FEMS Microbiology Reviews* 20(3): 591–604. doi:10.1016/S0168-6445(97)00036-3.

Brandl, H., R. Bosshard, and M. Wegmann. 1999. Computer-munching microbes: Metal leaching from electronic scrap by bacteria and fungi. *Process Metallurgy* 9(C): 569–576. doi:10.1016/S1572-4409(99)80146-1.

Burgstaller, W. and F. Schinner. 1993. Leaching of metals with fungi. *Journal of Biotechnology* 27(2): 91–116. doi:10.1016/0168-1656(93)90101-R.

Calgoro C. O., E. H. Tanabe, D. A. Bertuol, F. P. Cianga Silvas, D. C. R. Spinosa, and J. A. S. Tenorio. 2015. *Leaching Process*. Switzerland: Springer International Publishing.

Cánovas, D., R. Vooijs, H. Schat, and V. De Lorenzo. 2004. The role of thiol species in the hypertolerance of *Aspergillus* sp. P37 to arsenic. *Journal of Biological Chemistry* 279(49): 51234–51240. doi:10.1074/jbc.M408622200.

Cheikh, M., J.-P. Magnin, N. Gondrexon, J. Willison, and A. Hassen. 2010. Zinc and lead leaching from contaminated industrial waste sludges using coupled processes. *Environmental Technology* 31(14): 1577–1585. doi:10.1080/09593331003801548.

Chiang, H.-L., K.-H. Lin, M.-H. Lai, T.-C. Chen, and S.-Y. Ma. 2007. Pyrolysis characteristics of integrated circuit boards at various particle sizes and temperatures. *Journal of Hazardous Materials* 149(1): 151–159. doi:10.1016/j.jhazmat.2007.03.064.

Chmielewski, A. G., T. S. Urbański, and W. Migdał. 1997. Separation technologies for metals recovery from industrial wastes. *Hydrometallurgy* 45(3): 333–344. doi:10.1016/S0304-386X(96)00090-4.

Creczynski-Pasa, T. B. and R. V. Antônio. 2004. Energetic metabolism of *Chromobacterium violaceum*. *Genetics and Molecular Research* 3(1): 162–166.

Crundwell, F. K. 1988. The influence of the electronic structure of solids on the anodic dissolution and leaching of semiconducting sulphide minerals. *Hydrometallurgy* 21(2): 155–190. doi:10.1016/0304-386X(88)90003-5.

Cui, J. and E. Forssberg. 2003. Mechanical recycling of waste electric and electronic equipment: A review. *Journal of Hazardous Materials* 99(3): 243–263. doi:10.1016/S0304-3894(03)00061-X.

Cui, J. and L. Zhang. 2008. Metallurgical recovery of metals from electronic waste: A review. *Journal of Hazardous Materials* 158(2): 228–256. doi:10.1016/j.jhazmat.2008.02.001.

Dalmijn, W. L. and J. A. van Houwelingen. 1995. New developments in the processing of the non-ferrous metal fraction of car scrap. In *Proceedings of the Third International Symposium on Recycling of Metals and Engineered Materials, Point Clear*, P. B. Queneau, and R. D. Peterson (Eds.), pp. 739–750. Warrendale, PA: TMS.

Dalmijn, W. L. and J. A. van Houwelingen. 1996. Glass recycling in the Netherlands. *Glass* 73: 3.

Dalrymple, I., N. Wright, R. Kellner, N. Bains, K. Geraghty, M. Goosey, and L. Lightfoot. 2007. An integrated approach to electronic waste (WEEE) recycling. *Circuit World* 33(2): 52–58. doi:10.1108/03056120710750256.

Dascalescu, L., R. Morar, A. Iuga, A. Samuila, V. Neamtu, and I. Suarasan. 1994a. Charging of particulates in the corona field of roll-type electroseparators. *Journal of Physics D: Applied Physics* 27(6): 1242–1251. doi:10.1088/0022-3727/27/6/023.

Dascalescu, L., A. Samuila, A. Iuga, R. Morar, and I. Csorvassy. 1994b. Influence of material superficial moisture on insulation-metal electroseparation. *IEEE Transactions on Industrial Applications* 30: 844–849.

Deng, X., L. Chai, Z. Yang, C. Tang, Y. Wang, and Y. Shi. 2013. Bioleaching mechanism of heavy metals in the mixture of contaminated soil and slag by using indigenous *Penicillium chrysogenum* strain F1. *Journal of Hazardous Materials* 248: 107–114. doi:10.1016/j.jhazmat.2012.12.051.

de Vasconcelos, A. T. R., D. F. De Almeida, M. Hungria et al. 2003. The complete genome sequence of *Chromobacterium violaceum* reveals remarkable and exploitable bacterial adaptability. *Proceedings of the National Academy of Sciences of the United States of America* 100(20): 11660–11665.

Dobson, R. S. and J. E. Burgess. 2007. Biological treatment of precious metal refinery wastewater: A review. *Minerals Engineering* 20(6): 519–532. doi:10.1016/j.mineng.2006.10.011.

Donati, E. R. and W. Sand. 2007. *Microbial Processing of Metal Sulfides*. Netherlands: Springer.

Dorin, R. and R. Woods. 1991. Determination of leaching rates of precious metals by electrochemical techniques. *Journal of Applied Electrochemistry* 21(5): 419.

Druschel, G. and M. Borda. 2006. Comment on 'Pyrite dissolution in acidic media' by M. Descostes, P. Vitorge, and C. Beaucaire. *Geochimica et Cosmochimica Acta* 70(20): 5246–5250. doi:10.1016/j.gca.2005.07.023.

Druschel, G. K. 2002. Sulfur biogeochemistry: Kinetics of intermediate sulfur species reactions in the environment. PhD thesis, University of Wisconsin, Madison, WI.

Dutrizac, J. E. and R. J. C. MacDonald. 1974. Ferric ion as a leaching medium. *Minerals Science and Engineering* 6(2): 59–95. http://inis.iaea.org/Search/search.aspx?orig_q=RN:5143359.

Edwards, K. J., P. L. Bond, T. M. Gihring, and J. F. Banfield. 2000. An archaeal iron-oxidizing extreme acidophile important in acid mine drainage. *Science* 287(5459): 1796–1799.

Elliott, H. A. and N. L. Shastri. 1999. Extractive decontamination of metal-polluted soils using oxalate. *Water, Air, and Soil Pollution* 110(3–4): 335–346.

Erüst, C., A. Akcil, C. S. Gahan, A. Tuncuk, and H. Deveci. 2013. Biohydrometallurgy of secondary metal resources: A potential alternative approach for metal recovery. *Journal of Chemical Technology & Biotechnology* 88(12): 2115–2132. doi:10.1002/jctb.4164.

Evangelou, V. P. B. 1995. *Pyrite Oxidation and Its Control*. Boca Raton, FL: CRC Press/Taylor & Francis.

Faramarzi, M. A. and H. Brandl. 2006. Formation of water-soluble metal cyanide complexes from solid minerals by *Pseudomonas plecoglossicida*. *FEMS Microbiology Letters* 259(1): 47–52. doi:10.1111/j.1574-6968.2006.00245.x.

Faramarzi, M. A., M. Stagars, E. Pensini, W. Krebs, and H. Brandl. 2004. Metal solubilization from metal-containing solid materials by cyanogenic *Chromobacterium violaceum*. *Journal of Biotechnology* 113(1): 321–326. doi:10.1016/j.jbiotec.2004.03.031.

Feng, D. and J. S. J. Van Deventer. 2002. Leaching behaviour of sulphides in ammoniacal thiosulphate systems. *Hydrometallurgy* 63(2): 189–200. doi:10.1016/S0304-386X(01)00225-0.

Fomina, M. and G. M. Gadd. 2014. Biosorption: Current perspectives on concept, definition and application. *Bioresource Technology* 160: 3–14. doi:10.1016/j.biortech.2013.12.102.

Fonti, V., A. Dell'Anno, and F. Beolchini. 2016. Does bioleaching represent a biotechnological strategy for remediation of contaminated sediments? *Science of the Total Environment* 563–564: 302–319. doi:10.1016/j.scitotenv.2016.04.094.

Fowler, T. A. and F. K. Crundwell. 1998. Leaching of zinc sulfide by *Thiobacillus ferrooxidans*: Experiments with a controlled redox potential indicate no direct bacterial mechanism. *Applied and Environmental Microbiology* 64(10): 3570–3575.

Fowler, T. A. and F. K. Crundwell. 1999. Leaching of zinc sulphide by *Thiobacillus ferrooxidans*: Bacterial oxidation of the sulphur product layer increases the rate of zinc sulphide dissolution at high concentrations of ferrous iron. *Applied and Environmental Microbiology* 65(12): 5285–5292.

Fowler, T. A., P. R. Holmes, and F. K. Crundwell. 1999. Mechanism of pyrite dissolution in the presence of *Thiobacillus ferrooxidans*. *Applied and Environmental Microbiology* 65: 2987–2993.

Furuuchi, M. and K. Gotoh. 1992. Shape separation of particles. *Powder Technology* 73(1): 1–9. doi:10.1016/0032-5910(92)87001-Q.

Gadd, G. M. 2007. Geomycology: Biogeochemical transformations of rocks, minerals, metals and radionuclides by fungi, bioweathering and bioremediation. *Mycological Research* 111(1): 3–49. doi:10.1016/j.mycres.2006.12.001.

Gadd, G. M. 2009. Biosorption: Critical review of scientific rationale, environmental importance and significance for pollution treatment. *Journal of Chemical Technology & Biotechnology* 84(1): 13–28. doi:10.1002/jctb.1999.

Gadd, G. M. 2010. Metals, minerals and microbes: Geomicrobiology and bioremediation. *Microbiology* 156: 609–643.

Gesing, A. J., D. Reno, R. Grisier, R. Dalton, and R. Wolanski. 1998. Non-ferrous metal recovery from auto shredder residue using Eddy current separators. In *Proceedings of the TMS Annual Meeting*, B. Mishra (Ed.), San Antonio, TX, pp. 973–984. Warrendale, PA: TMS.

Ghorbani, Y., M. Oliazadeh, A. Shahverdian, R. Roohi, and A. Piraehgar. 2007. Use of some isolated fungi in biological leaching of aluminum from low grade bauxite. *African Journal of Biotechnology* 6(June): 1284–1288.

Giaveno, M. A., M. S. Urbieta, and E. Donati. 2011. Mechanisms of bioleaching: Basic understanding and possible industrial implications. In *Biohydrometallurgical Processes: A Practical Approach*, L. G. Santos Sobral, D. M. de Oliveira, C. E. Gomes de Souza (Eds.), pp 24–38. Rio de Janeiro, Brazil: CETEM, MCTI.

Gonen, F. and Z. Aksu. 2007. Copper(II) bioaccumulation properties of the yeast *C-tropicalis*: Effect of copper(II) on growth kinetics. *Journal of Biotechnology* 131: S165–S166.

González-Toril, E., E. Llobet-Brossa, E. O. Casamayor, R. Amann, and R. Amils. 2003. Microbial ecology of an extreme acidic environment, the Tinto River. *Applied and Environmental Microbiology* 69(8): 4853–4865. doi:10.1128/AEM.69.8.4853-4865.2003.

Gungor, A. and S. M. Gupta. 1998. Disassembly sequence planning for products with defective parts in product recovery. *Computers & Industrial Engineering* 35(1–2): 161–164. doi:10.1016/S0360-8352(98)00047-3.

Ha, V. H., J.-c. Lee, J. Jeong, H. T. Hai, and M. K. Jha. 2010. Thiosulfate leaching of gold from waste mobile phones. *Journal of Hazardous Materials* 178(1–3): 1115–1119. doi:10.1016/j.jhazmat.2010.01.099.

Hackl, R. P., D. B. Dreisinger, E. Peters, and J. A. King. 1995. Passivation of chalcopyrite during oxidative leaching in sulfate media. *Hydrometallurgy* 39(1–3): 25–48. doi:10.1016/0304-386X(95)00023-A.

Hagelüken, C. 2006. Improving metal returns and eco-efficiency in electronics recycling—A holistic approach for interface optimisation between pre-processing and integrated metals smelting and refining. In *Proceedings of the IEEE International Symposium on Electronics and the Environment*, Scottsdale, AZ, pp. 218–223.

Hallberg, K. B., B. M. Grail, C. A. du Plessis, and D. B. Johnson. 2011. Reductive dissolution of ferric iron minerals: A new approach for bio-processing nickel laterites. *Minerals Engineering* 24(7): 620–624. doi:10.1016/j.mineng.2010.09.005.

Hallberg, K. B. and D. B. Johnson. 2003. Novel acidophiles isolated from moderately acidic mine drainage waters. *Hydrometallurgy* 71(1): 139–148. doi:10.1016/S0304-386X(03)00150-6.

Hedley, N. and H. Tabachnick. 1958. *Chemistry of Cyanidation*. New York: American Cyanamid Company, Explosives and Mining Chemicals Department.

Helfferich, F. 1995. *Ion Exchange*. New York: Dover Publications, Inc.

Hetherington, P. 2006. Process intensification: A study of calcium carbonate precipitation methods on a spinning disc reactor. PhD thesis, Newcastle University, Newcastle upon Tyne, U.K.

Higashiyama, Y. and K. Asano. 1998. Recent progress in electrostatic separation technology. *Particulate Science and Technology* 16(1): 77–90. http://www.tandfonline.com/doi/abs/10.1080/02726359808906786#.WLFs5ocmJJY.mendeley.

Hilson, G. and A. J. Monhemius. 2006. Alternatives to cyanide in the gold mining industry: What prospects for the future? *Journal of Cleaner Production* 14(12): 1158–1167. doi:10.1016/j.jclepro.2004.09.005.

Hoberg, H. 1993. Applications of mineral processing in waste treatment and scrap recycling. In *Proceedings of the XVIII International Mineral Processing Congress*, Sydney, New South Wales, Australia, p. 27.

Hoffmann, J. E. 1992. Recovering precious metals from electronic scrap. *Journal of the Minerals, Metals and Materials Society* 44(7): 43–48.

Hong, K.-J., S. Tokunaga, and T. Kajiuchi. 2000. Extraction of heavy metals from MSW incinerator fly ashes by chelating agents. *Journal of Hazardous Materials* 75(1): 57–73. doi:10.1016/S0304-3894(00)00171-0.

Hoque, Md. E. and O. J. Philip. 2011. Biotechnological recovery of heavy metals from secondary sources—An overview. *Materials Science and Engineering C* 31(2): 57–66. doi:10.1016/j.msec.2010.09.019.

Horeh, N. B., S. M. Mousavi, and S. A. Shojaosadati. 2016. Bioleaching of valuable metals from spent lithium-ion mobile phone batteries using *Aspergillus niger*. *Journal of Power Sources* 320: 257–266. doi:10.1016/j.jpowsour.2016.04.104.

Iuga, A., L. Dăscălescu, R. Morar, I. Csorvassy, and V. Neamiu. 1989. Corona—Electrostatic separators for recovery of waste non-ferrous metals. *Journal of Electrostatics* 23(April): 235–243. doi:10.1016/0304-3886(89)90051-X.

Iuga, A., V. Neamtu, I. Suarasan, R. Morar, and L. Dascalescu. 1998. Optimal high-voltage energization of corona-electrostatic separators. *IEEE Transactions on Industrial Applications* 34: 286–293.

Johnson, D. B. 1998. Biodiversity and ecology of acidophilic microorganisms. *FEMS Microbiology Ecology* 27(4): 307–317. doi:10.1111/j.1574-6941.1998.tb00547.x.

Johnson, D. B. 2012. Reductive dissolution of minerals and selective recovery of metals using acidophilic iron- and sulfate-reducing acidophiles. *Hydrometallurgy* 127: 172–177. doi:10.1016/j.hydromet.2012.07.015.

Kelly, D. P. and A. P. Wood. 2000. Reclassification of some species of *Thiobacillus acidithiobacillus* gen. nov., *Halothiobacillus*. *International Journal of Systmatic and Evolutionary Microbiology* 50: 511–516.

Khaliq, A., M. A. Rhamdhani, G. Brooks, and S. Masood. 2014. Metal extraction processes for electronic waste and existing industrial routes: A review and Australian perspective. *Resources* 3: 152–179.

Kim, M. J., J. Y. Seo, Y. S. Choi, and G. H. Kim. 2016. Bioleaching of spent Zn–Mn or Ni–Cd batteries by *Aspergillus* species. *Waste Management* 51. 168–173. doi:10.1016/j.wasman.2015.11.001.

Kisielowska, E. and E. Kasinska-Pilut. 2005. Copper bioleaching from after-flotation waste using microfungi. *Acta Montanistica Slovaca* 10: 156–160.

Knowles, C. J. and A. W. Bunch. 1986. Microbial cyanide metabolism. *Advances in Microbial Physiology* 27: 73–111.

Kołodzziej, B. and Z. Adamski. 1984. A ferric chloride hydrometallurgical process for recovery of silver from electronic scrap materials. *Hydrometallurgy* 12(1): 117–127. doi:10.1016/0304-386X(84)90052-5.

Koyanaka, S., H. Ohya, S. Endoh, H. Iwata, and P. Ditl. 1997. Recovering copper from electric cable wastes using a particle shape separation technique. *Advanced Powder Technology* 8(2): 103–111. doi:10.1016/S0921-8831(08)60469-0.

Kratochvil, D. and B. Volesky. 1998. Advances in the biosorption of heavy metals. *Trends in Biotechnology* 16(7): 291–300. doi:10.1016/S0167-7799(98)01218-9.

Kuyucak, N. and B. Volesky. 1988. Biosorbents for recovery of metals from industrial solutions. *Journal of Biotechnology Letters* 10(2): 137–142.

Kuzugudenli, O. E. and C. Kantar. 1999. Alternatives to gold recovery by cyanide leaching. *Erzincan Üniversitesi Fen Bilimleri Enstitüsü Dergisi* 15(1–2): 119–127.

La Brooy, S. R., H. G. Linge, and G. S. Walker. 1994. Review of gold extraction from ores. *Minerals Engineering* 7(10): 1213–1241. doi:10.1016/0892-6875(94)90114-7.

Lee, J.-C., H. T. Song, and J.-M. Yoo. 2007. Present status of the recycling of waste electrical and electronic equipment in Korea. *Resources, Conservation and Recycling* 50(4): 380–397. doi:http://dx.doi.org/10.1016/j.resconrec.2007.01.010.

Lewis, A. E. 2010. Review of metal sulphide precipitation. *Hydrometallurgy* 104(2): 222–234. doi:10.1016/j.hydromet.2010.06.010.

Li, J., L. Huang, and X. Xu. 2011. Experimental research of leaching gold from waste printed circuit board by sodium hypochlorite method. *Chinese Journal of Environmental Engineering* 5: 453–456. doi:CNKI:SUN:HJJZ.0.2011-02-045.

Liu, R., J. Li, and Z. Ge. 2016. Review on *Chromobacterium violaceum* for gold bioleaching from e-waste. *Procedia Environmental Sciences* 31: 947–953. doi:10.1016/j.proenv.2016.02.119.

Liu, Y., L. L. Beer, and W. B. Whitman. 2012. Sulfur metabolism in archaea reveals novel processes. *Environmental Microbiology* 14(10): 2632–2644. doi:10.1111/j.1462-2920.2012.02783.x.

Lowson, R. T. 1982. Aqueous oxidation of pyrite by molecular oxygen. *Chemical Reviews* 82(5): 461–497. doi:10.1021/cr00051a001.

Lu, W., Y. Lu, F. Liu, K. Shang, W. Wang, and Y. Yang. 2011. Extraction of gold(III) from hydrochloric acid solutions by CTAB/n-heptane/iso-amyl alcohol/Na$_2$SO$_3$ microemulsion. *Journal of Hazardous Materials* 186c: 2166–2170. doi:10.1016/j.jhazmat.2010.12.059.

Luther III, G. W. 1987. Pyrite oxidation and reduction: Molecular orbital theory considerations. *Geochimica et Cosmochimca Acta* 51: 3193–3199.

Mack, C., B. Wilhelmi, J. R. Duncan, and J. E. Burgess. 2007. Biosorption of precious metals. *Biotechnology Advances* 25(3): 264–271. doi:10.1016/j.biotechadv.2007.01.003.

Malik, A. 2004. Metal bioremediation through growing cells. *Environment International* 30(2): 261–278. doi:10.1016/j.envint.2003.08.001.

Mangold, S., J. Valdés, D. S. Holmes, and M. Dopson. 2011. Sulfur metabolism in the extreme acidophile *Acidithiobacillus caldus*. *Frontiers in Microbiology* 2(Feb): 1–18. doi:10.3389/fmicb.2011.00017.

Mao, J., S. W. Won, K. Vijayaraghavan, and Y.-S. Yun. 2009. Surface modification of *Corynebacterium glutamicum* for enhanced Reactive Red 4 biosorption. *Bioresource Technology* 100: 1463–1466. doi:10.1016/j.biortech.2008.07.053.

Marsden, J. and I. House. 1992. *The Chemistry of Gold Extraction*. London, UK: Ellis Horwood Ltd.

Mathieu, G. I., R. Provencher, and J. G. Tellier. 1990. Mechanical sorting of aluminum metal from spent potlining. In *Proceedings of the 119th TMS Annual Meeting*, D. R. Gaskell (Ed.), Anaheim, CA, pp. 361–367. Warrendale, PA: TMS.

McGuire, M. M., K. J. Edwards, J. F. Banfield, and R. J. Hamers. 2001. Kinetics, surface chemistry, and structural evolution of microbially mediated sulfide mineral dissolution. *Geochimica et Cosmochimica Acta* 65(8): 1243–1258. doi:10.1016/S0016-7037(00)00601-3.

Meier-Staude, R. and R. Koehnlechner. 2000. Elektrostatische trennung von leiter/nichtleitergemischen in der betrieblichen praxis (Electrostatic separation of conductor/non-conductor mixtures in operational practice). *Aufbereitungstechnik* 41: 118–123.

Meyer, M. 1995. Development and realization of shredder fluff recycling. In *Proceedings of the Third International Symposium on Recycling of Metals and Engineered Materials, Point Clear*, P. B. Queneau, and R. D. Peterson (Eds.), pp. 765–776. Warrendale, PA: TMS.

Moses, C. O., D. K. Nordstrom, J. S. Herman, and A. L. Mills. 1987. Aqueous pyrite oxidation by dissolved oxygen and by ferric iron. *Geochimica et Cosmochimica Acta* 51: 1561–1571.

Motokawa, Y. and G. Kikuchi. 1971. Glycine metabolism in rat liver mitochondria. *Archives of Biochemistry and Biophysics* 146(2): 461–466. doi:10.1016/0003-9861(71)90149-4.

Mustin, C., P. de Donato, J. Berthelin, and P. Marion. 1993. Surface sulphur as promoting agent of pyrite leaching by *Thiobacillus ferrooxidans*. *FEMS Microbiology Reviews* 11: 71–78.

Naja, G., C. Mustin, B. Volesky, and J. Berthelin. 2008. Biosorption study in a mining wastewater reservoir. *Journal of Industrial Microbiology* 34: 14–27.

Nnorom, I. C. and O. Osibanjo. 2009. Toxicity characterization of waste mobile phone plastics. *Journal of Hazardous Materials* 161(1): 183–188. doi:10.1016/j.jhazmat.2008.03.067.

Nordstrom, D. K. 1982. Aqueous pyrite oxidation and the consequent formation of secondary iron minerals. In *Acid Sulfate Weathering, Pedogeochemistry and Relationship to Manipulation of Soil Minerals*, L. R. Hossner, J. A. Kittrick, and D. F. Fanning (Eds.), pp. 37–55. Madison, WI: Soil Science Society of America Press.

Norrgran, D. A. and J. A. Wernham. 1991. Recycling and secondary recovery applications using an eddy-current separator. *Minerals and Metallurgical Processing* 8: 184–187.

Norris, P. R., N. P. Burton, and N. A. M. Foulis. 2000. Acidophiles in bioreactor mineral processing. *Extremophiles* 4(2): 71.

Ohmura, N., K. Sasaki, N. Matsumoto, and H. Saiki. 2002. Anaerobic respiration using Fe(3+), S(0), and H(2) in the chemolithoautotrophic bacterium *Acidithiobacillus ferrooxidans*. *Journal of Bacteriology* 184(8): 2081–2087. doi:10.1128/JB.184.8.2081.

Okibe, N. and D. B. Johnson. 2004. Biooxidation of pyrite by defined mixed cultures of moderately thermophilic acidophiles in pH-controlled bioreactors: Significance of microbial interactions. *Biotechnology and Bioengineering* 87(5): 574–583. doi:10.1002/bit.20138.

Osorio, H., S. Mangold, Y. Denis, I. Ñancucheo, M. Esparza, D. B. Johnson, V. Bonnefoy, M. Dopson, and D. S. Holmesa. 2013. Anaerobic sulfur metabolism coupled to dissimilatory iron reduction in the extremophile *Acidithiobacillus ferrooxidans*. *Applied and Environmental Microbiology* 79(7): 2172–2181. doi:10.1128/AEM.03057-12.

Pant, D., D. Joshi, M. K. Upreti, and R. K. Kotnala. 2012. Chemical and biological extraction of metals present in e waste: A hybrid technology. *Waste Management* 32(5): 979–990. doi:10.1016/j.wasman.2011.12.002.

Paretsky, V. M., N. I. Antipov, and A. V. Tarasov. 2004. Hydrometallurgical method for treating special alloys, jewelry, electronic and electrotechnical scrap. In *Proceedings of the Minerals, Metals & Materials Society (TMS) Annual Meeting*, Charlotte, NC, pp. 713–721.

Peters, R. W. 1999. Chelant extraction of heavy metals from contaminated soils. *Journal of Hazardous Materials* 66(1): 151–210. doi:10.1016/S0304-3894(99)00010-2.

Preston, D. and K. A. Natarajan. 2004. Bacterial leaching. *Resonance* 9(8): 27–34. doi:10.1007/BF02837575.

Pronk, J. T., J. C. de Bruyn, P. Bos, and J. G. Kuenen. 1992. Anaerobic growth of *Thiobacillus ferrooxidans*. *Applied Environmental Microbiology* 58: 2227–2230.

Qu, Y. and B. Lian. 2013. Bioleaching of rare earth and radioactive elements from red mud using *Penicillium tricolor* RM-10. *Bioresource Technology* 136: 16–23. doi:10.1016/j.biortech.2013.03.070.

Qu, Y., B. Lian, B. Mo, and C. Liu. 2013. Bioleaching of heavy metals from red mud using *Aspergillus niger*. *Hydrometallurgy* 136: 71–77. doi:10.1016/j.hydromet.2013.03.006.

Quatrini, R., C. Appia-Ayme, Y. Denis, E. Jedlicki, D. S. Holmes, and V. Bonnefoy. 2009. Extending the models for iron and sulfur oxidation in the extreme acidophile *Acidithiobacillus ferrooxidans*. *BMC Genomics* 10: 394. doi:10.1186/1471-2164-10-394.

Quatrini, R., C. Appia-Ayme, Y. Denis et al. 2006. Insights into the iron and sulfur energetic metabolism of *Acidithiobacillus ferrooxidans* by microarray transcriptome profiling. *Hydrometallurgy* 83(1): 263–272. doi:10.1016/j.hydromet.2006.03.030.

Quinet, P., J. Proost, and A. Van Lierde. 2005. Recovery of precious metals from electronic scrap by hydrometallurgical processing routes. *Minerals and Metallurgical Processing* 22(1): 17–22.

Rawlings, D. E. 2002. Heavy metal mining using microbes. *Annual Review of Microbiology* 56(1): 65–91.

Ren, W.-X., P.-J. Li, Y. Geng, and X.-J. Li. 2009. Biological leaching of heavy metals from a contaminated soil by *Aspergillus niger*. *Journal of Hazardous Materials* 167(1): 164–169. doi:10.1016/j.jhazmat.2008.12.104.

Reuter, M., C. Hudson, A. Van Schaik, K. Heiskanen, C. Meskers, and C. Hageluken. 2013. Metal recycling: Opportunities, limits, infrastrucuture. Geneva, Switzerland: United Nations Environment Programme.

Rimstidt, J. D. and D. J. Vaughan. 2003. Pyrite oxidation: A state-of-the-art assessment of the reaction mechanism. *Geochimica et Cosmochimica Acta* 67(5): 873–880. doi:10.1016/S0016-7037(02)01165-1.

Ritcey, G. M. 2006. Solvent extraction in hydrometallurgy: Present and future. *Tsinghua Science and Technology* 11(2): 137–152. doi:10.1016/S1007-0214(06)70168-7

Rohwerder, T., T. Gehrke, K. Kinzler, and W. Sand. 2003. Bioleaching review part A: Progress in bioleaching: Fundamentals and mechanisms of bacterial metal sulfide oxidation. *Applied Microbiology and Biotechnology* 63: 239–248.

Rohwerder, T. and W. Sand. 2003. The sulfane sulfur of persulfides is the actual substrate of the sulfur-oxidizing enzymes from Acidithiobacillus and Acidiphilium spp. *Microbiology* 149: 1699–1710.

Rohwerder, T. and W. Sand. 2007. Mechanisms and biochemical fundamentals of bacterial metal sulfide oxidation. In *Microbial Processing of Metal Sulfides*, E. R. Donati, and W. Sand (Eds.), pp. 35–58. Dordrecht, the Netherlands: Springer.

Rossi, G. 1993. Biodepyritization of coal: Achievements and problems. *Fuel* 72: 1581–1592.

Roto, P. 1998. Metal processing and metal working industry. In *Encyclopedia of Occupational Health and Safety*, J. M. Stellman (Ed.). Geneva, Switzerland: CIS International Labour Office. Accessed February 14, 2017, http://www.ilocis.org/documents/chpt82e.htm.

Rowe, O. F., J. Sánchez-España, K. B. Hallberg, and D. B. Johnson. 2007. Microbial communities and geochemical dynamics in an extremely acidic, metal-rich stream at an abandoned sulfide mine (Huelva, Spain) underpinned by two functional primary production systems. *Environmental Microbiology* 9(7): 1761–1771. doi:10.1111/j.1462-2920.2007.01294.x.

Sadegh Safarzadeh, M., M. S. Bafghi, D. Moradkhani, and M. Ojaghi Ilkhchi. 2007. A review on hydrometallurgical extraction and recovery of cadmium from various resources. *Minerals Engineering* 20(3): 211–220. doi:10.1016/j.mineng.2006.07.001.

Saidan, M., B. Brown, and M. Valix. 2012. Leaching of electronic waste using biometabolised acids. *Chinese Journal of Chemical Engineering* 20(3): 530–534. doi:10.1016/S1004-9541(11)60215-2.

Sand, W., T. Gehrke, R. Hallman, and A. Schippers. 1995. Sulfur chemistry, bio¢lm, and the (in)direct attack mechanism—A critical evaluation of bacterial leaching. *Applied Microbiology and Biotechnology* 43(6): 961.

Sand, W., T. Gehrke, P.-G. Jozsa, and A. Schippers. 1997. *Biotechnology Comes of Age*, p. 366. Glenside, Adelaide, Australia: Australian Mineral Foundation, Glenside.

Sand, W., T. Gehrke, P.-G. Jozsa, and A. Schippers. 2001. (Bio)chemistry of bacterial leaching—Direct vs. indirect bioleaching. *Hydrometallurgy* 59(2–3): 159–175. doi:10.1016/S0304-386X(00)00180-8.

Santhiya, D. and Y. P. Ting. 2006. Use of adapted *Aspergillus niger* in the bioleaching of spent refinery processing catalyst. *Journal of Biotechnology* 121(1): 62–74. doi:10.1016/j.jbiotec.2005.07.002.

Schippers, A. 2004. Biogeochemistry of metal sulfide oxidation in mining environments, sediments, and soils. *Geological Society of America Special Papers* 379: 49–62.

Schippers, A. 2007. Microorganisms involved in bioleaching and nucleic acid-based molecular methods for their identification and quantification. In *Microbial Processing of Metal Sulfides*, E. R. Donati, and W. Sand (Eds.), pp. 3–33. Dordrecht, the Netherlands: Springer.

Schippers, A. and B. B. Jørgensen. 2002. Biogeochemistry of pyrite and iron sulfide oxidation in marine sediments. *Geochimica et Cosmochimica Acta* 66: 85–92.

Schippers, A., P. G. Jozsa, and W. Sand. 1996. Sulfur chemistry in bacterial leaching of pyrite. *Applied and Environmental Microbiology* 62(9): 3424–3431.

Schippers, A., T. Rohwerder, and W. Sand. 1999. Intermediary sulfur compounds in pyrite oxidation: Implications for bioleaching and biode-pyritization of coal. *Applied Microbiology and Biotechnology* 52: 104–110.

Schippers, A. and W. Sand. 1999. Bacterial leaching of metal sulfides proceeds by two indirect mechanisms via thiosulfate or via polysulfides and sulfur. *Applied and Environmental Microbiology* 65(1): 319–321.

Schubert, H. 1994. Wirbelstromsortierung—Grundlagen, scheider, anwendungen (Eddy current separation—Foundations, separators, application). *Aufbereitungstechnik* 35: 553–562.

Schubert, H. G. and G. Warlitz. 1994. Sorting metal/non-metal mixtures using a corona electrostatic separator. *Aufbereitungstechnik* 35: 449–456.

Shamsuddin, M. 1986. Metal recovery from scrap and waste. *Journal of Metals* 38: 24–31.

Shen, R., G. Zhang, M. Dell'Amico, P. Brown, and O. Ostrovski. 2007. A feasibility study of recycling of manganese furnace dust. *INFACON* XI: 507–519.

Sheng, P. P. and T. H. Etsell. 2007. Recovery of gold from computer circuit board scrap using aqua regia. *Waste Management & Research: The Journal of the International Solid Wastes and Public Cleansing Association, ISWA* 25(4): 380–383. doi:10.1177/07342 42X07076946.

Shuey, S. A. and P. Taylor. 2005. Review of pyrometallurgical treatment of electronic scrap. *Mining Engineering* 57(4): 67–70.

Singer, P. C. and W. Stumm. 1970. Acidic mine drainage: The rate determining step. *Science* 167: 1121–1123.

Singh, S. and S. Cameotra. 2015. Anaerobic bioleaching by acidophilic bacterial strains. In *Environmental Microbial Biotechnology, Soil Biology*, L. B. Sukla et al. (Eds.). Switzerland: Springer International Publishing.

Smart, R. S. C., M. Jasieniak, K. E. Prince, and W. M. Skinner. 2000. SIMS studies of oxidation mechanisms and polysulfide formation in reacted sulfide surfaces. *Minerals Engineering* 13(8–9): 857–870. doi:10.1016/S0892-6875(00)00074-1.

Stahl, I. and P.-M. Beier. 1997. Sorting of plastics using the electrostatic separation process. In *Proceedings of the XX International Mineral Processing Congress*, Vol. 5, Hoberg, H., and H. Von Blottnitz (Eds.). Aachen, Germany: GDMB, Clausthal-Zellerfeld, pp. 395–401.

Sum, E. Y. L. 1991. The recovery of metals from electronic scrap. *Journal of the Minerals, Metals and Materials Society* 43(4): 53–61.

Syed, S. 2016. Silver recovery aqueous techniques from diverse sources: Hydrometallurgy in recycling. *Waste Management* 50: 234–256. doi:10.1016/j.wasman.2016.02.006.

Tavlarides, L. L., J. H. Bae, and C. K. Lee. 1985. Solvent extraction, membranes, and ion exchange in hydrometallurgical dilute metals separation. *Separation Science and Technology* 22(2–3): 581–617. doi:10.1080/01496398708068970.

Thomas, J. E., C. F. Jones, W. M. Skinner, and R. S. C. Smart. 1998. The role of surface sulfur species in the inhibition of pyrrhotite dissolution in acid conditions. *Geochimica et Cosmochimica Acta* 62(9): 1555–1565. doi:10.1016/S0016-7037(98)00087-8.

Thomas, J. E., W. M. Skinner, and R. S. C. Smart. 2001. A mechanism to explain sudden changes in rates and products for pyrrhotite dissolution in acid solution. *Geochimica et Cosmochimica Acta* 65(1): 1–12. doi:10.1016/S0016-7037(00)00503-2.

Thurston, R. S., K. W. Mandernack, and W. C. Shanks. 2010. Laboratory chalcopyrite oxidation by *Acidithiobacillus ferrooxidans*: Oxygen and sulfur isotope fractionation. *Chemical Geology* 269(3): 252–261. doi:10.1016/j.chemgeo.2009.10.001.

Tributsch, H. 2001. Direct versus indirect bioleaching. *Hydrometallurgy* 59(2): 177–185. doi:10.1016/S0304-386X(00)00181-X.

Tributsch, H. and J. C. Bennett. 1981a. Semiconductor-electrochemical aspects of bacterial leaching. 1. Oxidation of metal sulphides with large energy gaps. *Journal of Chemical Technology and Biotechnology* 31(1): 565–577. doi:10.1002/jctb.503310186.

Tributsch, H. and J. C. Bennett. 1981b. Semiconductor-electrochemical aspects of bacterial leaching. Part 2. Survey of rate-controlling sulphide properties. *Journal of Chemical Technology and Biotechnology* 31(1): 627–635. doi:10.1002/jctb.503310186.

Van Der Valk, H. J. L., B.C. Braam, and W.L. Dalmijn. 1986. Eddy-current separation by permanent magnets part I: Theory. *Resources and Conservation* 12(3–4): 233–252. doi:10.1016/0166-3097(86)90014-3.

Valenzuela, L., A. Chi, S. Beard, A. Orell, N. Guiliani, J. Shabanowitz, D. F. Hunt, and C. A. Jerez. 2006. Genomics, metagenomics and proteomics in biomining microorganisms. *Biotechnology Advances* 24(2): 197–211. doi:10.1016/j.biotechadv.2005.09.004.

Vaughan, D. J. and J. R. Craig. 1978. *Mineral Chemistry of Metal Sulfides*. Cambridge, U.K.: Cambridge University Press.

Vera, M., A. Schippers, and W. Sand. 2013. Progress in bioleaching: Fundamentals and mechanisms of bacterial metal sulfide oxidation—Part A. *Applied Microbiology and Biotechnology* 97(17): 7529–7541. doi:10.1007/s00253-013-4954-2.

Volesky, B. 1990. Biosorption by fungal biomass. In *Biosorption of Heavy Metals*, B. Volesky (Ed.), pp. 139–172. Boca Raton, FL: CRC Press.

Volesky, B. 2003. Biosorption process simulation tools. *Hydrometallurgy* 71(1): 179–190. doi:10.1016/S0304-386X(03)00155-5.

Volesky, B. 2007. Biosorption and me. *Water Research* 41(18): 4017–4029. doi:10.1016/j.watres.2007.05.062.

Welch, S. A., W. W. Barker, and J. F. Banfield. 1999. Microbial extracellular polysaccharides and plagioclase dissolution. *Geochimica et Cosmochimica Acta* 63(9): 1405–1419. doi:10.1016/S0016-7037(99)00031-9.

Wentzien, S., W. Sand, A. Albertsen, and R. Steudel. 1994. Thiosulfate and tetrathionate degradation as well as biofilm generation by *Thiobacillus intermedius* and *Thiobacillus versutus* studied by microcalorimetry, HPLC, and ion-pair chromatography. *Archives of Microbiology* 161: 116–125.

Wernham, J. A., J. A. Marin, and D. E. Heubel. 1993. Aluminum removal from recycled pet. In *Proceedings of the First International Conference on Processing Materials for Properties*, Honolulu, HI, pp. 759–762. Warrendale, PA: TMS.

Williamson, M. A. and J. D. Rimstidt. 1994. The kinetics and electrochemical rate-determining step of aqueous pyrite oxidation. *Geochimica et Cosmochimica Acta* 58(24): 5443–5454. doi:10.1016/0016-7037(94)90241-0.

Wills, B. A. 1988. *Mineral Processing Technology*, 4th ed., pp. 377–381. Oxford, U.K.: Pergamon Press.

Wilson, R. J., T. J. Veasey, and D. M. Squires. 1994. The application of mineral processing techniques for the recovery of metal from post-consumer wastes. *Minerals Engineering* 7(8): 975–984. doi:10.1016/0892-6875(94)90027-2.

Wu, H. Y. and Y. P. Ting. 2006. Metal extraction from municipal solid waste (MSW) incinerator fly ash—Chemical leaching and fungal bioleaching. *Enzyme and Microbial Technology* 38(6): 839–847. doi:10.1016/j.enzmictec.2005.08.012.

Wu, J. B., L. J. Qiu, L. B. Chen, and D. H. Chen. 2009. Gold and silver selectively leaching from printed circuit boards scrap with acid thiourea solution. *Non-Ferrous Metals* 61: 90–93. doi:CNKI:SUN:YOUS.0.2009-04-020.

Xin, B., B. Chen, N. Duan, and C. Zhou. 2011. Extraction of manganese from electrolytic manganese residue by bioleaching. *Bioresource Technology* 102(2): 1683–1687.

Xu, T. J. and Y. P. Ting. 2009. Fungal bioleaching of incineration fly ash: Metal extraction and modeling growth kinetics. *Enzyme and Microbial Technology* 44(5): 323–328. doi:10.1016/j.enzmictec.2009.01.006.

Xu, Y. and M. A. A. Schoonen. 2000. The absolute energy positions of conduction and valence bands of selected semiconductiong minerals. *American Mineralogist* 85: 543–556.

Yang, B. 1994. Ion exchange in organic extractant system. *Ion Exchange Adsorption* 10: 168–179.

Yang, H., J. Liu, and J. Yang. 2011. Leaching copper from shredded particles of waste printed circuit boards. *Journal of Hazardous Materials* 187(1): 393–400. doi:10.1016/j.jhazmat.2011.01.051.

Yang, J., Q. Wang, Q. Wang, and T. Wu. 2008. Comparisons of one-step and two-step bioleaching for heavy metals removed from municipal solid waste incineration fly ash. *Environmental Engineering Science* 25(5): 783–789. doi:10.1089/ees.2007.0211.

Yang, J., Q. Wang, Q. Wang, and T. Wu. 2009. Heavy metals extraction from municipal solid waste incineration fly ash using adapted metal tolerant *Aspergillus niger*. *Bioresource Technology* 100(1): 254–260. doi:10.1016/j.biortech.2008.05.026.

Yannopoulos, J. C. 1991. *The Extractive Metallurgy of Gold*. New York: Van Nostrand Reinhol.

Yilmazer, P. and N. Saracoglu. 2009. Bioaccumulation and biosorption of copper(II) and chromium(III) from aqueous solutions by *Pichia stipitis* yeast. *Journal of Chemical Technology & Biotechnology* 84(4): 604–610. doi:10.1002/jctb.2088.

Zhang, J.-Z. and F. J. Millero. 1993. The products from the oxidation of H_2S in seawater. *Geochimica et Cosmochimica Acta* 57(8): 1705–1718. doi:10.1016/0016-7037 (93)90108-9.

Zhang, S. and E. Forssberg. 1998. Optimization of electrodynamic separation for metals recovery from electronic scrap. *Resources, Conservation and Recycling* 22(3): 143–162. doi:10.1016/S0921-3449(98)00004-4.

Zhong, F., D. Li, and J. Wei. 2006. Experimental study on leaching gold in printed circuit boards scrap with thiourea. *Journal of Non Ferrous Metal Recycling and Utilization* 6: 25–27. doi:CNKI:SUN:HJJZ.0.2011-02-045.

4 Recycling of Electronic Waste

4.1 INTRODUCTION

The advancements in electronic technology, on the one hand, have magnified the manufacturing of electronic devices and their use worldwide and, on the other, have minimized their life span, leading them concomitantly to end up as waste. One of the fastest-growing wastes in municipal waste comes from the category of electronic waste (e-waste) or wastes from electrical and electronic equipment (WEEE) (Bertram et al. 2002). A United Nations Environment Program (UNEP) report estimated that the electrical waste and e-waste generated globally is around 50 million tons annually (Huisman et al. 2007, UNEP 2009). The same report projected that by the next decade, there would be a 500% surge in e-waste production in developing nations. Among e-waste, printed circuit boards (PCBs) have quite diverse composition, containing polymers, ceramics, and metals. The metal content is around 28%–30% (copper: 10%–20%, lead: 1%–5%, nickel: 1%–3%, precious metals such as silver, platinum, and gold are also present in the electronic scrap to a total of 0.3%–0.4%). Many other elements (Ga, In, Ti, Si, Ge, As, Sb, Se, and Te) may be found in chips, with An, Pb, and Cd in solder joints; Ga, Si, Se, and Ge in semiconductors; and tantalum in capacitors (Li et al. 2004). Constituents of the other remaining important materials are plastics (19%), bromine (4%), and glass and ceramics (49%). Besides these inorganic elements, the other important organic compounds such as isocyanates, phosgene, and acrylic and phenolic resins are also found in circuit boards (Ludwig et al. 2003, Ilyas et al. 2010). Ceramics present in PCBs include silica, alumina, alkaline earth oxides, mica, and barium titanate (Li et al. 2004).

Recycling of such e-waste is important from both aspects of waste treatment and recovery of valuable metals. It has been claimed that the purity level of precious metals in waste PCBs (WPCBs) is more than 10 times higher than that of rich-content minerals, making them a potentially economically viable mineral resource (Huang et al. 2009). Various mechanical, pyrometallurgical, and hydrometallurgical methods are used for recycling these wastes. Mechanical separation can just be useful as a pretreatment method (Syed 2012). Pyrometallurgy, based on incineration, smelting in a plasma arc furnace or blast furnace, drossing, sintering, melting, and high-temperature reactions in a gas phase, is used for the recovery of nonferrous metals as well as precious metals from WPCBs (Huang et al. 2009). Hydrometallurgical processing of PCB waste usually includes both chemical leaching and electrochemical processing (Li et al. 2004). This involves a series of acid or caustic (cyanide, halide, thiourea, thiosulfate) leaching of the waste. The obtained solutions are subjected to precipitation of impurities, solvent extraction, adsorption, and ion exchange in order to purify and concentrate metals (Cui and Zhang 2008, Huang et al. 2009). But these methods cause secondary pollution, as halogenated flame retardants in the smelter feed can lead to the formation of dioxins and furans (Menad et al. 1998) or produce high volumes of effluents and also are not economical due to high energy consumption (Krebs et al. 1997).

Process-integrated biotechnology is characterized by the application of biocatalysts (e.g., microorganisms, enzymes) in an industrial process and the substitution of existing processes. Biotechnology not only reduces the cost but also reduces the environmental footprint of a given process. Regarding sustainability, biological processes can contribute, to a large extent, to future technologies, including waste treatment (Wang et al. 2009). Bioleaching technology offers many advantages over conventional methods, including its relative simplicity, low operating costs, less exacting operational requirements, low energy input, reduced need for skilled labor, and, most importantly, environmental friendliness. However, it requires a longer period of operation compared with other methods, such as chemical leaching (Huang et al. 2009, Amiri et al. 2011b). Bioleaching has been used for the recovery of precious metals and copper from ores for many years. However, limited researches were carried out on the bioleaching of metals from e-waste (Cui and Zhang 2008).

Indeed, bioleaching enables the recycling of metals by a process close to natural biogeochemical cycles of waste materials, thus reducing the demand for natural resources such as ores, energy, and landfill space (Krebs et al. 1997). Undoubtedly, leaching metals from waste WPCBs by using microorganisms will be a promising alternative in comparison with traditional methods. For these reasons, techniques facilitating and improving the bioleaching process should be of interest. Additionally, there is a need for research on seeking the available source of active bioleaching biomass (Huang et al. 2009, Lee and Pandey 2012). This chapter focuses on the biohydrometallurgical recycling of metals from e-waste.

4.2 BIOLEACHING OF METALS FROM E-WASTE

4.2.1 SOLDER

Solders are an indispensable part of any electronic device as they provide both electrical connection and mechanical support in the packaging modules (Ma and Suhling 2009). Various types of solders, tin–copper (Sn–Cu), tin–copper–silver (Sn–Cu–Ag), and tin–lead (Sn–Pb) solders, are used extensively in electronic devices. Sufficient attention should be paid on the possible spreading of these heavy metals leached from solders in a natural environment. Thus, there is a need to recycle the waste lead and the lead-free solders as they contain materials that are valuable and recyclable as well as toxic.

Considering the toxicity of metals toward microbial cells, Hocheng et al. (2014) used a two-step process for effective bioleaching of metals from solders using culture supernatants of *Aspergillus niger* and *Acidithiobacillus ferrooxidans*. Faster metal dissolution (99%) was achieved in 60 h, at 200 rpm shaking speed, 30°C temperature, by using 100 mL of *A. niger* culture supernatant as compared with *At. ferrooxidans* culture supernatant. Metals were removed faster from Sn–Cu–Ag solder than from Sn–Cu and Sn–Pb solders by *A. niger* culture supernatant. Bioleaching of metals by *A. niger* culture supernatant is based on the acidolysis/complexolysis process carried out by an organic acid. An acidic environment created by citric acid present in *A. niger* culture supernatant favors the dissolution of metals. Citric acid provides both a source of protons, which protonate the anionic component, and an organic acid anion (Morley et al. 1996). The protonation of oxygen atoms occurs around the surface of metallic compounds. The protons and oxygen associated with water displace the metal from the surface (Mulligan et al. 2004, Anjum et al. 2012). In the case of *At. ferrooxidans*, the bioleaching process is carried out based on the oxidation mechanism. *At. ferrooxidans* uses Fe^{2+} as an energy source and produces Fe^{3+}, which is a strong oxidant. It can oxidize metals and convert them into a soluble form (Jadhav et al. 2013). Scanning electron microscopy of the surface of solders during the bioleaching process shows the deterioration of the Sn–Cu–Ag solder surface, indicating the removal of metals from the solder surface (Figure 4.1).

Further, to achieve selective precipitation of metals, hydrogen sulfide (H_2S) gas, sodium chloride (NaCl), and sodium hydroxide (NaOH) were used. The addition of NaOH and NaCl in the bioleached solution from Sn–Cu–Ag solder precipitated tin (85%) and silver (80%), respectively. Passing of H_2S gas at pH 8.1 selectively precipitated lead (57.18%) from the Sn–Pb bioleached solution.

The results of the present study suggest that metal dissolution based on the acidolysis/complexolysis mechanism is more efficient as compared with the oxidation mechanism. One possible reason behind this is the presence of metals mainly as oxides, carbonates, and silicates rather than as sulfides in industrial solid waste materials. It is easier to leach metals via acid generated by microorganisms rather than

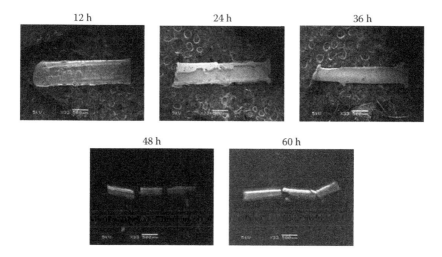

FIGURE 4.1 Scanning electron microscopy (SEM) analysis of the Sn–Cu–Ag solder surface during bioleaching by *Aspergillus niger* culture supernatant. (With kind permission from Springer Science+Business Media: *Appl. Biochem. Biotechnol.*, Microbial leaching of waste solder for recovery of metal, 173, 2014, 193, Hocheng, H., Hong, T., and Jadhav, U.)

via ferric ion (Vestola et al. 2010). Although chemical leaching of solders using various chemicals such as NaOH, sodium persulfate, nitric acid, and organic solvents is effective, these chemicals are harmful to the environment (Mecucci and Scott 2002, Yoo et al. 2012). Thus, the investigated method puts forth an effective environment-friendly alternative for recycling of waste solders.

4.2.2 Printed Circuit Boards (PCBs)

PCBs constitute a major part of e-waste which are being produced enormously and increasing day by day. Treating e-waste before disposal not only removes the toxic substances but also lowers the disposal costs and further is also the attractive secondary metal source. Table 4.1 below shows various microorganisms involved in bioleaching of metals from different e-wastes.

4.2.2.1 Microorganisms for Metal Recovery from PCBs

The study carried out by Brandl et al. (2001) suggests that the production of inorganic and organic acids by bacteria (*Thiobacillus thiooxidans, T. ferrooxidans*) and fungi (*A. niger, Penicillium simplicissimum*) led to the mobilization of metals from electronic scrap. Dust collected from shredding processes of electronic scrap was used in the process. The addition of a greater amount of scrap resulted in an increase in the initial pH due to the alkaline nature of scrap. To avoid toxicity, a two-step process was introduced, where biomass was produced in the absence of electronic scrap as the first step, and subsequently, different scrap concentrations were added to the cultures and incubated further for the bioleaching process. One-step leaching experiments using fungi demonstrated that at a concentration of scrap >10 g/L in the medium, microbial growth was inhibited. In a two-step leaching process, however, concentrations of up to 100 g/L of electronic scrap could be easily treated after acclimatization of fungi. Both fungal strains were able to mobilize Cu and Sn by 65%, and Al, Ni, Pb, and Zn by more than 95%. The two-step leaching process was demonstrated using a commercial gluconic acid (NaglusoTM, 2.5 M) produced by *A. niger*.

At scrap concentrations of 5–10 g/L, a consortium of *Thiobacillus* strains were able to leach more than 90% of the available Cu, Zn, Ni, and Al. Pb precipitated as $PbSO_4$, while Sn precipitated probably as SnO. This was confirmed by applying a computer program named MICROQL to model the speciation of dissolved components under the given conditions. The authors further suggests that a two-step leaching process will be more appropriate for increasing the leaching efficiency from the point of view of industrial application. The authors suggest the following advantages for a two-step process:

1. Biomass is not in direct contact with the metal-containing waste and might be recycled.
2. Waste material is not contaminated by microbial biomass.
3. Acid formation can be optimized in the absence of waste material.
4. Higher waste concentrations can be applied as compared with the one-step process, resulting in increased metal yields.

TABLE 4.1

Bioleaching of Metals from Electronic Waste by Various MicroOrganisms

Type of Waste	Microorganisms	Leaching of Metals (%)	Reference
Electronic scrap	*Thiobacillus (T.) thiooxidans* +	Cu, Zn, Al, and Ni (90)	Brandl et al. (2001)
	T. ferrooxidans, Aspergillus (A.) niger + *Penicillium simplicissimum*	Cu, Zn, Al, and Ni (100)	
Electronic scrap	*Chromobacterium (C.) violaceum, Pseudomonas (P.) fluorescens, P. plecoglossicida*	Au (68.5)	Brandl et al. (2008)
Electronic scrap	*Acidithiobacillus* sp. + *Leptospirillum* sp.	Cu and Ni (100)	Vestola et al. (2010)
Electronic scrap	*Sulfobacillus (S.) thermosulfidooxidans* + acidophilic heterotroph (code A1TSB)	Cu (89), Zn (83), Ni (81), and Al (79)	Ilyas et al. 2007
Electronic scrap	*S. thermosulfidooxidans* + *Thermoplasma acidophilum*	Cu (86), Zn (80), Ni (74), and Al (64)	Ilyas et al. (2010)
Electronic scrap	*S. thermosulfidooxidans* + *S. acidophilus*	Cu (85), Zn (74), Ni (78), and Al (68)	Ilyas et al. (2013)
Printed circuit boards (PCBs)	*S. thermosulfidooxidans*	Cu (95), Zn (96), Ni (94), and Al (91)	Ilyas and Lee (2014)
Electronic scrap material (ESM)	*C. violaceum*		Natarajan and Ting (2014)
	Wild type	Au (11)	
	Mutated strain at pH 9	Au (18)	
	pH 9.5	Au (22.5)	
	pH 10	Au (19)	
ESM	*C. violaceum*		Natarajan et al. (2015)
	Wild type	Au (11.3)	
	pBAD	Au (29.6)	
	pTAC	Au (24.6)	
Electronic waste	*C. violaceum*	Cu (95.7) and Au (30)	Natarajan and Ting (2015)

(Continued)

TABLE 4.1 (*Continued*)
Bioleaching of Metals from Electronic Waste by Various MicroOrganisms

Type of Waste	Microorganisms	Leaching of Metals (%)	Reference
Electronic waste	*Acidithiobacillus* (*At.*) *caldus* BRGM3, *Leptospirillum* (*L.*) *ferriphilum* BRGM1, *S. benefaciens* BRGM2 and *Ferroplasma* (*Fp.*) *acidiphilum* BRGM4, *At. ferrooxidans* T ESM14882, and *Acidiphilium* sp. SJH		Bryan et al. (2015)
Printed wire boards (PWBs)	*At. ferrooxidans*	Cu (99.0)	Wang et al. (2009)
	At. thiooxidans	Cu (74.9)	
	Consortium	Cu (99.9)	
PCBs	*Acidithiobacillus* + *Gallionella* + *Leptospirillum* sp.	Cu (95)	Xiang et al. (2010)
PCBs	*Acidithiobacillus* + *Gallionella* + *Leptospirillum* sp.	Cu (96.8), Zn (91.6), and Al (88.2)	Zhu et al. (2011)
PCBs	Mixed consortium	Cu (23.97), Sn (59.96), Pb (9.30), and Fe (5.92)	Adhapure et al. (2013)
Waste printed circuit boards (WPCBs)	*At. ferrooxidans*	Cu (96.8), Zn (83.8), and Al (75.4)	Yang et al. (2014)
PCBs	*P. chlororaphis*	Cu (52.3), Ag (12.1), and Au (8.2)	Ruan et al. (2014)
Computer printed circuit boards (CPCBs)	*At. ferrooxidans*	Cu and Ni (100)	Arshadi and Mousavi (2014)
PCBs	*Bacillus* (*B.*) *megaterium*	Au (36.81)	Arshadi and Mousavi (2015a)
	At. ferrooxidans + *B. megaterium*	Au (63.8)	
Mobile phone printed circuit boards (MPPCBs)	*At. ferrooxidans*	Cu (70)	Arshadi and Mousavi (2015b)
MPPCBs	*B. megaterium*	Cu (72) and 65 g Au/ton	Arshadi et al. (2016)
PCBs	*At. ferrooxidans* strain Z1	Cu (92.57), Al (85.24), and Zn (95.18)	Yang et al. (2017)

(Continued)

TABLE 4.1 (*Continued*)

Bioleaching of Metals from Electronic Waste by Various MicroOrganisms

Type of Waste	Microorganisms	Leaching of Metals (%)	Reference
PCBs	*At. ferrooxidans*	Cu (80)	Li et al. (2015)
	C. violaceum	Au (70.6)	
PCBs	*L. ferriphilum*	Cu (94.08), Zn (99.80), and Ni (99.80)	Shah et al. (2015)
PCBs	*At. ferrooxidans*	Cu (94.1)	Liang et al. (2016)
PCBs	*S. thermosulfidooxidans*	Cu (85)	Rodrigues et al. (2015)
PCBs	*At. ferrooxidans*	Cu (95.30), Zn (91.68), Mn (90.32), Al (86.31), and Ni (59.07)	Nie et al. (2015)
WPCBs	*At. ferrooxidans*	Cu (94.8)	Chen et al. (2015)
PCBs	*At. ferrivorans* and *At. thiooxidans* (DSM 9463), *P. putida* (WSC361), and *P. fluorescens* (E11.3)		Isıldar et al. (2016)
Scrap TV circuit boards (STVBs)	MES1 (*At. ferrooxidans*, *L. ferrooxidans*, *At. thiooxidans*)	Cu (89)	Bas et al. 2013
MPPCBs	*C. violaceum*	Cu (24.6)	Chi et al. (2011)
		Au (11.31)	
WPCBs	*C. violaceum*	Cu (37) and Au (13)	Tran et al. (2011)
Waste MPPCBs	*C. violaceum*	Au (20–30)	Tay et al. (2013)
Gold-plated finger integrated circuits found in computer motherboards (GFICMs) and MPPCBs	*A. niger* MXPE6 + *A. niger* MX7	Au (87 for PCb and 28) Cu (0.2 for PCB and 29)	Madrigal-Arias et al. (2015)
Waste electric cables	*At. ferrooxidans*	Cu (90)	Lambert et al. (2015)

Choi et al. (2004) investigated the potential use of *At. ferrooxidans* for bioleach-ing of copper contained in PCBs of waste computers. PCB shreds containing a high amount of copper were selected after disposal of shreds containing plastic. The authors suggest that $Fe_2(SO_4)_3$ formed by *At. ferrooxidans* during the leaching process oxidizes elemental copper from PCBs to the cupric ion (Equation 4.1) as follows (Choi et al. 2004):

$$Fe_2(SO_4)_3 + Cu \rightarrow Cu^{2+} + 3SO_4^{2-} \qquad (4.1)$$

It is clear that under conditions of the initial addition of Fe^{2+} ion, the concentration of copper in the precipitate was even higher than that remaining in the solution. Therefore, after the bioleaching of PCBs shreds by *At. ferrooxidans*, subsequent treatment of reaction precipitate in a proper way should be sought in an effort to develop a more efficient method for the recovery of copper in practical processes. The addition of 7 g/L of ferrous ion was suggested to be optimum, depending upon the total amount of copper both in the solution and in the precipitate, and the recov-ery efficiency of copper against the input was 24%.

Further, to increase the solubility of copper and enhance the efficiency of the bioleaching process, citric acid was added as a complexing agent. The total leached copper that remained dissolved in the absence of citric acid was increased from 37 wt% to more than 80 wt% after the addition of citric acid. This demonstrates that the addition of a complexing agent such as citric acid to the bioleaching solution can increase the solubility of leached metal ions.

Yang et al. (2009) investigated the use of the mesophilic bacteria *At. ferrooxidans* for copper bioleaching from PCBs. Results show the influence of Fe^{3+} and pH on copper solubilization. The authors show the importance of Fe^{3+} in the copper leach-ing mechanism as follows:

$$Fe^{2+} + O_2 + H^+ \xrightarrow{\text{A. ferroxidans}} Fe^{3+} + H_2O \qquad (4.2)$$

$$Cu^0 + 2Fe_{(aq)}^{3+} \rightarrow 2Fe_{(aq)}^{2+} + Cu_{(aq)}^{2+\Delta G^\theta} \qquad (4.3)$$

The procedure for the solubilization of copper is straightforward, as the Gibbs free energy of the reaction is a subtractive value of −82.90 kJ/mol, which means that the reaction (4.3) can take place under thermodynamic conditions. Higher copper extrac-tion was achieved by a greater concentration of Fe^{3+} (6.66–2.36 g/L) and a lower pH range (1.5–2.0), thus suggesting them to be the two main variables involved in copper mobilization.

Wang et al. (2009) isolated *At. ferrooxidans* and *At. thiooxidans* from an acidic mine drainage and evaluated their capacity for bioleaching of printed wire boards (PWBs). Both microorganisms alone and in the consortium were studied for their bioleachability. The authors crushed the electronic scrap and sieve-fractioned it to obtain sieve fractions of different sizes. Results indicated that the leaching percent-age for all the cultures increased with a decrease in the size of the sieve fraction sample; the reason behind could be that the ratio area increases with the decrease in the sample size and this increased ratio area is favorable for microbial attachment. Under a 7.8 g/L concentration of PWBs, the percentages of copper solubilized were 99.0%, 74.9%, and 99.9% for the 0.5–1.0 mm sieve fractions of the sample at 9 days of leaching time using a pure culture of *At. ferrooxidans*, a pure culture of *At. thiooxidans*, and a mixed culture, respectively, while the percentages of copper, lead, and zinc solubilized were all more than 88.9% for the <0.35 mm sieve fractions of the sample at 5 days of leaching time using the same three cultures. Variations in pH and redox potential of leaching solution with time implied that oxidation of Fe^{2+} to Fe^{3+} in the culture medium in the presence of *At. ferrooxidans* led to the mobilization of metals (Wang et al. 2009).

Time and the amount of PCB addition adversely affect both bacterial growth and metal bioleaching. Thus, Liang et al. (2010) utilized the strategy of multiple PCB additions, that is, increasing the PCB dose with time, which not only improved metal recovery but also minimized growth inhibition of *At. thiooxidans* and *At. ferrooxidans*. A fine powder of PCB was added (4 g/L at 48 h, 6 g/L at 96 h, and 8 g/L at 144 h) to separate bacterial cultures. As a result, the percentages of Cu, Ni, Zn, and Pb leached by *At. thiooxidans* were 78%, 73%, 75%, and 71%, and corresponding values were 80%, 73%, 76%, and 72% for *At. ferrooxidans* after 240 h of cultivation. Also, it worked great with the mixed culture, extracting Cu, Ni, Zn, and Pb at 94%, 89%, 90%, and 86%, respectively. Moreover, the increased redox potential and lowered pH value were elucidated to be the mechanism involved in achieving greater metal bioleaching in a mixed culture (Liang et al. 2010).

Xiang et al. (2010) used an enriched bacterial consortium for the solubilization of copper from PCBs and determined the optimum conditions for the same. The bacterial consortium was enriched from a natural acid mine drainage (AMD) collected from a local pyrite mine in Guangdong, China. It consisted of bacteria mainly from genera *Acidithiobacillus* and *Gallionella* and a small amount of *Leptospirillum* sp. This consortium effectively solubilized copper from PCBs. The extraction of copper was attributed to ferrous ion–oxidizing bacteria. The leaching rate of copper was higher at lower PCB dosages, with initial pH around 1.5 and Fe^{2+} concentration of 9 g/L. Optimized conditions resulted in 95% recovery of copper and also remarkably shortened the leaching period from about 12 to 5 days (Xiang et al. 2010).

In continuation of this study, Zhu et al. (2011) used metal concentrates of PCBs to study bioleaching, from which the nonmetallic part of PCBs was removed. The powdered PCB was subjected to hydraulic sorting so as to remove nonmetallic components, mainly plastics, from the PCB powder, which can impart bacterial toxicity. The influence of the initial pH, Fe^{2+} concentration, the dosage of metal concentrate, particle size, and inoculation quantity on the bioleaching process was studied. The optimum conditions were found to be pH 2.0, 12 g/L initial Fe^{2+} concentration, 10% inoculation quantity, and 60–80 # mesh PCB metal concentrate. Zhu et al. (2011) further proposed a two-step metal bioleaching under these optimum conditions. In the first step, they inoculated a 10% inoculum in the medium and precultured without PCB metal concentrate. When the Fe^{2+} concentration of the medium came down to 8 g/L, PCB metal concentrate was added under the obtained optimum conditions as the second step. This resulted in a sharp decrease in the time period required for leaching, taking only 45 h to leach 96.8% of copper.

Also, 88.2% aluminum and 91.6% zinc were recovered in 98 h. Thus, the removal of the nonmetallic part from the PCB resulted in a reduction in the leaching time period from 5 days to just 45 h, suggesting the nonmetallic PCB part to be a limiting factor of the bioleaching process.

Bioleaching of metals from electronic scrap was studied using a mixed culture of metal-adapted *S. thermosulfidooxidans* and an unidentified acidophilic heterotroph (code A1TSB) (Ilyas et al. 2007). Prior to bioleaching studies, the authors treated the electronic scrap with a high-density, saturated NaCl solution so as to remove toxic components, that is, plastics and organic compounds, without changing the physicochemical properties of the metallic part. The microorganisms were acclimatized to tolerate the increasing metal concentration by repeated subculturing on metal-containing media, as unadapted cells showed drastic metal bioleaching rates. Further, Ilyas et al. (2007) proposed the addition of sulfur to the medium, which not only increased the rate of metal solubilization but also stabilized the pH. Also, sulfur acted as an additional energy source for the growth of microorganisms. In this study, it was found that the pure culture of acidophilic heterotroph was unable to leach metals from the scrap; rather, it enhanced the metal leaching rate due to its synergistic effects on the growth of *S. thermosulfidooxidans*. At a scrap concentration of 10 g/L, a mixed consortium of the metal-adapted cultures was able to leach more than 81% Ni, 89% Cu, 79% Al, and 83% Zn. Although Pb and Sn were also leached, they were detected in the precipitate formed during bioleaching (Ilyas et al. 2007).

Ilyas et al. (2010) took the shake flask bioleaching experiments to the next level of column studies, which ought to be the main step toward industrial application. As electronic scrap is alkaline in nature, the column of scrap was washed with distilled water, followed by sulfuric acid (H_2SO_4), to stabilize the pH. After stabilization, preleaching was stopped and bioleaching was carried out using the mixed, adapted consortium of *Sulfobacillus thermosulfidooxidans* and *Thermoplasma acidophilum*. During the whole leaching process, about 80% Zn, 64% Al, 86% Cu, and 74% Ni were leached. These findings may facilitate an industrial-scale implementation of this process for recycling of metals from electronic scrap.

Further, the authors focused on applying additional pretreatment conditions and investigating the effect of additional energy sources using consortia of moderately thermophilic bacteria (iron and sulfur oxidizers) in extracting metals from electronic scrap (Ilyas et al. 2013). Different consortia of moderately thermophilic bacteria, *S. thermosulfidooxidans* with *Thermoplasma acidophilum* and *S. thermosulfidooxidans* with *S. acidophilus*, were cultured with additional energy sources (FeS_2, S_0, $FeS_2 + S_0$) and washing charge material as a pretreatment. At a scrap concentration of 10%, an adapted consortium of *S. thermosulfidooxidans* and *Thermoplasma acidophilum*, containing $FeS_2 + S_0$ (1%), extracted approximately 85% Cu, 75% Al, 80% Ni, and 80% Zn from pretreated electronic scrap. However, a consortium of *S. thermosulfidooxidans* and *S. acidophilus*, containing $FeS_2 + S_0$, extracted 90% Cu, 80% Al, 82% Ni, and 85% Zn. During column bioleaching studies of 165 days, approximately 74% Zn, 68% Al, 85% Cu, and 78% Ni were leached, as shown in Figure 4.2. Such results have the potential to be implemented on an industrial scale.

Very less information is available for the utilization of sulfur sources, such as the uptake of sulfur, its mobilization, and the possible role of soluble intermediates

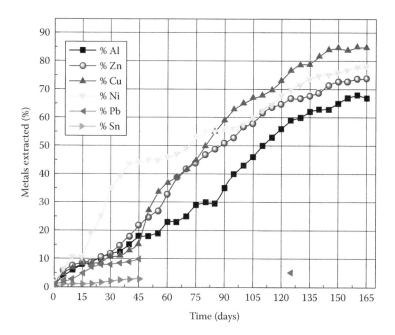

FIGURE 4.2 Metals extraction for column bioleaching experiments. (Reprinted from Ilyas, S. et al., *Hydrometallurgy*, 131, 138, 2013. With permission.)

during the mobilization and oxidation processes (Franz et al. 2009, He et al. 2009). Taking these points into account, Ilyas and Lee (2014) carried out the bioleaching of WPCBs in a stirred-tank reactor and investigated the effect of aeration, O_2, CO_2, agitation, various sulfur types and dosages, and pulp density on the growth of *S. thermosulfidooxidans* and bioleaching efficiency. Moreover, sulfur bioavailability, metabolism, and oxidation degree were also evaluated.

For the bioleaching experiments, 2 L baffles glass reactors containing impellers, a thermostatic bath, and gas streams (individual and mixed) were employed. Bioleaching results revealed that medium-1, supplemented with biogenic S_0, showed 91% Al, 95% Cu, 96% Zn, and 94% Ni recovery at 10% pulp density, using 25% O_2 + 0.03% CO_2 enriched air at 2.5% S_0 dosage in a stirred-tank reactor for 15 days. However, 74% Al, 81% Cu, 83% Zn, and 78% Ni were recovered from medium-2, supplemented with technical sulfur and using 30% O_2 + 2% CO_2 enriched air at 3.5% S_0 dosage and 10% pulp density.

The acid required for the maximal bacterial productivity and metal bioleaching was compensated by increasing the S_{0b} dosage from 1.5% to 2.5% in medium-1 and increasing the S_{0t} dosage from 1.5% to 3.5% in medium-2; however, further increase in their concentration only contributed to acidification. Biogenic sulfur, compared with technical sulfur, was oxidized rapidly due to its greater availability and hydrophobicity. Prior to its transport into the cytoplasm and oxidation, *S. thermosulfidooxidans* grown on different S_0 sources converted both S_{0b} and S_{0t} using amphiphilic compounds, depending on their bioavailability, as an initial mobilization step.

FTIR (Fourier transform infrared) and XANES (X-ray absorption near-edge structure spectroscopy) analyses indicate a significant modification of the spectra of technical sulfur due to the addition of several new groups ($-NH$, $-OH$, $-S(=O)_2-$, $-SO_2-O-$, $-CH_3$, or $-CH_2$) and the presence of polythionates, C$-$S$-$H, sulfanes, and sulfonate species in both media at different degrees and time periods, whereas the spectra of biogenic sulfur suggested insignificant modification.

The bacterial cell density was highest (5.5–6.0 × 10^7 cells/mL) with S_{0b}, along with high metal extraction (94%) at higher pulp densities (10%–14%), as compared with S_{0t}. Hence, S_{0b} can be used as a low-cost energy source instead of S_{0t}. Moreover, it was observed that PEP carboxylase activity (5.1 nmol/min mg protein) was significantly higher in medium-1, and PEP carboxykinase and pyruvate carboxylase activities (1.10 and 0.5 nmol/min mg protein, respectively) were higher in medium-2, suggesting different preferential modes of growth. An economical feasibility study (Opex level) indicated that the results of the leaching process might have the potential to be implemented at an operational scale (Ilyas and Lee 2014).

Bas et al. (2013) studied the bioleaching of copper from low-grade scrap TV circuit boards (STVBs) with low iron content using a mixed mesophilic culture of acidophilic bacteria. STVBs received after heat treatment for solder removal were crushed into powder and used for bioleaching. In view of the low iron content of STVBs, pyrite (containing 42.2% Fe, 0.6% Cu, and 44.6% S) was added (up to 50 g/L) as a supplementary source of iron and sulfur. A mixed culture of mesophilic bacteria, MES1 (*At. ferrooxidans*, *L. ferrooxidans*, *At. thiooxidans*), was used in this study.

Many studies have already demonstrated the beneficial effect of the external addition of Fe(II)/Fe(III) on the rate and extent of copper extraction (Choi et al. 2004, Yang et al. 2009, Xiang et al. 2010, Zhu et al. 2011). The rate of copper bioleaching from e-waste is apparently controlled intimately by the availability (i.e., initial concentration) and the rate of bio-oxidation of ferrous ion (i.e., the rate of generation of ferric ion) in the bioleaching environment. Also, the adverse effect of the limited availability of soluble iron at the onset of bioleaching and/or low iron content of the e-waste sample used is known (Brandl et al. 2001, Ilyas et al. 2007, 2010, Willscher et al. 2007). All these studies infer that an external supply of iron and acid is required for efficient bioleaching of e-waste with low iron content, in particular. In this study, the authors observed the discernible slow bioleaching rate of copper, owing to the low iron content of STVBs. The external addition and/or increase in the concentration of Fe(II), from zero to 8 g/L, markedly enhanced the dissolution rate of copper from 35% to 89%. Oxidation of Fe(II) and Cu is an acid-consuming reaction. Thus, the increasing concentration of Fe(II) resulted in increased acid consumption up to 50%.

To overcome this problem and considering the inherent characteristics of the bioleaching process (i.e., acid consuming and need for iron supplement), Bas et al. (2013) proposed a novel approach of pyrite addition as a source of iron and acid, since bio-oxidation of pyrite is an acid-generating reaction with concurrent release of iron (Equations 4.4 and 4.5):

$$4FeS_2 + 15O_2 + 2H_2O \xrightarrow{\text{bacteria}} 4Fe^{3+} + 8SO_4^{2-} + 4H^+ \qquad (4.4)$$

$$FeS_2 + 14Fe^{3+} + 8H_2O \rightarrow 15Fe^{2+} + 2SO_4^{2-} + 16H^+ \quad (4.5)$$

Thus, copper extraction was observed to increase from 24% to 84% in the presence of 50 g/L pyrite concentrate. Furthermore, a significant decrease (62%) in the consumption of acid was also noted.

In bioleaching systems, inoculum size is an important parameter which directly affects the rate and extent of bioleaching (Ilyas et al. 2010, Xu et al. 2010), since the number of active bacteria increases with an increase in the concentration of the inoculum. An increase in the size of the inoculum from 10% to 50% v/v was observed to improve the bioleaching rate of copper. This enhancement in the dissolution rate can be attributed to the rapid conversion of Fe(II) into Fe(III) because of the increased size of the bacterial population. The increased availability of Fe(II), which was carried over by the inoculum, could have also contributed to the improvement in the bioleaching rate. Thus, the authors highlight the practical importance of the availability of iron (and hence, the iron content of wastes) for the successful development of a bioleaching process for e-waste and the potential for utilization of pyrite as a cheap source of iron and sulfur in bioleaching of e-waste (Bas et al. 2013).

Adhapure et al. (2013) applied a consortium from bauxite and pyrite ore samples for bioleaching of metals from PCBs. The employed consortium was obtained by applying a top–down method. Various process parameters such as PCB concentration, pH, and ferrous ion concentration were studied. Nearly 96.93% and 93.33% of Cu and Zn were solubilized at 10 g/L PCB, respectively, whereas only 10.26% Ni was solubilized at 30 g/L PCB. Further, only a 0.58% Pb solubilization was obtained at 20 g/L PCB. X-ray diffraction (XRD) analysis of the precipitate formed during bioleaching showed the presence of Sn (59.96%), Cu (23.97%), Pb (9.30%), and Fe (5.92%). Moreover, the authors suggest that the consortium can be useful in the commercial bioleaching process under nonsterile conditions.

Three different modes of copper leaching, namely abiotic chemical leaching using inorganic H_2SO_4, indirect leaching using bacterially generated H_2SO_4, and direct leaching using the acidophilic bacteria *At. thiooxidans*, were performed to study the mechanism that influences metal solubilization (Hong and Valix 2014). The chemical method involved using H_2SO_4 reagent, while spent acid leaching involved the use of biogenic H_2SO_4 that has been harvested from the fermentation of elemental sulfur nutrient and direct leaching involved the addition of waste in a growing culture.

The yield of bacterially generated H_2SO_4 used for both indirect and direct leaching was 14.9 g/100 mL grown in a medium containing 25 g/100 mL of elemental sulfur and basalt salts for 14 days at 30°C. This acid was diluted to achieve various pH for leaching tests. The variables tested were solution pH, temperature, time, pulp density, and copper concentration in the waste. Increasing the acid concentration, copper concentration in the waste, higher temperature, and prolonged leaching favored higher yields and higher copper selectivity. In the case of chemical leaching, both passivation and galvanic coupling reduced the Cu yield and diminished the leaching. Passivation of copper with chemical leaching appears to result from the oxidation of cuprous ion, promoted by oxygen and peroxide dissociation. Passivation by copper oxidation and galvanic coupling effects can be overcome in wastes containing a high percentage of Cu (>90%) by using a low pH (<1.0), a high temperature (90°C), a low pulp density (10 g/100 mL), and long periods of leaching. The effect of galvanic coupling, however, is prevalent and is more difficult to control as the percentage of copper decreases.

Direct leaching was performed with 10 g/100 mL of waste, with Cu-rich waste being added to the microbial culture once the pH reached 1.0. For indirect leaching, the biogenic acid was harvested and used in leaching the waste. Leaching using biogenic acid or direct leaching with microorganisms has added complexities. It is apparent that 10 g/100 mL of waste does not pose any toxic effects on *At. thiooxidans* and is the optimal pulp density to achieve maximum copper dissolution. However, the presence of incompletely oxidized sulfide compounds and sulfates leads to the formation of CuS and copper sulfate precipitates, both of which appear to contribute to copper passivation. However, a growth medium containing partially oxidized sulfide compounds promoted copper surface passivation, resulting in lower Cu recovery (60%) relative to abiotic leaching (98%) (Hong and Valix 2014).

Mäkinen et al. (2015) conducted a study to determine the use of froth flotation and autotrophic acid bioleaching in the treatment and metal recovery from PCBs. Froth flotation, which is a separation technology widely applied, for example, in the mining industry, is a very interesting method also for PCB treatment in order to separate valuable metals from plastics and other nonmetallic substances. In the flotation process, the solid material is crushed to liberate its compounds. Then, the crushed material is mixed with a liquid to produce a slurry, which is agitated and gas is introduced into the system. Gas bubbles attach on the surfaces of hydrophobic particles and lift them up to form a froth layer, which is collected. Material that is not afloat remains in the solution during agitation and can be collected later from the bottom of the floatation apparatus (Yarar 2000, Heiskanen 2014). However, PCB froth, a "waste" fraction produced by flotation, still contains residual metals and other harmful elements in quantities that may prevent further utilization or disposal strategies according to legislation.

Flotation, conducted after crushing and sieving of PCBs, produced two fractions from the crushed PCBs: metal-rich concentrate and metal-poor froth. The metal-rich concentrate can be treated using pyrometallurgical methods. The metal-poor PCB froth still contained significant concentrations of copper and was therefore treated through autotrophic acid bioleaching. Thus, autotrophic acid bioleaching was utilized to remove residual metals, especially copper, from the froth. The mixed acidophilic culture used, enriched from a sulfide ore mine site, contained *At. ferrooxidans, At. thiooxidans/albertensis, At. caldus, L. ferrooxidans, S. thermosulfidooxidans, S. thermotolerans*, and some members of the *Alicyclobacillus* genus (Halinen et al. 2009). The mixed culture was adapted and used for bioleaching studies. It was observed that a high initial Fe^{2+} concentration (7.8 g/L) and a low pH (1.6), as well as a low PCB froth concentration (50 g/L), were beneficial to the rapid and selective dissolution of copper. With these parameters, copper solubilization of 99% was reached in 3 days, with Cu (6.8 g/L) and Fe (7.0 g/L) being the only major metallic species in solution. Also Ni, Zn, Co, and Mn were efficiently leached, but concentrations remained low in solution due to these metals' scarcity in PCBs. Tentatively, it can be estimated that froth flotation and bioleaching can be utilized in PCB treatment.

This study illustrates that the main barriers to the industrial application of this process were batch-wise operation and insufficient optimization of biological reactions. In light of these concerns, factors such as the amount of biologically produced H_2SO_4 and Fe^{3+} and their rate of production must be taken into account (Mäkinen et al. 2015).

In another study, Yang et al. (2017) explored the ability of *At. ferrooxidans* strain Z1, a new isolate from the mixed culture of acidophilic bacteria enriched from natural AMD as a pure culture, to recover copper from metal concentrates of WPCBs. Various process parameters influencing the process, namely initial pH, initial Fe(II) concentration, dosage of metal concentrates, inoculation quantity, and particle size, were studied and optimized. WPCBs were shredded, powdered, and subjected to hydraulic sorting to remove the nonmetallic part and obtain the metal concentrate, which was used in bioleaching studies. A two-step bioleaching method was employed. The initial pH of the solution and the Fe^{2+} concentration remarkably affected metal recovery. Under the optimum conditions of an initial pH of 2.25, an initial Fe(II) concentration of 9 g/L, a metal concentrate dosage of 12 g/L, an inoculation quantity of 10%, and a particle size of 0.178–0.250 mm, 92.57% Cu was leached within 78 h, whereas 85.24% aluminum and 95.18% zinc were leached after 183 h. Thus, the strain was found to effectively recover copper from PCBs.

Recycling of WPCBs by bioheap leaching could be of great importance in industrial operations. Column leaching is used as a simulating model for heap or dump leaching processes, which gives information on what could be expected in heap or dump leaching and how leaching conditions can be optimized (Muñoz et al. 1995, Qiu et al. 2011, Ilyas et al. 2013). Also, the information of kinetics is very important to optimize the leaching parameters and to improve column bioleaching performance (Olson et al. 2003, Rohwerder et al. 2003). Hence, Chen et al. (2015) explored the ability of *At. ferrooxidans* to extract copper from WPCBs and analyzed the kinetics of column bioleaching.

A maximum copper extraction of 94.8% was obtained using column bioleaching after 28 days. It was also observed that Cu oxidation was inevitably influenced by variations in Fe^{3+} concentration, so this fact should be considered in conventional kinetic models of the bioleaching process. Further, results revealed that the rate of copper dissolution is controlled by external diffusion, rather than by internal diffusion, and is unaffected by the precipitation layer, because of iron hydrolysis and formation of jarosite precipitate at the surface of the material. The addition of H_2SO_4 to maintain the pH of the solution at 2.25 had an insignificant effect on the size and morphology of the precipitate, and thus, the kinetics of column bioleaching remained unaffected.

In the bioleaching process, the formation of jarosite precipitate can be restrained by adding dilute H_2SO_4 and maintaining the acidic condition of the leaching medium. In such way, the Fe^{2+}–Fe^{3+} cycle process can be continued to create a favorable condition for Cu bioleaching. Experimental results show the feasibility of column Cu bioleaching from WPCBs by using *At. ferrooxidans*, and an increase in the concentration of Fe^{3+} and the velocity of the leaching solution cycling may increase the kinetics of copper bioleaching (Chen et al. 2015).

Rodrigues et al. (2015) explored a new approach for copper bioleaching from PCBs using moderate thermophiles in a rotating-drum reactor. A rotating-drum reactor could be used to treat material (waste) at increased pulp densities and reduce global energy consumption when compared with stirred-tank reactors. Liu et al. (2007) suggest that such a reactor will ensure a reduced impact on microbial cells due to the lower degree of collisions between PCB particles. Moreover, this configuration allows the use of a high solid load without negatively influencing the bio-oxidation of Fe^{2+}, which tends to be the case when impeller-driven reactors are used (Jin et al. 2013).

Nevertheless, very few studies have demonstrated the use of high solid concentrations, likely due to the inhibitory effect of fine PCB particles on bacterial growth, which is may be due to the stimulation of quick release of harmful factors from the ground PCB. To avoid this, in the present work, the authors used relatively coarse PCB sheets. Owing to this, the overall cost associated with the grinding process was reduced.

Initially, shake flask studies were carried out to evaluate the effect of particle size (−208 μm + 147 μm), ferrous ion concentration (1.25–10.0 g/L), and pH (1.5–2.5) on copper leaching using mesophilic and moderately thermophilic microorganisms. Nearly 94% and 99% of Cu was recovered at 30°C and 50°C using mesophilic and moderately thermophilic cultures, respectively, at a low solid density of 10 g/L. Although jarosite precipitation was observed in these conditions, it did not affect the rate of Cu dissolution. A parallel increase in Cu extraction was observed with initial ferrous ion concentration only during the first 2 days; however, 5.0 g/L was found to be the optimum concentration. On the other hand, it was found that a pulp density of up to 25.0 g/L could be achieved with the use of coarse PCB particles (20 mm) at 50°C in shake flasks with a 5.0 g/L initial ferrous ion concentration.

Under these conditions and when PCBs were subjected to a preweakening step (via a jaw crusher), followed by lacquer coating removal, nearly 76% copper was extracted in 8 days, also suggesting a limited release of inhibiting elements in solution. Bioleaching with moderate thermophiles at 50°C and 5.0 g/L ferrous ion concentration has enabled faster copper extraction than the one done at 30°C, but care should be exercised if the solution potential is chosen to monitor bacterial growth since the available metallic copper rapidly reduces ferric iron to its divalent state. Further large-scale experiments were carried out using coarser nonground PCBs and *S. thermosulfidooxidans* was selected due to the faster leaching kinetics fostered by higher temperatures using a rotating-drum reactor, as shown in Figure 4.3.

The reactor consisted of a perforated drum (with 10 mm openings) driven by a motor and fitted with a gas sparger. The perforated drum was immersed in a fixed cylindrical compartment containing a liquid medium/pregnant leach solution. The temperature of the reactor was maintained at 50°C ± 1°C. The experiments were performed at a rotation speed of 80 rpm using 20 mm long PCB sheets (25.0 g/L pulp density), immersed in 12 L of a medium containing 10% (v/v) of the inoculum. As compared with shake flasks, leaching in the aerated rotating-drum reactor was found to be efficient and about 85% Cu was extracted in 8 days (44% for abiotic). Also, acid consumption was more than two times lower for biotic leaching than for abiotic leaching at the same initial concentration of ferrous ion (5.0 g/L). Scanning electron microscopy (SEM) and energy dispersive X-ray spectroscopy analyses of samples before and after leaching suggest that metal dissolution from the internal layers was restricted by the fact that the metal surface was not entirely available and accessible to the solution in the case of the 20 mm long sheets.

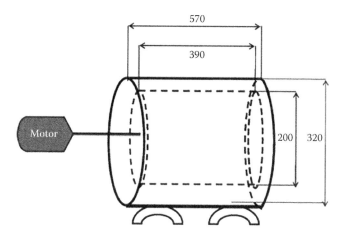

FIGURE 4.3 Schematics of a rotating-drum reactor (dimensions in millimeters). (Reprinted from Rodrigues, M.L.M. et al., *Waste Manage.*, 41, 148, 2015. With permission.)

These results suggest a promising use of this technology for the recovery of copper from PCB waste and could be integrated into a global copper recycling process from such sources (Rodrigues et al. 2015).

A two-step process was developed to enhance the biorecovery of Cu, Zn, and Ni from WPCBs using ferric sulfate generated by a *Leptospirillum ferriphilum*–dominated consortium and the factors influencing the process were investigated (Shah et al. 2015). A two-step process was employed, where the consortium was grown in the absence of e-waste. When more than 90%–95% Fe^{2+} was oxidized by the microbes, e-waste was added in the second step.

The bioleaching system, with a pH of 2.0, showed the maximum dissolution of Cu (88%), Zn (99.80%), and Ni (97.13%) at 10 g/L pulp density. A pH value below this had an inhibitory effect on the Fe^{3+} regeneration capability of the consortium and thus metal extraction, whereas a pH value above this decreased metal extraction due to metal precipitation and low solubility of Fe^{3+} iron.

Pretreatment of PCBs prior to bioleaching was helpful in removing the toxic non-metallic part, which can hamper bacterial activity. PCB pretreatment with acidified distilled water (pH 5) and NaCl solution increased metal extraction by 3.8%–7.98%, resulting in 94.08%, 99.80%, and 97.99% extraction of Cu, Zn, and Ni, respectively. Particle size had an evident effect on metal extraction and resulted in a twofold increase in metal extraction of Cu and Zn at a particle size of 75 mm, whereas for Ni, it was 1680 mm, giving 97.35%–99.80% metal extraction. The addition of a 0.1% chelating agent (ethylene diamine disuccinic acid and oxalic acid) increased Cu and Zn extraction by 2.76- to 3.12-folds and also reduced the operation time from 6–8 days to as low as 2–4 days. An Fe^{3+} iron concentration of 16.57 g/L exhibited maximum dissolution, with 62.45% Cu, 88.28% Zn, and 63.88% Ni solubilization within 10–12 days of reaction time, at 75 g/L pulp density. Further increase in Fe^{3+} iron concentration decreased metal extraction, may be due to an increased jarosite precipitation, where the precipitate covered the WPCB, making it nonpolarized and thus slowing down the rate of metal extraction. Maximum Zn was solubilized first, followed by Cu and Ni. The dissolution of Ni, compared with that of Cu and Zn, followed a different trend in experiments involving various chelating agents and particle sizes.

A high pulp density of PCB addition in parts resulted in the increased of extraction of Cu, Zn, and Ni from 118.74 to 182.99, 19.76 to 32.70, and 1.83 to 2.59 mg/day, respectively, compared with computer PCBs (CPCBs) addition in one lot. A further experiment with optimized parameters increased the extraction rate of Cu, Zn, and Ni to 236.8, 33.0, and 3.29 mg/day, respectively. This developed two-step process can be applied on a large scale for rapid extraction of multiple metals from e-waste, thereby transforming the problem of pollution into a profitable metal resource (Shah et al. 2015).

The effects of dissolved oxygen (DO) levels in the culture medium on cell growth and copper extraction from WPCBs were investigated in the case of *At. ferrooxidans* by Liang et al. (2016). *At. ferrooxidans*, as a chemolithotrophic aerobic bacterium, can obtain energy by the oxidation of ferrous ions (Fe^{2+}) to ferric ions (Fe^{3+}) and use molecular oxygen (O_2) as a terminal electron acceptor, which indicates that Fe^{2+} oxidation was markedly affected by the DO status of the medium. So, maintaining DO at a suitable level to enhance Fe^{2+} use efficiency is very important for bacterial growth and metal recovery from WPCBs.

The experiment to study the effects of DO levels on bacterial growth and copper extraction was carried out in a 5 L fermenter with a working volume of 3 L. Aeration was maintained at 1 vvm (air volume/culture volume/min), an agitation speed of 50–200 rpm to control and maintain DO levels at 5%–25% of the saturation value, a temperature of 30°C, and a culture period of 200 h. The culture period was divided into two stages of cell growth and copper extraction.

Primarily, a lower DO level was adopted to satisfy bacterial growth and avoid excessive Fe^{2+} oxidation, whereas at a later stage, a higher DO level was used to enhance copper extraction. The shift time of DO was determined via simulating the Gauss function. Controlling 10% DO for initial 64 h with a 18 g/L pulp density and later switching it to 20% resulted in the final copper recovery of 94.1%, increased by 37.6% and 48.3% compared with constant DO of 10% and 20% operations. Moreover, the leaching period was shortened to 60 h from 108 h. Thus, the application of the DO shifting strategy is helpful in enhancing copper extraction from PCBs on a large scale (Liang et al. 2016).

Nie et al. (2015) immobilized *At. ferrooxidans* on cotton gauze in a two-step reactor to study bioleaching parameters of metal concentrates from WPCBs. The ability of *At. ferrooxidans* to oxidize ferrous ion plays a critical role which can affect the whole bioleaching rate. The rate of Fe^{2+} oxidation could be improved by immobilization *of At. ferrooxidans*. Reports also suggest that *At. ferrooxidans* cells have a natural tendency of immobilization on different surfaces, which makes it a potential candidate for cellular immobilization (Giaveno et al. 2008). Immobilized *At. ferrooxidans* can attain high cell concentrations and higher cellular metabolic activity inside the reactor and thus will reduce the loss of biomass, further enhancing the bioleaching efficiency. Many matrices are available for immobilization of cells. Cotton gauze is a natural material of textile fibers that has a high water absorption capacity, high

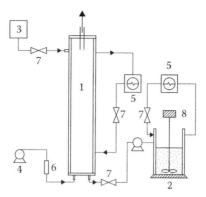

FIGURE 4.4 Schematic diagram of a bioleaching reactor: 1, bio-oxidation reactor; 2, leaching reactor; 3, reservoir; 4, air pump; 5, constant-temperature water tank; 6, gas flowmeter; 7, valve; 8, motor stirrer. (Reprinted from Nie, H. et al., *Appl. Biochem. Biotechnol.*, 177, 675, 2015. With permission.)

mechanical strength in the wet state, ultrafine network structure, and good storage stability, and is cheap and also nontoxic to microorganisms, making it a suitable candidate for microbial immobilization.

The bioleaching reactor was composed of the bio-oxidation reactor and the leaching reactor, as shown in Figure 4.4. The bio-oxidation reactor had a working volume of 500 mL, an outlet for effluent and compressed air at the bottom, and an inlet for fresh medium and exhaust air at the top. It was aerated at a rate of 1 L/min. It was used for continuous oxidation of ferrous ion by immobilized cells.

The temperature of the bioleaching reactor was maintained at 30°C and an agitation speed of 300 rpm. The leaching reactor was used for extraction of metals from WPCBs.

The average ferrous ion oxidation rate was found to be 0.54 g/(L·h), and a ferrous ion oxidation ratio of 96.90% was obtained after 12 h at an aeration rate of 1 L/min in the bio-oxidation reactor. The retention time at 6 h was optimum for the dissolution of copper in this system. After 96 h, the highest leaching efficiency of copper reached 91.68% under the conditions of the content of the metal powder 12 g/L, the retention time 6 h, and the aeration rate 1 L/min. The bioleaching efficiency of copper could be above 91.12% under repeated continuous batch operation. SEM analyses of the metal powder revealed the smoother surface of the leaching residue as compared with raw powder. Leaching efficiencies of 95.32%, 91.68%, 90.32%, 86.31%, and 59.07% were obtained for Cu, Zn, Mn, Al, and Ni, respectively. Moreover, no remarkable fluctuation was observed in repeated batches of copper extraction.

XRD analyses showed jarosite as a reddish-brown precipitate formed on the cotton gauze in the bio-oxidation reactor and the leaching reactor. Often, jarosite precipitation is described as an unwanted phenomenon in the process of bioleaching because it limits the amount of biomass retention due to kinetic barriers. On the other hand, the jarosite precipitate participates in biofilm formation and is a suitable support for bacterial adsorption due to its porous and loose nature, and it also reinforces cotton gauze (Pogliani and Donati 2000, Zhu et al. 2013). The use of the jarosite precipitate as a biomass matrix has good prospects in the practical use in the future (Nie et al. 2015).

Bryan et al. (2015) studied the addition of pyrite as a source of lixiviant during bioleaching of e-waste. The mix consortia used in the study comprised *At. caldus* BRGM3, *L. ferriphilum* BRGM1, *S. benefaciens* BRGM2 and *Ferroplasma acidiphilum* BRGM4, *At. ferrooxidans* T DSM14882, and *Acidiphilium* sp. SJH. Bioleaching experiments were carried out using a two-step process and by supplementing pyrite in the culture medium. The use of pyrite by microbes generates a lixiviant (ferric iron and proton acidity), which leaches metals from PCBs and maintains a low solution pH. But PCB addition may hamper this production and reduce metal recovery due to its toxicity toward microbes. Hence, to investigate it, the authors studied the influence of elemental metal addition on pyrite leaching. It was found that Cu, Cr, Ni, Sn, and Zn had an immediate inhibitory effect on pyrite oxidation, but after the adaptation phase, leaching recovered. Terminal restriction fragment length polymorphism profiles before and after pyrite leaching suggested the dominance of iron-oxidizing *L. ferriphilum* rather than the acquisition or emergence of more tolerant, but different, organisms. The microbial consortium was typical of most moderately thermophilic mineral bioleaching systems.

The addition of 1% (w/v) ground PCB initially inhibited bioleaching, which recovered rapidly, but an increasing pulp density $\geq 5\%$ significantly decreased culture viability and leaching. The increasing pulp density decreased copper recovery. But at the same time, significant Cu leaching was observed, suggesting the role of the abiotic process, as the leaching culture was no longer viable and also because the leaching kinetics was linear, thus underscoring the need for proper abiotic controls in these types of tests.

Further, it was observed that the high elemental iron content created a lag phase in copper dissolution possibly due to displacement reactions. Thus, this factor must be dealt with during further leaching experiments. Also, acid consumption by PCBs was overcome by acid production, suggesting the need for optimization of pyrite oxidation. Thus, the authors have highlighted areas to be considered during future studies and provided a base to establish a proof of concept of using pyrite as a substrate for e-waste bioleaching (Bryan et al. 2015).

Lambert et al. (2015) studied the leaching kinetics of copper from waste electric cables and reagent consumption during chemical leaching and bioleaching by *At. ferrooxidans*. The residues obtained after recycling of waste electric cables can act as a source of metal. Cu dissolution was observed both in the absence and in the presence of bacteria in an H_2SO_4 medium with ferric iron; that is, more than 90% of Cu was recovered in both bioleaching and chemical leaching in the presence of Fe^{3+}. Efficient Cu leaching kinetics was obtained for the biotic test at 35°C and 7.0 g/L initial Fe^{3+} concentration, whereas for the abiotic test, it was at 50°C and 6.4 g/L Fe^{3+} concentration. These results suggest the essential role of Fe^{3+} in the kinetics of copper leaching in both biological and chemical systems. The presence of Fe^{3+} may enhance the oxidation of copper and thus improve its dissolution into the solution. It seems that Fe^{3+} is more active during copper oxidation than O_2, which does it at a slower rate. Further results revealed that a 2 g/L Fe^{3+} concentration showed good copper leaching kinetics (90% Cu after 3 days), and increasing this concentration to 6 g/L may bring about a three-time decrease in the leaching period.

Faster leaching kinetics was observed with an increase in temperature but also resulted in an elevation in acid consumption to maintain pH, probably due to a more pronounced degradation of plastics in cable scrap. A temperature of 35°C seems to be the best compromise between efficient copper leaching kinetics and limited acid consumption. Even though 90% Cu recovery was achieved in 1 day for both tests, it was observed that the lower-temperature test enables a lower H_2SO_4 consumption (0.93 mol H_2SO_4/mol Cu^{2+} against 2.76 mol H_2SO_4/mol Cu^{2+} for the test at 50°C).

At. ferrooxidans–assisted copper leaching was faster due to ferric iron regeneration, but at the same time, this increase in kinetics was relatively limited, which was affirmed by measuring the reaction rate for both mechanisms. The chemical mechanism of copper dissolution is too fast compared with the bio-oxidation of Fe^{2+} to see a positive effect of bacterial presence. The copper leaching rate varies between 0.001 and 0.011 mol/L h, while the maximal Fe^{2+} oxidation rate was 0.002 mol/L h. So the regeneration of Fe^{3+} by bacterial action should only partially compensate the Fe^{3+} consumed by copper. Note that the Fe^{2+} oxidation rate obtained in this work is very low compared with velocities obtained in other studies. The bacterial consortium is probably not the best choice to have a positive effect on copper leaching through the bio-oxidation of Fe^{2+}.

Finally, the pre-exponential constant A, the activation energy Ea, and the kinetic order b of the kinetics law were calculated. The activation energy of copper dissolution was calculated to be 20.4 kJ/mol. Very close values of the optimized parameters (A, Ea, and b) were obtained for biotic and abiotic tests. This confirms that the biological mechanism has a low positive effect compared with the chemical mechanism on copper leaching kinetics (Lambert et al. 2015).

4.2.2.2 Recovery of Gold

Ruan et al. (2014) explored the ability of a new strain, *Pseudomonas chlororaphis* (PC), to recover gold, silver, and copper from metallic particles of crushed WPCBs. The *Pseudomonas* strain employed in the study was isolated from reed roots from a mining region on the basis of its ability to produce CN^-. Eleven strains were selected depending on the color reaction. Genetic analysis of the strain was carried out using the 16S rDNA sequence technique and the category was distinguished using a phylogenetic tree. Although *Chromobacterium violaceum* is known for CN^- production and metal recovery, it is not feasible for industrial application due to its obligatory culture conditions. However, a *Pseudomonas* strain with CN^- production ability is suitable for industrial application due to its common culture conditions and strong acclimation ability.

Mixed metallic particles were obtained from corona electrostatic separation of crushed WPCBs, wherein metallic and nonmetallic particles were separated. Of all the isolated strains, PC produced the highest concentration of CN^- (7.11 mg/L) and hence was selected for further studies. The influence of various process parameters such as pH, temperature, shaking speed, and additives on CN^- production was evaluated. According to the published literature, a pH range of 7–10 was selected for the study (Chi et al. 2011). Nearly 7.1 and 6.88 mg/L of CN^- was produced at pH 7 and 10, respectively, which suggests a very less influence of pH on CN^- production by

Pseudomonas, whereas a temperature of 60°C and a shaping speed of 60 rpm were found to be optimum. In the case of additives, methionine (2 g/L) exerted a positive influence on CN⁻ production and glycine (4.4 g/L) boosted this positive influence. Dissolution of gold, silver, and copper by the produced CN⁻ was indicated by the presence of Au^+, Ag^+, and Cu^{2+}. Results of SEM analyses suggested that PC cells and their secretions adsorbed metal ions. However, the increased concentration of metal ions in the solution destroyed the bacterial growth. About 8.2% gold, 12.1% silver, and 52.3% copper in metallic particles from crushed WPCBs were dissolved into the solution by PC (Ruan et al. 2014).

Gold biorecovery from electronic scrap material (ESM) was explored using *C. violaceum*, along with the effect of pretreatment of ESM and its pulp density (Natarajan and Ting 2014). The higher gold content in ESM (10–1000 g gold/ton e-waste; Cui and Zhang 2008) as compared with ores (0.5–13.5 g gold/ton gold ore; Korte et al. 2000) renders it a more economical and highly attractive alternative source for gold than ores. Furthermore, there is also a pressing need to discover alternative sources of gold due to the depletion of natural resources and the global increase in the demand for gold.

C. violaceum produces cyanide (as secondary metabolite) in the early stationary phase as cyanide ion (CN⁻) and the nondissociated form of hydrogen cyanide (HCN) (Lawson et al. 1999). At physiological pH, cyanide is present mainly as HCN (pKa 9.3) (Kita et al. 2006). Cyanogenic microorganisms form water-soluble metal–cyanide complexes from metal-containing solids such as PCB scrap under alkaline pH (Faramarzi et al. 2004). As the pKa of HCN is 9.3, conducting the gold dissolution reaction under alkaline conditions increases the total cyanide ions available for bioleaching. The process of gold–cyanide complex formation in the presence of oxygen, known as gold cyanidation, is summarized in Elsner's equation (Kita et al. 2006).

Au dissolution in a cyanide solution consists of an anodic (4.6) and a cathodic (4.7) reaction and is summarized by Elsner's equation (4.8) as follows (Hedley and Tabachnick 1958):

$$4Au + 8CN^- \rightarrow 4Au(CN)_2^- + 4e^- \tag{4.6}$$

$$O_2 + 2H_2O + 4e^- \rightarrow 4OH^- \tag{4.7}$$

$$4Au + 8CN^- + O_2 + 2H_2O \rightarrow 4Au(CN)_2^- + 4OH^- \tag{4.8}$$

The challenge for such a reaction is bacterial growth under alkaline conditions. In this respect, the authors carried out a mutation of the bacteria to grow under alkaline conditions to increase the concentration of cyanide ions available for gold bioleaching. Wild-type *C. violaceum* was exposed to 100 mM of the mutagen *N*-nitroso-*N*-ethyl urea (ENU) at pH 9, 9.5, and 10 as the selection pressure so as to obtain cells capable of growing at alkaline pH.

A high Cu concentration in ESM interferes with gold cyanidation. Also, other metals such as nickel, iron, silver, and zinc also form stable complexes with cyanide. Thus, ESM was pretreated with nitric acid to remove these base metals (nearly 80% Cu removal) and lessen the competition for gold cyanidation. To circumvent the toxic effect of ESM, bacteria were cultured without ESM till they reached a significant cell density and cyanide production has reached its peak. After this was attained, ESM was added to the cultures for bioleaching studies using wild-type and mutated strains at different pulp densities and then under the alkaline pH range.

The mutation markedly enhanced bioleaching under alkaline conditions through increased availability of CN for gold dissolution. An increase in the pulp density decreased gold recovery due to a higher metal concentration and its inhibitive effects on bacterial growth and cyanide production. Bioleaching with *C. violaceum* mutated to grow at pH 9.5 resulted in greater gold recovery compared with the unadapted strain and mutated strains at pH 9 and pH 10. It was noted that at 0.5% pulp density of pretreated ESM, mutated bacteria attained a gold biorecovery of 18% at pH 9, 22.5% at pH 9.5, and 19% at pH 10, while that of the unadapted strain (at pH 7) yielded only a 11% recovery. Results showed that gold bioleaching efficiency from electronic scrap was enhanced under alkaline conditions with mutated bacteria compared with bioleaching at physiological pH (around 7) of *C. violaceum* (Natarajan and Ting 2014).

In continuation of the earlier study, Natarajan et al. (2015) investigated and compared gold recovery by wild-type and two genetically engineered strains (pBAD and pTAC) with an additional cyanide-producing operon of *C. violaceum*. ESM was pretreated to remove metals competing for metal cyanide complexation with gold. The effect of pulp density on the leaching performance by the various strains was also analyzed.

The cyanide lixiviant that complexes with gold is derived from the secondary metabolite HCN produced by oxidative decarboxylation of glycine, a reaction that is catalyzed by the enzyme HCN synthase in *C. violaceum* (Knowles and Bunch 1986). This enzyme is encoded by the hcnABC operon cluster (Laville et al. 1998). Regulation of this operon under quorum control restricts its widespread use in metal recovery, as the amount of lixiviant produced is limited to 20 mg/L of cyanide at the onset of the stationary phase (in a bacterial culture with approximately 1×10^{16} colony-forming units/mL) (Natarajan and Ting 2014). To overcome the limited cyanogenic capability of wild-type *C. violaceum*, two metabolically engineered strains, pBAD hcn (induced by L-Arabinose) and pTAC hcn (induced by IPTG), were constructed with an additional copy of the cyanide-producing operon (hcnABC) to produce more cyanide lixiviant.

Two metabolically engineered *C. violaceum* strains (0.002% L-Arabinose and 1 mM IPTG for pBAD and pTAC, respectively) were constructed by site-specific integration of the duplicated hcnABC operon under transcriptional control of the exogenous promoters pBAD (L-Arabinose inducible) and pTAC (Isopropyl thiogalactoside [IPTG] inducible) to produce more cyanide lixiviant. The genetic modifications were done as per the method of Tay et al. (2013). Two versions of engineered strains were produced, each requiring a specific inducer, L-(+)-Arabinose and IPTG. The inducer binds an upstream regulatory element of the promoter, leading to its activation and subsequent binding to the promoter sequence. As a result, other elements required for transcription will bind to the promoter, including the polymerase, hence initiating transcription of the gene. The two engineered strains were named as *C. violaceum* pBAD hcnABC and *C. violaceum* pTAC hcnABC. Choi et al. (2005)

developed this method of site-specific genomic integration of an inducible cyano-genic operon using Tn7-mediated transposition.

During the exponential phase, low cyanide production was obtained, which peaked toward the early stationary phase for both wild-type and engineered strains. The inducers for the engineered strains were added during the mid-logarithmic growth phase, that is, 12 and 16 h after inoculation for the pBAD and pTAC strains, respectively. Cyanide production increased for both engineered strains 4–5 h after the addition of the inducer. In the absence of the inducer, both engineered strains produced similar cyanide levels as the wild-type strain, indicating the need for induction to activate the additional cyanide-producing operon in the two strains. A comparison of lixiviant profiles of wild-type and engineered strains exhibited that the pBAD (induced with 0.002% L-Arabinose) and pTAC strains (induced with 1 mM IPTG) produced peak concentrations (at 30 h after inoculation) of 34.5 and 31 mg/L of cyanide, respectively, showing significant increases over the wild-type peak concentration of 20 mg/L of cyanide.

A two-step bioleaching protocol was followed, in that ESM was added at different pulp densities to the culture when the maximum cyanide production was attained in the early stationary phase. Gold recovery from ESM waste after 8 days of bioleaching (Figure 4.5) demonstrated a significant increase for both strains of engineered *C. violaceum*, as compared with wild-type bacteria. At 0.5% w/v pulp density, the engineered strains achieved the highest gold recovery of 29.6% (pBAD) and 24.6% (pTAC), while the wild-type strain showed a modest recovery of 11.3%. The enhanced gold leaching performance by the pBAD strain over two other strains

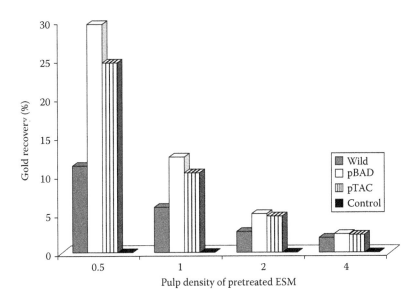

FIGURE 4.5 Gold recovery of pretreated electronic waste material obtained using wild-type, pBAD (induced with 0.002% Arabinose), and pTAC (induced with 1 mM IPTG) strains. (Reprinted from Natarajan, G. et al., *Minerals Eng.*, 75, 32, 2015. With permission.)

could be due to a higher cyanide production by the pBAD strain. As the pulp density increased, gold recovery generally decreased due to the higher toxicity of metals within ESM and its inhibitive effects on bacterial growth and cyanide production.

The authors' findings highlight the utility of lixiviant metabolic engineering in the construction of enhanced bioleaching microbes for the recovery of precious metals such as Au from e-waste. As e-waste generation continues to increase, the success in biological processing of wastes depends on the integration of developments in microbiological and hydrometallurgical engineering into a single system. Since many factors affect the overall productivity in gold recovery from e-waste, subsequent optimization studies with the most efficient genetically engineered strain will need to be investigated (Natarajan et al. 2015).

In a further study, Natarajan and Ting (2015) studied gold recovery from e-waste using spent medium with pH modification.

Since the leaching of metal ions in the solution and the formation of the metal–cyanide complex may be toxic to bacteria and adversely affect cyanide production, an approach to improve gold recovery without affecting bacterial growth is to decouple growth/cyanide production from bioleaching. Further, as base metals (mainly copper) present in ESM (and at a higher concentration than in gold) are capable of forming soluble metal–cyanide complexes, bioleaching of gold may be enhanced through pretreatment of ESM to reduce competition for cyanide ions from these metals (Natarajan and Ting 2014). Hence, in this present study, pretreated ESM is used for bioleaching.

Two-step bioleaching and spent-medium leaching approaches were studied. In two-step bioleaching, ESM was added to the bacterial culture after the latter had reached a significant cell density and bacterial cyanide production had reached its peak (during the early stationary phase). In spent-medium leaching, ESM was added to bacterial cell–free metabolites, which were obtained after centrifuging and filtering the cells.

Cyanide production by pure and mixed cultures of cyanogenic bacteria (*C. violaceum*, *P. fluorescens*, and *Pseudomonas aeruginosa*) was compared so as to select the highest cyanide-producing bacteria. ESM was pretreated in order to remove copper and other base metals. The highest cyanide concentration of 20 mg/L was observed in the pure culture of *C. violaceum* during the early stationary phase (20 h after inoculation). At a pulp density of 0.5% w/v, *C. violaceum* showed the highest gold bioleaching (11.3%), followed by the mixed culture of *C. violaceum* and *P. aeruginosa* (10.2%) in two-step bioleaching. At the same pulp density of ESM, spent-medium bioleaching achieved gold recovery of 18%.

Gold and copper recovery was higher in spent-medium leaching as compared with two-step leaching. The authors attribute the following results for the same: (1) In the absence of bacteria, oxygen is utilized in gold complex formation in spent-medium leaching. (2) The biogenic cyanide in spent-medium leaching is only completely utilized in the leaching of gold, whereas in two-step leaching, it is also consumed during its conversion into beta cyano-alanine during the mid and late stationary phases, as bacterial growth/cyanide production is not decoupled from the bioleaching process. (3) Metals are biosorbed and bioaccumulated during two-step leaching, thus reducing the concentration of gold in the bioleached solution.

Further, bacterial growth and gold complexation were decoupled by using spent medium and adjusting its pH to 10. This approach increased the availability of cyanide ions for gold bioleaching, recovering 30% Au and 95.7% Cu at 0.5% w/v pulp density. Thus, the pH modification of spent medium further improved metal solubilization and yielded greater metal recovery (compared with two-step bioleaching) (Natarajan and Ting 2015).

Li et al. (2015) studied the bioleaching of gold from PCBs using *C. violaceum*. Many metals and metalloids form a complex with cyanide, so the high amount of base metals dissolved by cyanide complexation during bioleaching hinders gold recovery from the leachate (Pham and Ting 2009). Thus, the base metal content of the sample must be reduced using suitable methods before proceeding to gold recovery. Hence, in the present study, e-waste was pretreated by *At. ferrooxidans* to remove copper before the bioleaching of gold.

An orthogonal experiment was conducted to optimize the process parameters to enhance Cu recovery. Prior to the gold bioleaching experiments, PCBs were subjected to leaching by *At. ferrooxidans* for 7 days. This pretreated powder was used for gold bioleaching by *C. violaceum*. The effects of various parameters such as aeration, pretreatment, particle size, and different levels of nutritive salts on gold leaching efficiency were studied. pH value was found to be the most significant factor, whereas particle size was the least influencing factor. The order of the influencing factors was as follows: pH > added amount of the powder > the amount of bacteria > reaction temperature > particle size.

The pH increased slightly during the generation of OH$^-$. The DO concentration of the solution decreased to a minimal level after 24 h without any oxygen supplement as a result of bacterial respiration and the reaction of gold and cyanide. The use of H_2O_2 to supplement DO did not increase the CN$^-$ level significantly (Kita et al. 2006, Chi et al. 2011). Hence, a homemade sterile oxygenator was used to supplement DO in this case, which increased the gold leaching efficiency. After an initial drop, the pH gradually increased in the next few days and reached 9.0–9.5, as the reaction between gold and cyanide generates OH$^-$.

Using untreated PCBs, only 19.8% gold was recovered, may be due to the presence of other compounds toxic to microorganisms. Nearly 80% copper and other base metals were removed by the pretreatment using *At. ferrooxidans*, thus increasing the gold/copper ratio in the residual solid.

About 40.1% of the gold leaching efficiency was obtained at a cyanide content of 2.155 mg after 7 days. Using the optimum particle size obtained with 200 # mesh, the gold leaching efficiency was increased to 46.3%; however, there was not much difference in the cyanide content. The addition of NaCl (1.7×10^{-1} mol/L) and $MgSO_4 \cdot 7H_2O$ (4×10^{-3} mol/L), resulted in a maximum gold leaching efficiency of 70.6% and 52.4% and a cyanide content of 6.425 and 4.918 mg were obtained, respectively. The obtained gold leaching efficiency of 70.6% in this study is higher than in previous reports (Li et al. 2015).

Madrigal-Arias et al. (2015) employed different strains of *A. niger* to recover gold, copper, and nickel from gold-plated finger integrated circuits found in computer motherboards (GFICMs) and MPPCBs. Au bioleaching ranged from 42% to 1% for *A. niger* strain MXPE6; using a combination of *A. niger* MXPE6 + *A.*

niger MX7, Au bioleaching was 87% and 28% for PCBs and GFICMs, respectively. Results indicate that by using a fungal consortium, the recovery of Au from GFICMs or PCBs is significantly increased when compared with single fungal inoculation. In contrast, bioleaching of Cu by *A. niger* MXPE6 was 24% and 5%; using a combination of both strains, the values were 0.2% and 29% for PCBs and GFICMs, respectively. Fungal Ni leaching was only found for PCBs, with no significant differences among treatments.

Waste mobile phone PCBs (MPPCBs) are a rich source of copper and precious metals such as gold and silver, which may be processed to ensure resource recycling. *C. violaceum* generates cyanide from glycine during growth and the early stationary phase. Such cyanogenic bacteria can be explored for gold extraction. However, the consumption of DO by bacteria during its growth inhibits metal leaching, thus hampering the leaching process. Thus, Chi et al. (2011) demonstrated the effect of the absence/presence of H_2O_2 to supplement DO during metal extraction. They used *C. violaceum* to leach gold and copper from the waste MPPCBs containing 34.5% Cu and 0.025% Au in a YP (yeast extract and polypeptone with glycine) medium. Adding 0.004% H_2O_2 increased DO without seriously affecting bacterial growth and improved copper leaching from 11.4% to 24.6% at pH 10.0 and gold recovery only marginally from 10.8% to 11.31% at pH 11.0. Significant copper leaching as compared with gold leaching may be the result of the high copper content present in the sample consuming cyanide produced at the higher DO level. The authors further suggest the use of a suitable method to decrease the copper content prior to gold bioleaching to improve gold recovery (Madrigal-Arias et al. 2015).

Isıldar et al. (2016) examined copper and gold recovery from PCBs using a two-step bioleaching approach. Attributing to the previous studies that showed improved gold recovery upon prior recovery of copper (Pham and Ting 2009), the authors proposed a new two-step approach based on the different chemical properties and leaching mechanisms of base and precious metals. *Acidophilic* strains of *At. ferrivorans* and *At. thiooxidans* (DSM 9463) and cyanide-producing strains *Pseudomonas putida* (WSC361) and *Pseudomonas fluorescens* (E11.3) were used in the study. In the preliminary step, *At. ferrivorans* and *At. thiooxidans* were grown and used for copper leaching from PCBs. In the second step, *P. fluorescens* and *P. putida* were grown to a point of maximum cyanide production and subsequently used for bioleaching of Cu-leached PCBs.

To avoid the inhibitory effect of PCBs on bacterial growth, cultures were grown in the absence of waste material to allow bacterial establishment and generation of optimal bioleaching conditions. Nearly 94%, 89%, and 98% Cu were recovered by pure cultures of *At. ferrivorans*, *At. thiooxidans*, and a mixture of both, respectively, at 1% (10 g/L) pulp density, pH 1.0–1.6, and ambient temperature (23°C ± 2°C) in 7 days As a result of an increase in pH, bioleaching decreased with an increase in pulp density. Low pH and high oxidation-reduction potential during the growth of acidophiles suggest the involvement of acidolysis and redoxolysis mechanisms.

A coculture of iron- and sulfur-oxidizing acidophiles showed higher copper recovery than pure cultures, indicating a twofold advantage of the involvement of both leaching mechanisms, redoxolysis and acidolysis, as well as a positive cooperative bioleaching mechanism when acidophiles coexist in the bioleaching coculture.

P. putida produced higher cyanide 21.5 (±1.5) mg/L as compared with 15.5 (±2.4) mg/L by *P. fluorescens* under optimum glycine concentration. Glycine had an inhibitory effect on cyanide production by *P. fluorescens* at concentrations above 7.5 g/L. Highest gold recovery of 44% was obtained at 0.5% pulp density, in alkaline conditions at pH 7.3–8.6, and 30°C in 2 days. However, complete recovery of gold was observed with chemical cyanide leaching at 100 mg/L (3.84 mM) cyanide concentration. The authors suggest the need for further studies to enhance bacterial cyanide production and increase the chemical stability of the free cyanide in the solution. This study provided a proof of concept of a two-step approach in metal bioleaching from PCBs by bacterially produced lixiviants (Isıldar et al. 2016).

4.2.3   KINETICS OF METAL RECOVERY

Taking into account the importance of H⁺ consumption during bioleaching and to analyze the leaching mechanism, Yang et al. (2014) investigated the kinetics of metal recovery and the relationship between H⁺ consumption and metal recovery from WPCBs using *At. ferrooxidans*. The bioleaching efficiency decreased rapidly as the WPCB concentration increased from 15 to 35 g/L. When the WPCB concentration was 15 g/L, Cu (96.8%), Zn (83.8%), and Al (75.4%) were recovered after 72 h by *At. ferrooxidans*. Experimental results indicated that the metal recovery rate was significantly influenced by acid. In the bioleaching process, metal recovery is related to the metal's reactivity, and the stronger the metal's reactivity, the faster and easier the bioleaching. In addition, the alkaline substance of WPCBs (direct consumption) and the oxidation reaction of Fe^{2+} (indirect consumption) consume H⁺. Thus, adding acid can maintain the pH of the leaching solution and contribute to improving the leaching efficiency indirectly by ensuring that the iron cycle proceeds well. Moreover, the bioleaching kinetics of metals fits the second-order model (with the addition of acid) well. However, the kinetics would change from the second-order model to the shrinking-core model because the amount of precipitate increases with time (without the addition of acid).

4.2.4   PROCESS OPTIMIZATION FOR METAL RECOVERY

There are a lot of factors that influence the biorecovery percentage. These parameters and their interaction play a great role in bioleaching recovery and a small improvement in them can be critical for metal dissolution. So process optimization is a crucial aspect in this field (Amiri et al. 2011a). Process optimization is necessary to attain maximum metal recovery with the minimum number of experiments. Statistical methods have the advantage of being rapid and can identify interactions among factors (Amiri et al. 2012). Response surface methodology (RSM) is used to model, analyze, and define the interaction of independent factors and optimize processes (Amani et al. 2012). Only a few reports are available on the use of RSM for bioleaching.

Bioleaching of Cu and Ni from CPCB waste was studied using adapted *At. ferrooxidans* (Arshadi and Mousavi 2014). The adaptation phase began at 1 g/L CPCB powder with 10% inoculation and the final pulp density was reached at 20 g/L after about 80 days. Four effective factors, including initial pH, particle size, pulp density, and initial Fe^{3+} concentration, were optimized to achieve maximum simultaneous recovery of Cu and Ni. Their interactions were also identified using a central composite design in RSM (CCD-RSM). The suggested optimal conditions were initial pH 3, initial Fe^{3+} 8.4 g/L, pulp density 20 g/L, and particle size 95 μm. Analysis of variance results showed that the most effective factor for Cu extraction was pulp density, while for Ni, it was pH. Nearly 100% of Cu and Ni were simultaneously recovered under optimum conditions in 20 days.

In continuation of this, simultaneous gold and copper recovery from PCBs was evaluated using CCD-RSM by Arshadi and Mousavi (2015a). The potential of *B. megaterium* for simultaneous recovery of copper and gold from PCBs was assessed. The effect of glycine concentration, initial pH, pulp density, and particle size as well as their interactions in the bioleaching process was evaluated using CCD-RSM. The PCB powder was suspended in brine (100 g NaCl/L) for 2 h to remove plastics from the upper solution. Maximum simultaneous gold and copper recovery occurred at an initial pH of 10, a pulp density of 2 g/L, a particle size obtained with mesh # 100, and a glycine concentration of 0.5 g/L, which led to a 36.81% gold extraction. The conducted experiment under these optimal conditions, maximizing gold and minimizing copper as a target, resulted in only a 14.02% gold recovery (Arshadi and Mousavi 2015a).

In continuation of this study, Arshadi and Mousavi (2015b) examined simultaneous Cu and Ni extraction from MPPCBs by employing *At. thiooxidans*. Adapted *At. ferrooxidans* was employed to maximize simultaneous Cu and Ni recovery from MPPCBs and identify the behavior and interaction of the influencing parameters on e-waste bioleaching by applying RSM. It took 55 days for *At. ferrooxidans* to get adapted to the increasing concentrations of PCBs and finally reach the threshold of 20 g PCB powder per liter of the medium. Initial pH, initial Fe^{3+} concentration, pulp density, and particle size were the four factors selected under a multiobjective optimization strategy using CCD-RSM to maximize simultaneous extraction of Cu and Ni from MPPCBs. CCD was used to build second- and third-order response surface models and fit a polynomial model, and also decrease the minimum number of experiments required for modeling.

The quadratic model suggested that particle size had a significant effect on Cu recovery, whereas, for Ni recovery, it was initial pH. However, recovery of Cu was more efficient as compared with Ni, which was 70%. But the decrease in pH increased Ni recovery, suggesting acidic pH being favorable for Ni recovery. Further, according to the model, best condition for maximum simultaneous recovery of Cu and Ni at 0.965 desirability was for an initial pH of 1, an initial Fe^{3+} concentration of 4.28 g/L, a pulp density of 8.50 g/L, and a particle size of 114.02 lm (obtained with # 100 mesh). Results of the experiments at optimum parameters revealed a total extraction of both metals. Moreover, the variation in Fe^{3+} concentration under optimum conditions showed most of the recovery occurring within 17 days.

During this study, the authors found that the gold contained in MPPCBs that *At. ferrooxidans* are unable to recover is about 200 times that in a typical gold mine. Thus, the authors suggest further studies for gold extraction from MPPCBs using cyanogenic bacteria.

The analysis of samples showed that copper makes about 40% (w/w) of the total weight of samples, while gold makes about 0.018% (w/w). It means that the amount of cyanide needed for the total extraction of Cu is much higher than Au. The higher amount of Cu has greater inhibitory effects, as it forms a complex with cyanide at a faster rate than does Au and thus inhibits gold recovery. To decrease this copper, the authors used the strategy of using *At. ferrooxidans* to treat this sample at 20 g/L pulp density, pH 3, and Fe^{3+} concentration of 8.4 g/L. After 20 days of this treatment, the sediment was subjected to bioleaching of gold using *B. megaterium* at an initial pH of 10, a pulp density of 2 g/L, a particle size obtained with mesh # 100, and a glycine concentration of 0.5 g/L. The sample without pretreatment after 2 days showed a gold recovery of 34.36%, which almost became constant with time, while for CPCBs with pretreatment and two-step bioleaching in 6 days, it reached a maximum recovery of 63.8%.

The factors affecting the simultaneous recovery of copper and gold from discarded MPPCBs using *Bacillus megaterium* and their interactions were studied using RSM by Arshadi et al. (2016). RSM is a group of mathematical and statistical techniques which is used in the development of an adequate functional relationship between interest responses and several variables that affect the response. First- and second-order models are usually employed in RSM. RSM is useful in establishing a relationship between the responses and the variables, predicting responses' values for a given set of conditions, and in determining the significance of the factors through hypothesis testing. Also, it helps in determining the optimum variables' condition for having the maximum or the minimum value of response over a certain region. Among RSM principles, CCD is the most popular among all second-order design methods. The total number of experiments in CCD is equal to $n = 2k + 2k + n_0$, where k and n_0 are the numbers of variables and center points, respectively (Khuri and Mukhopadhyay 2010).

Due to the presence of precious metals in e-waste, especially PCBs, there is a great motive for their recovery instead of simply disposing of them. The purity of these valuable metals has been found to be 10 times greater than that of rich-content minerals (Hadi et al. 2015).

B. megaterium, a cyanogenic bacterium, was employed in the present study due to its ability to produce cyanide and recover gold (Arshadi and Mousavi 2015c). Various important parameters such as initial pH, initial glycine concentration, and pulp density were selected for statistical evaluation by RSM so as to reach the maximum simultaneous extraction efficiency of both metals.

Experimental design was based on CCD with three factors, five level. The most effective parameter in Au recovery from MPPCBs was pulp density, with a linear negative effect; that is, increasing the pulp density rapidly decreased Au recovery. At lower pulp densities, regardless of the value of the initial pH, Au recovery increased, whereas for CPCBs, the initial pH and pulp density had a sharp interaction, which led to different results. These results arise from the different natures of the samples; CPCBs are highly alkaline, while MPPCBs are neutral.

To maximize Au recovery, the optimal conditions suggested by the models were an initial pH of 10, a pulp density of 8.13 g/L, and a glycine concentration of 10 g/L. Under these optimal conditions, approximately 72% Cu and 65 g Au/ton MPPCBs, which is seven times greater than the recovery from gold mines, were extracted.

Higher initial pH and pulp density enhanced Cu recovery. At higher initial pH, the produced cyanide exists in the stable form of CN, and thus the amount of the free and available cyanide in the solution increases. Increasing the pulp density causes an increment in the number of particles and also the available surface for cyanide to form complexes with Cu. This, consequently, improves Cu recovery.

The effect of Cu removal on Au recovery was also investigated, wherein first Cu was totally extracted using *At. ferrooxidans* and then the remaining sediment was used in further study. Cu elimination from MPPCBs had an insignificant effect on Au recovery. It was found that when the ratio of Cu to Au is high, Cu elimination can considerably improve Au recovery. *B. megaterium* could extract the total Au from MPPCBs containing 130 g Au/ton. Further, it was found that there is a big difference in the bioleaching results of MPPCBs and CPCBs.

MPPCBs contain 1800 g/ton and 6.65% w/w of Au and Cu, respectively. The amount of Au in CPCBs is nearly 8.82 times less than that of Au in MPPCBs, whereas that of Cu in CPCBs is about 6 times greater than in MPPCBs. First, nearly 8.82 times of cyanide is required to recover Au from MPPCBs, similar to that of CPCBs; however, bacterial cyanide production is limited and unable to reach that point. So the expected gold recovery from MPPCBs is low when compared with CPCBs. Further, the amount of Cu is less, while that of Au is much higher in MPPCBs than in CPCBs, and consequently, the inhibitory effect of Cu is low. And thus, the elimination of Cu does not significantly impact Au recovery, whereas for CPCBs, it has a significant effect and nearly 64% of Au was recovered in the study. These findings are thus a golden key in bioleaching of PCBs of e-waste (Arshadi et al. 2016).

4.2.5 Use of Biosurfactants in Metal Recovery

There are some suggestions in the literature that the addition of surfactants, complexing agents, or other additives may positively influence bioleaching using mesophilic bacteria (Cui and Zhang 2008, Lan et al. 2009). Karwowska et al. (2014) studied the role of biosurfactant-producing bacteria in bioleaching of metals from PCBs. They evaluated the role of the sulfur-oxidizing *Acidithiobacillus* sp. alone and in consortium with a mixture of biosurfactant-producing strains *B. subtilis* PCM 2021 and *B. cereus* PCM 2019.

Two bioleaching media containing a mixture of municipal activated sludge and municipal wastewater with 1% sulfur and 1% sulfur plus biosurfactant were used for the experiments. It was revealed that zinc was removed effectively both in a traditional solution acidified by microbial oxidation of sulfur and when using a microbial culture containing sulfur-oxidizing and biosurfactant-producing bacteria. The average process efficiency was 48% for Zn dissolution. Cadmium removal was similar in both media, with the highest metal release of 93%. For nickel and copper, a better effect was obtained in the acidic medium, with a process efficiency of 48.5% and 53%, respectively. Chromium was the only metal that was removed more effectively in the bioleaching medium containing both sulfur-oxidizing and biosurfactant-producing bacteria. Lead was removed from PCBs at a very low efficiency (below 0.5%). In this research, the bioleaching activity of sulfur-oxidizing bacteria was supported with the addition of biosurfactants produced by *B. cereus* and *B. subtilis* strains.

The use of biosurfactant-producing bacteria is statistically significant to the bioleaching process for copper, lead, nickel, and chromium. Zinc was removed effectively in both bioleaching media, but faster in the acidic environment.

The results of the experiment concerning the influence of temperature and aeration mode on the bioleaching efficiency revealed the different susceptibilities of metals to the bioleaching process in both tested cultures under applied process conditions. Aerating the culture medium with compressed air increased the release of all metals in the medium with sulfur and biosurfactant, and of Ni, Cu, Zn, and Cr in the acidic medium. Increasing the temperature of the medium (to 37°C) had a more significant impact in the acidic environment than in the neutral environment.

4.2.6 LEACHING PROCESS USING ENZYMES

The application of enzymes in bioremediation, also termed as white biotechnology, is an emerging technology (Alcalde et al. 2006). There are several benefits of using enzymes for environmental applications. Enzymes can function both under mild conditions, replacing harsh conditions and harsh chemicals, and under extreme conditions, thus saving energy and preventing pollution. The application of enzymes avoids the generation of secondary waste, since they are highly specific. Enzymes are readily absorbed back into nature and do not create disposal problems in bioremediation processes, as in biomass produced using living cells (Alcalde et al. 2006, Chaudhuri et al. 2013).

The use of enzymes for bioleaching of metals will help develop novel processes with better affinity, capacity, and selectivity. Enzymes can be used both in place of hazardous chemicals and for in situ production of such chemicals. H_2O_2 is one such chemical. H_2O_2 is required to carry out the Fenton process. The use of glucose oxidase for H_2O_2 production reduces the risk of its transportation and handling during the bio-Fenton process. This makes the bioleaching process safer. The use of enzymes in the bio-Fenton process reduces the process time, which in turn reduces the time required for the bioleaching process. Jadhav and Hocheng (2015) evaluated the ability of such enzymes as a new possible tool for environment-friendly leaching of metals from PCBs (enzymatic bio-Fenton process).

During the Fenton process, Fe(III) ions are produced along with hydroxyl radicals (Karimi et al. 2012, Sekaran et al. 2013). In the bio-Fenton reaction, H_2O_2 is produced by biological means. The reaction of H_2O_2 with Fe(II) produces Fe(III) and hydroxyl radicals (HO·). The authors used glucose oxidase for oxidation of glucose, thus producing gluconic acid and H_2O_2 (Equation 4.9) (Eskandarian et al. 2013):

$$\text{Glucose} + O_2 \xrightarrow{\text{Glucoseoxidase}} \text{Gluconic acid} + H_2O_2 \qquad (4.9)$$

The H_2O_2 released during this process reacted with Fe(II) and oxidized Fe(II) to Fe(III) according to Equation 4.10 (Karimi et al. 2012, Eskandarian et al. 2013):

$$\text{Fe(II)} + H_2O_2 \rightarrow \text{Fe(III)} \qquad (4.10)$$

The effect of incubation time on H_2O_2 revealed that an increase in incubation time increased H_2O_2 production, but the curve reached a plateau after 12 h, which may be ascribed to the decrease in substrate concentration or inactivation of the enzyme during the process. It was also found that within 90 min, 50 mM Fe(II) was converted into Fe(III). Effects of various parameters such as temperature, pH, and shaking speed on the conversion of Fe(II) into Fe(III) exhibited the optimum temperature to be 30°C, pH 5.5, and shaking speed 150 rpm. This enzymatic bio-Fenton process was further applied for bioleaching of metals. The results showed that metal recovery increased with an increase in the initial Fe(II) concentration. A maximum recovery of metals, 61% and 60% for Ti and Al, respectively, was achieved in 48 h at a 250 mM initial Fe(II) concentration. Metal recovery decreased with an increase in the concentration of the PCB powder. The authors also carried out the bioleaching process using glucose oxidase immobilized in calcium alginate beads. Almost similar metal recovery was achieved using both free and immobilized glucose oxidase enzyme. But as the number of bioleaching cycles increased, metal recovery decreased, as shown in Figure 4.6. Low stability and high membrane porosity of the immobilized calcium alginate beads and loss of enzyme activity may be the plausible reasons.

This study is a stepping stone to the development of enzymatic bioleaching, which will initiate the application of new enzymes for their possible use in the metal recovery process. Also, this study has opened up avenues to investigate many aspects of enzyme application in bioremediation, such as large-scale enzyme production using cheaper substrates so as to make the process cost-effective. Enzyme immobilization studies will facilitate the recycling of enzymes, broaden their applications, and

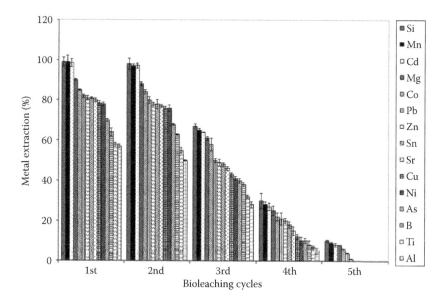

FIGURE 4.6 Effect of immobilization on bioleaching performance of glucose oxidase enzyme. (Reprinted from Jadhav, U. and Hocheng, H., *Clean Technol. Environ. Policy*, 17, 947, 2015. With permission.)

improve their stability (Kharrat et al. 2011, Zhou and Hartmann 2013, Zhao et al. 2014). The immobilization will protect the enzyme from heavy metals and will also contribute to reducing the cost of the process. Consequently, these studies will help initiate the development of a bioreactor system for bioleaching of metals using immobilized enzymes. Besides, techniques such as recombinant DNA technology, protein engineering, and computer-assisted rational enzyme design can be applied to enzyme-assisted environmental biotechnology (Ahuja et al. 2004).

4.3   CONCLUSION

The finding in this study renders the ability of biohydrometallurgical processing to be implemented at an industrial scale for recycling of metals from waste electronic materials. Preadaptation of microorganisms to high metal ion concentrations and using mixed consortia of different bacteria greatly enhance the metal solubilization rate, as these microorganisms can survive under high concentration of metal ions. Also, pretreatment of electronic scrap reduces the time required to stabilize the pH and provides optimum conditions for a bacterial attack on the scrap surface. For a commercially interesting process, a direct growth of microorganisms in the presence of electronic scrap is poorly suited and not advisable. Therefore, a two-step process seems appropriate to increase the leaching efficiency, for example, for an industrial application. In the first step, microorganisms will be grown in the absence of electronic scrap, followed by the second step, where the formed metabolites will be used for metal solubilization. Also, process optimization and the use of a combination of techniques such as mutation and genetic manipulation to obtain more effective microbial populations can overcome other hurdles and bring this technology at par with its counterparts. But studies also provide evidence for the precipitation of some metals such as Pb and Sn, along with other metal ions. Thus, some additional strategies must be developed to either prevent such precipitation or recover metals from the precipitate.

REFERENCES

Adhapure, N. N., S. S. Waghmare, V. S. Hamde, and A. M. Deshmukh. 2013. Metal solubilization from powdered printed circuit boards by microbial consortium from bauxite and pyrite ores. *Applied Biochemistry and Microbiology* 49(3): 256–262. doi:10.1134/S0003683813030034.

Ahuja, S. K., G. M. Ferreira, and A. R. Moreira. 2004. Utilization of enzymes for environmental applications. *Critical Reviews in Biotechnology* 24(2–3): 125–154. doi:10.1080/07388550490493726.

Alcalde, M., M. Ferrer, F. J. Plou, and A. Ballesteros. 2006. Environmental biocatalysis: From remediation with enzymes to novel green processes. *Trends in Biotechnology* 24(6): 281–287. doi:10.1016/j.tibtech.2006.04.002.

Amani, T., M. Nosrati, S. M. Mousavi, and R. K. Kermanshahi. 2012. Analysis of the syntrophic anaerobic digestion of volatile fatty acids using enriched cultures in a fixed-bed reactor. *Water Environment Research* 84: 460–472.

Amiri, F., S. M. Mousavi, and S. Yaghmaei. 2011a. Enhancement of bioleaching of a spent Ni/Mo hydroprocessing catalyst by *Penicillium simplicissimum*. *Separation and Purification Technology* 80(3): 566–576. doi:10.1016/j.seppur.2011.06.012.

Amiri, F., S. M. Mousavi, S. Yaghmaei, and M. Barati. 2012. Bioleaching kinetics of a spent refinery catalyst using *Aspergillus niger* at optimal conditions. *Biochemical Engineering Journal* 67: 208–217. doi:10.1016/j.bej.2012.06.011.

Amiri, F., S. Yaghmaei, S. M. Mousavi, and S. Sheibani. 2011b. Recovery of metals from spent refinery hydrocracking catalyst using adapted *Aspergillus niger*. *Hydrometallurgy* 109(1–2): 65–71. doi:10.1016/j.hydromet.2011.05.008.

Anjum, F., M. Shahid, and A. Akcil. 2012. Biohydrometallurgy techniques of low grade ores: A review on black shale. *Hydrometallurgy* 117: 1–12. doi:10.1016/j.hydromet.2012.01.007.

Arshadi, M. and S. M. Mousavi. 2014. Simultaneous recovery of Ni and Cu from computer-printed circuit boards using bioleaching: Statistical evaluation and optimization. *Bioresource Technology* 174: 233–242. doi:10.1016/j.biortech.2014.09.140.

Arshadi, M. and S. M. Mousavi. 2015a. Enhancement of simultaneous gold and copper extraction from computer printed circuit boards using *Bacillus megaterium*. *Bioresource Technology* 175: 315–324. doi:10.1016/j.biortech.2014.10.083.

Arshadi, M. and S. M. Mousavi. 2015b. Multi-objective optimization of heavy metals bioleaching from discarded mobile phone PCBs: Simultaneous Cu and Ni recovery using *Acidithiobacillus ferrooxidans*. *Separation and Purification Technology* 147: 210–219. doi:10.1016/j.seppur.2015.04.020.

Arshadi, M. and S. M. Mousavi. 2015c. Statistical evaluation of bioleaching of mobile phone and computer waste PCBs: A comparative study. *Advanced Materials Research* 1104(May): 87–92. doi:10.4028/www.scientific.net/AMR.1104.87.

Arshadi, M., S. M. Mousavi, and P. Rasoulnia. 2016. Enhancement of simultaneous gold and copper recovery from discarded mobile phone PCBs using *Bacillus megaterium*: RSM based optimization of effective factors and evaluation of their interactions. *Waste Management* 57: 158–167. doi:10.1016/j.wasman.2016.05.012.

Bas, A. D., H. Deveci, and E. Y. Yazici. 2013. Bioleaching of copper from low grade scrap TV circuit boards using mesophilic bacteria. *Hydrometallurgy* 138: 65–70. doi:10.1016/j.hydromet.2013.06.015.

Bertram, M, T. E. Graedel, H. Rechberger, and S. Spatari. 2002. The contemporary European copper cycle: Waste management subsystem. *Ecological Economics* 42(1): 43–57. doi:10.1016/S0921-8009(02)00100-3.

Brandl, H., R. Bosshard, and M. Wegmann. 2001. Computer-munching microbes: Metal leaching from electronic scrap by bacteria and fungi. *Hydrometallurgy* 59: 319–326. doi:10.1016/S1572-4409(99)80146-1.

Brandl, H., S. Lehmann, M. A. Faramarzi, and D. Martinelli. 2008. Biomobilization of silver, gold, and platinum from solid waste materials by HCN-forming microorganisms. *Hydrometallurgy* 94(1): 14–17. doi:10.1016/j.hydromet.2008.05.016.

Bryan, C. G., E. L. Watkin, T. J. McCredden, Z. R. Wong, S. T. L. Harrison, and A. H. Kaksonen. 2015. The use of pyrite as a source of lixiviant in the bioleaching of electronic waste. *Hydrometallurgy* 152: 33–43. doi:10.1016/j.hydromet.2014.12.004.

Chaudhuri, G., G. A. Shah, P. Dey, S. Ganesh, P. Venu-Babu, and R. Thilagaraj. 2013. Enzymatically mediated bioprecipitation of heavy metals from industrial wastes and single ion solutions by mammalian alkaline phosphatase. *Journal of Environmental Science and Health. Part A, Toxic/Hazardous Substances & Environmental Engineering* 48(1): 79–85. doi:10.1080/10934529.2012.707851.

Chen, S., Y. Yang, C. Liu, F. Dong, and B. Liu. 2015. Column bioleaching copper and its kinetics of waste printed circuit boards (WPCBs) by *Acidithiobacillus ferrooxidans*. *Chemosphere* 141: 162–168. doi:10.1016/j.chemosphere.2015.06.082.

Chi, T. D., J. C. Lee, B. D. Pandey, K. Yoo, and J. Jeong. 2011. Bioleaching of gold and copper from waste mobile phone PCBs by using a cyanogenic bacterium. *Minerals Engineering* 24(11): 1219–1222. doi:10.1016/j.mineng.2011.05.009.

Choi, M.-S., K.-S. Cho, D.-S. Kim, and D.-J. Kim. 2004. Microbial recovery of copper from printed circuit boards of waste computer by *Acidithiobacillus ferrooxidans*. *Journal of Environmental Science and Health, Part A* 39: 2976–2982.

Choi, K. H., J. B. Gaynor, K. G. White et al. 2005. A Tn7-based broad-range bacterial cloning and expression system. *Nature Methods* 2: 443–448.

Cui, J. and L. Zhang. 2008. Metallurgical recovery of metals from electronic waste: A review. *Journal of Hazardous Materials* 158(2–3): 228–256. doi:10.1016/j.jhazmat.2008.02.001.

Eskandarian, M., A. Karimi, and M. Shabgard. 2013. Studies on enzymatic biomachining of copper by glucose oxidase. *Journal of the Taiwan Institute of Chemical Engineers* 44(2): 331–335. doi:10.1016/j.jtice.2012.11.005.

Faramarzi, M. A., M. Stagars, E. Pensini, W. Krebs, and H. Brandl. 2004. Metal solubilization from metal-containing solid materials by cyanogenic *Chromobacterium violaceum*. *Journal of Biotechnology* 113(1): 321–326. doi:10.1016/j.jbiotec.2004.03.031.

Franz, B., T. Gehrke, H. Lichtenberg, J. Hormes, C. Dahl, and A. Prange. 2009. Unexpected extracellular and intracellular sulfur species during growth of *Allochromatium vinosum* with reduced sulfur compounds. *Microbiology* 155(8): 2766–2774. doi:10.1099/mic.0.027904-0.

Giaveno, A., L. Lavalle, E. Guibal, and E. Donati. 2008. Biological ferrous sulfate oxidation by *A. ferrooxidans* immobilized on chitosan beads. *Journal of Microbiological Methods* 72(3): 227–234. doi:10.1016/j.mimet.2008.01.002.

Hadi, P., C. Ning, W. Ouyang, M. Xu, C. S. K. Lin, and G. McKay. 2015. Toward environmentally-benign utilization of nonmetallic fraction of waste printed circuit boards as modifier and precursor. *Waste Management* 35: 236–246. doi:10.1016/j.wasman.2014.09.020.

Halinen, A.-K., N. Rahunen, A. H. Kaksonen, and J. A. Puhakka. 2009. Heap bioleaching of a complex sulfide ore: Part I: Effect of pH on metal extraction and microbial composition in pH controlled columns. *Hydrometallurgy* 98(1): 92–100. doi:10.1016/j.hydromet.2009.04.005.

He, H., C. G. Zhang, J. L. Xia et al. 2009. Investigation of elemental sulfur speciation transformation mediated by *Acidithiobacillus ferrooxidans*. *Current Microbiology* 58: 300–307.

Hedley, H. and H. Tabachnick. 1958. *Chemistry of Cyanidation*. Mineral Dressing Notes, No. 23. pp. 23–28, New York: American Cyanamid Co.

Heiskanen, K. 2014. Physical separation 101. In *Handbook of Recycling: State-of-the-Art for Practitioners, Analysts, and Scientists*, Worrell, E. and M. A. Reuter (Eds.). pp. 537–543, Elsevier.

Hocheng, H., T. Hong, and U. Jadhav. 2014. Microbial leaching of waste solder for recovery of metal. *Applied Biochemistry and Biotechnology* 173: 193–204.

Hong, Y. and M. Valix. 2014. Bioleaching of electronic waste using acidophilic sulfur oxidising bacteria. *Journal of Cleaner Production* 65: 465–472. doi:10.1016/j.jclepro.2013.08.043.

Huang, K., J. Guo, and X. Zhenming. 2009. Recycling of waste printed circuit boards: A review of current technologies and treatment status in China. *Journal of Hazardous Materials* 164(2): 399–408. doi:10.1016/j.jhazmat.2008.08.051.

Huisman, J., F. Magalini, R. Kuehr et al. 2007. 2008 Review of directive 2002/96 on waste electrical and electronic equipment. ENV.G.4/ETU/2006/0032. United Nations University, Bonn, Germany, p. 347.

Ilyas, S., M. A. Anwar, S. B. Niazi, and M. Afzal Ghauri. 2007. Bioleaching of metals from electronic scrap by moderately thermophilic acidophilic bacteria. *Hydrometallurgy* 88(1–4): 180–188. doi:10.1016/j.hydromet.2007.04.007.

Ilyas, S., R. Chi, H. N. Bhatti, M. A. Ghauri, and M. A. Anwar. 2010. Column bioleaching of metals from electronic scrap. *Hydrometallurgy* 101(3): 135–140. doi:10.1016/j.hydromet.2009.12.007.

Ilyas, S. and J. C. Lee. 2014. Bioleaching of metals from electronic scrap in a stirred tank reactor. *Hydrometallurgy* 149: 50–62. doi:10.1016/j.hydromet.2014.07.004.

Ilyas, S., J. C. Lee, and R. A. Chi. 2013. Bioleaching of metals from electronic scrap and its potential for commercial exploitation. *Hydrometallurgy* 131–132: 138–143. doi:10.1016/j.hydromet.2012.11.010.

Isıldar, A., J. van de Vossenberg, E. R. Rene, E. D. van Hullebusch, and P. N. L. Lens. 2016. Two-step bioleaching of copper and gold from discarded printed circuit boards (PCB). *Waste Management* 57: 149–157. doi:10.1016/j.wasman.2015.11.033.

Jadhav, U. and H. Hocheng. 2015. Enzymatic bioleaching of metals from printed circuit board. *Clean Technology and Environmental Policy* 17: 947–956.

Jadhav, U. U., H. Hocheng, and W.-H. Weng. 2013. Innovative use of biologically produced ferric sulfate for machining of copper metal and study of specific metal removal rate and surface roughness during the process. *Journal of Materials Processing Technology* 213(9): 1509–1515. doi:10.1016/j.jmatprotec.2013.03.015.

Jin, J., S.-y. Shi, G.-l. Liu, Q.-h. Zhang, and W. Cong. 2013. Comparison of Fe^{2+} oxidation by *Acidithiobacillus ferrooxidans* in rotating-drum and stirred-tank reactors. *Transactions of Nonferrous Metals Society of China* 23(3): 804–811. doi:10.1016/S1003-6326(13)62532-7.

Karimi, A., M. Aghbolaghy, A. Khataee, and S. S. Bargh. 2012. Use of enzymatic bio-Fenton as a new approach in decolorization of malachite green. *The Scientific World Journal* 2012(1): 691569. doi:10.1100/2012/691569.

Karwowska, E., D. Andrzejewska-Morzuch, M. Łebkowska, A. Tabernacka, M. Wojtkowska, A. Telepko, and A. Konarzewska. 2014. Bioleaching of metals from printed circuit boards supported with surfactant-producing bacteria. *Journal of Hazardous Materials* 264: 203–210. doi:10.1016/j.jhazmat.2013.11.018.

Kharrat, N., Y. B. Ali, S. Marzouk, Y.-T. Gargouri, and M. Karra-Châabouni. 2011. Immobilization of *Rhizopus oryzae* lipase on silica aerogels by adsorption: Comparison with the free enzyme. *Process Biochemistry* 46(5): 1083–1089. doi:10.1016/j.procbio.2011.01.029.

Khuri, A. I. and S. Mukhopadhyay. 2010. Response surface methodology. *Wiley Interdisciplinary Reviews: Computational Statistics* 2(2): 128–149. doi:10.1002/wics.73.

Kita, Y., H. Nishikawa, and T. Takemoto. 2006. Effects of cyanide and dissolved oxygen concentration on biological Au recovery. *Journal of Biotechnology* 124(3): 545–551. doi:10.1016/j.jbiotec.2006.01.038.

Knowles, C. J. and A. W. Bunch. 1986. Microbial cyanide metabolism. *Advances in Microbial Physiology* 27: 73–111. doi:10.1016/S0065-2911(08)60304-5.

Korte, F., M. Spiteller, and F. Coulston. 2000. The cyanide leaching gold recovery process is a nonsustainable technology with unacceptable impacts on ecosystems and humans: The disaster in Romania. *Ecotoxicology and Environmental Safety* 46(3): 241–245. doi:10.1006/eesa.2000.1938.

Krebs, W., C. Brombacher, P. P. Bosshard, R. Bachofen, and H. Brandl. 1997. Microbial recovery of metals from solids. *FEMS Microbiology Reviews* 20(3): 605–617. doi:10.1016/S0168-6445(97)00037-5.

Lambert, F., S. Gaydardzhiev, G. Léonard, G. Lewis, P. F. Bareel, and D. Bastin. 2015. Copper leaching from waste electric cables by biohydrometallurgy. *Minerals Engineering* 76: 38–46. doi:10.1016/j.mineng.2014.12.029.

Lan, Z., Y. Hu, and W. Qin. 2009. Effect of surfactant OPD on the bioleaching of marmatite. *Minerals Engineering* 22(1): 10–13. doi:10.1016/j.mineng.2008.03.002.

Laville, J., C. Blumer, C. Von Schroetter, V. Gaia, G. Défago, C. Keel, and D. Haas. 1998. Characterization of the hcnABC gene cluster encoding hydrogen cyanide synthase and anaerobic regulation by ANR in the strictly aerobic biocontrol agent. *Pseudomonas fluorescens* CHA0. *Journal of Bacteriology* 180(12): 3187–3196.

Lawson, E. N., M. Barkhuizen, and D. W. Dew. 1999. Gold solubilisation by the cyanide producing bacteria *Chromobacterium violaceum*. In *Process Metallurgy*, R. Amils and A. Ballester (Eds.), pp. 239–246. Amsterdam, the Netherlands: Elsevier.

Lee, J. and B. D. Pandey. 2012. Bio-processing of solid wastes and secondary resources for metal extraction—A review. *Waste Management* 32: 3–18.

Li, J., C. Liang, and C. Ma. 2015. Bioleaching of gold from waste printed circuit boards by *Chromobacterium violaceum*. *Journal of Material Cycles and Waste Management* 17(3): 529–539. doi:10.1007/s10163-014-0276-4.

Li, J., P. Shrivastava, Z. Gao, and H. C. Zhang. 2004. Printed circuit board recycling: A state of-the-art survey. *IEEE Transactions on Electronics Packaging Manufacturing* 27: 33–42.

Liang, G., P. Li, W. Liu, and B. Wang. 2016. Enhanced bioleaching efficiency of copper from waste printed circuit boards (PCBs) by dissolved oxygen-shifted strategy in *Acidithiobacillus ferrooxidans*. *Journal of Material Cycles and Waste Management* 18(4): 742–751. doi:10.1007/s10163-015-0375-x.

Liang, G., Y. Mo, and Q. Zhou. 2010. Novel strategies of bioleaching metals from printed circuit boards (PCBs) in mixed cultivation of two acidophiles. *Enzyme and Microbial Technology* 47(7): 322–326. doi:10.1016/j.enzmictec.2010.08.002.

Liu, G., J. Yin, and W. Cong. 2007. Effect of fluid shear and particles collision on the oxidation of ferrous iron by *Acidithiobacillus ferrooxidans*. *Minerals Engineering* 20(13): 1227–1231. doi:10.1016/j.mineng.2007.06.002.

Ludwig, C., S. Hellweg, and S. Stucki. 2003. *Municipal Solid Waste Management: Strategies and Technologies for Sustainable Solutions*, pp. 320–322. Berlin, Germany: Springer.

Ma, H. and J. C. Suhling. 2009. A review of mechanical properties of lead-free solders for electronic packaging. *Journal of Material Science* 44: 1141–1158.

Madrigal-Arias, J. E., R. Argumedo-Delira, A. Alarcón, M. R. Mendoza-López, O. García-Barradas, J. S. Cruz-Sánchez, R. Ferrera-Cerrato, and M. Jiménez-Fernández. 2015. Bioleaching of gold, copper and nickel from waste cellular phone PCBs and computer goldfinger motherboards by two *Aspergillus niger* strains. *Brazilian Journal of Microbiology: [Publication of the Brazilian Society for Microbiology]* 46(3): 707–713. doi:10.1590/S1517-838246320140256.

Mäkinen, J., J. Bachér, T. Kaartinen, M. Wahlström, and J. Salminen. 2015. The effect of flotation and parameters for bioleaching of printed circuit boards. *Minerals Engineering* 75: 26–31. doi:10.1016/j.mineng.2015.01.009.

Mecucci, A. and K. Scott. 2002. Leaching and electrochemical recovery of copper, lead and tin from scrap printed circuit boards. *Journal of Chemical Technology & Biotechnology* 77(4): 449–457. doi:10.1002/jctb.575.

Menad, N., B. Björkman, and E. G. Allain. 1998. Combustion of plastics contained in electric and electronic scrap. *Resources, Conservation and Recycling* 24(1): 65–85. doi:10.1016/S0921-3449(98)00040-8.

Morley, G. F., J. A. Sayer, S. J. Wilkinson, M. M. Gharieb, and G. M. Gadd. 1996. Fungal sequestration, solubilization and transformation of toxic metals. In *Fungi and Environmental Change*, Frankland, J. C., Magan, N., and G. M. Gadd, (Eds.), pp. 235–256. Cambridge, U.K.: Cambridge University Press.

Mulligan, C. N., M. Kamali, and B. F. Gibbs. 2004. Bioleaching of heavy metals from a low-grade mining ore using *Aspergillus niger*. *Journal of Hazardous Materials* 110(1): 77–84. doi:10.1016/j.jhazmat.2004.02.040.

Muñoz, J. A., M. L. Blázquez, A. Ballester, and F. González. 1995. A study of the bioleaching of a Spanish uranium ore. Part III: Column experiments. *Hydrometallurgy* 38(1): 79–97. doi:10.1016/0304-386X(94)00038-5.

Natarajan, G., S. B. Tay, W. S. Yew, and Y.-P. Ting. 2015. Engineered strains enhance gold biorecovery from electronic scrap. *Minerals Engineering* 75: 32–37. doi:10.1016/j.mineng.2015.01.002.

Natarajan, G. and Y.-P. Ting. 2014. Pretreatment of e-waste and mutation of alkali-tolerant cyanogenic bacteria promote gold biorecovery. *Bioresource Technology* 152: 80–85. doi:10.1016/j.biortech.2013.10.108.

Natarajan, G. and Y. P. Ting. 2015. Gold biorecovery from e-waste: An improved strategy through spent medium leaching with pH modification. *Chemosphere* 136: 232–238. doi:10.1016/j.chemosphere.2015.05.046.

Nie, H., N. Zhu, Y. Cao, Z. Xu, and P. Wu. 2015. Immobilization of *Acidithiobacillus ferrooxidans* on cotton gauze for the bioleaching of waste printed circuit boards. *Applied Biochemistry and Biotechnology* 177(3): 675–688. doi:10.1007/s12010-015-1772-2.

Olson, G. J., J. A. Brierley, and C. L. Brierley. 2003. Bioleaching review part B. *Applied Microbiology Biotechnology* 63: 249–257.

Pham, V. A. and Y. P. Ting. 2009. Gold bioleaching of electronic waste by cyanogenic bacteria and its enhancement with bio-oxidation. *Advanced Materials Research* 71–73(May): 661–664. doi:10.4028/www.scientific.net/AMR.71-73.661.

Pogliani, C. and E. Donati. 2000. Immobilisation of *Thiobacillus ferrooxidans*: Importance of jarosite precipitation. *Process Biochemistry* 35(9): 997–1004. doi:10.1016/S0032-9592(00)00135-7.

Qiu, G., Q. Li, R. Yu et al. 2011. Column bioleaching of uranium embedded in granite porphyry by a mesophilic acidophilic consortium. *Bioresource Technology* 102(7): 4697–4702. doi:10.1016/j.biortech.2011.01.038.

Rodrigues, M. L. M., V. A. Leão, O. Gomes, F. Lambert, D. Bastin, and S. Gaydardzhiev. 2015. Copper extraction from coarsely ground printed circuit boards using moderate thermophilic bacteria in a rotating-drum reactor. *Waste Management* 41: 148–158. doi:10.1016/j.wasman.2015.04.001.

Rohwerder, T., T. Gehrke, K. Kinzler, and W. Sand. 2003. Bioleaching review. Part A. *Applied Microbiology Biotechnology* 63: 239–248.

Ruan, J., X. Zhu, Y. Qian, and J. Hu. 2014. A new strain for recovering precious metals from waste printed circuit boards. *Waste Management* 34(5): 901–907. doi:10.1016/j.wasman.2014.02.014.

Sekaran, G., S. Karthikeyan, C. Evvie, R. Boopathy, and P. Maharaja. 2013. Oxidation of refractory organics by heterogeneous Fenton to reduce organic load in tannery wastewater. *Clean Technology and Environmental Policy* 15: 245–253.

Shah, M. B., D. R. Tipre, M. S. Purohit, and S. R. Dave. 2015. Development of two-step process for enhanced biorecovery of Cu-Zn-Ni from computer printed circuit boards. *Journal of Bioscience and Bioengineering* 120(2): 167–173. doi:10.1016/j.jbiosc.2014.12.013.

Syed, S. 2012. Recovery of gold from secondary sources—A review. *Hydrometallurgy* 115: 30–51. doi:10.1016/j.hydromet.2011.12.012.

Tay, S. B., G. Natarajan, M. N. B. A. Rahim, H. T. Tan, M. C. M. Chung, Y. P. Ting, and W. S. Yew. 2013. Enhancing gold recovery from electronic waste via lixiviant metabolic engineering in *Chromobacterium violaceum*. *Scientific Reports* 3: 2236. doi:10.1038/srep02236.

Tran, C. D., J. Lee, B. D. Pandey, J. Jeong, K. Yoo, and T. H. Huynh. 2011. Bacterial cyanide generation in the presence of metal ions (Na^+, Mg^{2+}, Fe^{2+}, Pb^{2+}) and gold bioleaching from waste PCBs. *Journal of Chemical Engineering of Japan* 44: 692–700.

UNEP. 2009. Sustainable innovation and technology transfer industrial sector studies, recycling from e-waste to resources. Geneva, Switzerland: United Nations Environment Programme & United Nations University.

Vestola, E. A., M. K. Kuusenaho, H. M. Närhi, O. H. Tuovinen, J. A. Puhakka, J. J. Plumb, and A. H. Kaksonen. 2010. Acid bioleaching of solid waste materials from copper, steel and recycling industries. *Hydrometallurgy* 103(1–4): 74–79. doi:10.1016/j.hydromet.2010.02.017.

Wang, J., J. Bai, J. Xu, and B. Liang. 2009. Bioleaching of metals from printed wire boards by *Acidithiobacillus ferrooxidans* and *Acidithiobacillus thiooxidans* and their mixture. *Journal of Hazardous Materials* 172(2): 1100–1105. doi:10.1016/j.jhazmat.2009.07.102.

Willscher, S., M. Katzschner, K. Jentzsch, S. Matys, and H. Pöllmann. 2007. Microbial leaching of metals from printed circuit boards. *Advanced Materials Research* 20–21: 99–102. doi:10.4028/www.scientific.net/AMR.20-21.99.

Xiang, Y., P. Wu, N. Zhu, T. Zhang, W. Liu, J. Wu, and P. Li. 2010. Bioleaching of copper from waste printed circuit boards by bacterial consortium enriched from acid mine drainage. *Journal of Hazardous Materials* 184(1): 812–818. doi:10.1016/j.jhazmat.2010.08.113.

Xu, Z., T. Yang, L. Yang, and Y. Li. 2010. Bioleaching metals from waste printed circuit boards and the shapes of microorganisms. In *Proceedings of XXVth International Mineral Processing Congress (IMPC)*, September 6–10, Brisbane, Queensland, Australia, pp. 4001–4007.

Yang, C., N. Zhu, W. Shen, T. Zhang, and P. Wu. 2017. Bioleaching of copper from metal concentrates of waste printed circuit boards by a newly isolated *Acidithiobacillus ferrooxidans* strain Z1. *Journal of Material Cycles and Waste Management* 19(1): 247–255. doi:10.1007/s10163-015-0414-7.

Yang, T., Z. Xu, J. Wen, and L. Yang. 2009. Factors influencing bioleaching copper from waste printed circuit boards by *Acidithiobacillus ferrooxidans*. *Hydrometallurgy* 97(1): 29–32. doi:10.1016/j.hydromet.2008.12.011.

Yang, Y., S. Chen, S. Li, M. Chen, H. Chen, and B. Liu. 2014. Bioleaching waste printed circuit boards by *Acidithiobacillus ferrooxidans* and its kinetics aspect. *Journal of Biotechnology* 173(1): 24–30. doi:10.1016/j.jbiotec.2014.01.008.

Yarar, B. 2000. *Flotation. Ullmann's Encyclopedia of Industrial Chemistry.* Weinheim, Germany: Wiley-VCH.

Yoo, K., J.-c. Lee, K.-s. Lee, B.-S. Kim, M.-s. Kim, S.-k. Kim, and B. D. Pandey. 2012. Recovery of Sn, Ag and Cu from waste Pb-free solder using nitric acid leaching. *Materials Transactions* 53(12): 2175–2180. doi:10.2320/matertrans.M2012268.

Zhao, B., X. Liu, Y. Jiang, L. Zhou, Y. He, and J. Gao. 2014. Immobilized lipase from *Candida* sp. 99–125 on hydrophobic ailicate: Characterization and applications. *Applied Biochemistry and Biotechnology* 173: 1802–1814.

Zhou, Z. and M. Hartmann. 2013. Progress in enzyme immobilization in ordered mesoporous materials and related applications. *Chemical Society Reviews* 42: 3894–3912.

Zhu, J., M. Gan, D. Zhang, Y. Hu, and L. Chai. 2013. The nature of schwertmannite and jarosite mediated by two strains of *Acidithiobacillus ferrooxidans* with different ferrous oxidation ability. *Materials Science and Engineering: C* 33(5): 2679–2685. doi:10.1016/j.msec.2013.02.026.

Zhu, N., Y. Xiang, T. Zhang, P. Wu, Z. Dang, P. Li, and J. Wu 2011. Bioleaching of metal concentrates of waste printed circuit boards by mixed culture of acidophilic bacteria. *Journal of Hazardous Materials* 192(2): 614–619. doi:10.1016/j.jhazmat.2011.05.062.

5 Recycling of Energy Storage Wastes

5.1 INTRODUCTION

The tremendous use of batteries and their ubiquitous applications in electronic equipment have raised a serious problem due to their toxicity, abundance, and permanence in the environment (Li and Xi 2005), which has substantially risen their amount of waste. Therefore, recycling of waste batteries is an important subject not only from the viewpoint of treatment of hazardous waste but also with respect to recovery of valuable metals (Huang et al. 2010). Increased recycling of batteries is advantageous to the environment in terms of both retrieving metals and avoiding the production of virgin metals. Moreover, it is financially remarkable as compared with disposal. The release of metals into the environment is a major issue due to the probable harm to humans as well as to nature. For these reasons, methods have been developed to recover metals.

Both physical and chemical methods are used for recycling batteries. Conventional methods such as pyro- and hydrometallurgical processes are expensive, cause secondary pollution, and have high energy consumption and low efficiency, which urges the turn toward low-cost, environmentally friendly, and highly efficient biohydrometallurgical process. Moreover, most of the reported or patented processes for recycling of spent batteries are in the pilot or laboratory stage (Li et al. 2009).

This chapter proposes various dimensions of metal recovery from energy storage wastes such as batteries, button cells, etc. Various types of batteries are used widely, which include lithium-ion batteries (LIBs), nickel–cadmium (Ni–Cd) batteries, nickel–metal hydride (Ni–MH) batteries, zinc–manganese (Zn–Mn) batteries (ZMBs), silver oxide batteries, etc.

5.2 BIOLEACHING OF METALS FROM ENERGY STORAGE WASTES

5.2.1 BATTERY

5.2.1.1 Lithium-Ion Batteries

Mishra et al. (2008) studied the bioleaching of spent LIBs containing $LiCoO_2$ using chemolithotrophic and acidophilic bacteria *Acidithiobacillus ferrooxidans*, which utilized elemental sulfur and ferrous ion as the energy source. Bacterial metabolites such as sulfuric acid (H_2SO_4) and ferrous ion were involved in the dissolution of metals from the spent batteries. Preliminary results showed a faster dissolution of cobalt as compared with lithium. The effect of the initial Fe(II) concentration suggested that higher Fe(II) concentrations resulted in decreased dissolution due to coprecipitation of Fe(III) with other metals in the residue.

The higher solid/liquid ratio (w/v) of spent batteries also affected metal dissolution by preventing cell growth, owing to its toxicity toward cells.

Figure 5.1 shows the energy dispersive x-ray analysis (EDX) mapping of the leach residues after leaching. Both the leach residues and the original sample showed the same distribution of lithium. However, a little distribution of cobalt on the surface of the leach residues indicated that it has been leached into the solution during the leaching process. Thus, the mapping demonstrated the dissolution rate of lithium to be slower than that of cobalt.

Xin et al. (2009) explored the bioleaching ability of a mixed culture of sulfur-oxidizing bacteria (SOB) and iron-oxidizing bacteria (IOB) using spent LIBs and interpreted the mechanism for Co and Li bioleaching. Spent lithium-ion secondary batteries were cut up and active cathode and anode materials were mixed and powdered. Both the SOB and IOB were screened from an old mining site (Anshan Mountain, Liaoning Province, North China). The dissolution of Co and Li from spent lithium-ion secondary batteries was found to be exclusively dependent on a

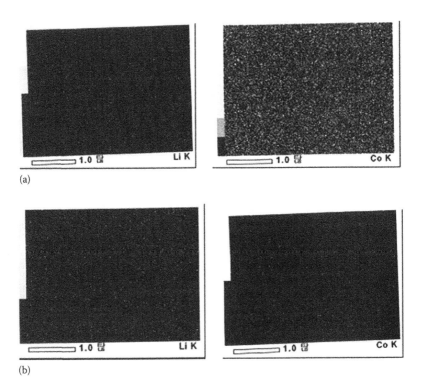

FIGURE 5.1   (a) SEM image of the raw sample and responding x-ray energy dispersive spectrum element distribution of Li and Co. (b) SEM image of the leach residues and responding x-ray energy dispersive spectrum element distribution of Li and Co. (Reprinted from *Waste Manage.*, 28, Mishra, D., Kim, D.J., Ralph, D.E., Ahn, J.G., and Rhee, Y.H., Bioleaching of metals from spent lithium ion secondary batteries using *Acidithiobacillus ferrooxidans*, 333–338, Copyright 2008, with permission from Elsevier.)

noncontact mechanism. The highest release of Li appeared at the lowest pH of 1.54, with elemental sulfur as an energy source, while the lowest occurred at the highest pH of 1.69, with FeS_2 as an energy source. The authors suggested acid dissolution as the main mechanism for Li bioleaching independent of energy sources.

In contrast, the highest release of Co appeared at higher pH and varied oxidation-reduction potential (ORP), with elemental sulfur and FeS_2 ($S + FeS_2$) as energy sources, and the lowest occurred at almost unchanged ORP, with elemental sulfur as an energy source. Bioleaching of Co was suggested to be a combined action of acid dissolution and reduction of Fe^{2+}, which was produced via a physical–chemical reaction between FeS_2 and Fe^{3+}. The dissolution mechanisms of Co differed with different energy sources. Co^{2+} was released by acid dissolution after insoluble Co^{3+} was reduced into soluble Co^{2+} by Fe^{2+} in both FeS_2 and $FeS_2 + S$ systems. The proposed mechanism was further confirmed from the results of the experiments on bioprocess-stimulated chemical leaching and from the changes in the structure and composition of bioleaching residues characterized by X-ray photoelectron spectroscopy (XPS), scanning electron microscopy (SEM), and EDX. Maximum leaching efficiencies of more than 90% for Co and 80% for Li were obtained. The diverse leaching bacteria and energy sources result in the production of diverse metabolic products and consequently introduce diverse leaching mechanisms leading to efficient bioleaching of metals (Xin et al. 2009).

Many reports state that the metal dissolution rate in the bioleaching process can be improved by using catalytic ions in the solution (Muñoz et al. 2007a,b,c), which holds great significance for optimizing the bioleaching process (Ballester et al. 1990, 1992). Taking this into account, Zeng et al. (2012) studied the influence of copper ions on bioleaching of cobalt from spent LIBs (mainly $LiCoO_2$) using *At. ferrooxidans* and also searched for an appropriate method to solve the problem of low efficiency.

Cobalt recovery increased from 43.1% in the absence of copper ions within 10 days to 99.9% at a copper concentration of 0.75 g/L after 6 days, indicating the intensified oxidation rate of $LiCoO_2$ and cobalt leaching from spent LIBs by copper ions, as shown in Figure 5.2. Moreover, these results were confirmed using EDX, x-ray diffraction (XRD), and SEM analyses.

Further, the authors proposed a probable mechanism for the catalytic effect of copper ions on bioleaching of Co from spent LIBs. The possible catalytic mechanisms are proposed in Equations 5.1 through 5.3 (Zeng et al. 2012):

$$Cu^{2+} + 2LiCoO_2 \rightarrow CuCo_2O_4 + 2Li^+ \tag{5.1}$$

$$CuCo_2O_4 \rightarrow 6Fe^{3+} \rightarrow 6Fe^{2+} + Cu^{2+} + 2O_2 + 2Co^{2+} \tag{5.2}$$

$$4Fe^{2+} + O_2 + 4H^+ \xrightarrow{\text{bacteria}} 4Fe^{3+} + 2H_2O \tag{5.3}$$

The main requirement for the metal ion to exhibit catalytic activity appears to be the formation of such intermediate products ($CuCo_2O_4$), which are oxidized by the leaching solution, reproducing the catalytic ion. First, $CuCo_2O_4$ was formed on the particle surface when the copper ion was added during the bioleaching process (Equation 5.1).

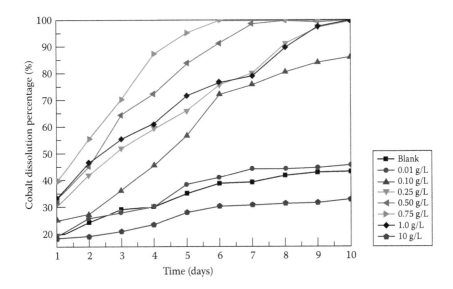

FIGURE 5.2 Dissolution percentage of the copper-catalyzed process under different Cu^{2+} concentrations. (Reprinted from *J. Hazard. Mater.*, 199–200, Zeng, G., X. Deng, S. Luo, X. Luo, and J. Zou, A copper-catalyzed bioleaching process for enhancement of cobalt dissolution from spent lithium–ion batteries, 164–169, Copyright 2012, with permission from Elsevier.)

Based on the reaction, Cu^{2+} was regenerated through the oxidation of $CuCo_2O_4$ by ferric iron (Equation 5.2). Then, ferric ion was produced from the ferrous ion according to Equation 5.3. The equation explains that $LiCoO_2$ underwent a cationic interchange reaction with copper ions to form $CuCo_2O_4$ on the surface of the sample, which could be easily dissolved by Fe^{3+}, thus accelerating Co recovery.

Thus, the primary leaching efficiency problem of Co recovery from LIBs was solved by applying low-cost copper ions as a catalyst, which is very important for recycling of spent LIBs.

Further, Zeng et al. (2013) continued the study investigating the influence of silver ions (Ag^+) on bioleaching of cobalt from spent LIBs. The pH of the solution with Ag^+ was found to be higher than that without Ag^+, and it reached a maximum at an Ag^+ concentration of 0.02 g/L. The authors further divide the $LiCoO_2$ leaching reaction as acid consumption (Equations 5.4 and 5.5), acid generation (Equations 5.6 and 5.7), ferric ion generation (Equation 5.5), and ferric ion consumption (Equations 5.6 and 5.7) reactions:

$$2LiCoO_2 \rightarrow 4H^+ \rightarrow 2Co^{2+} + 2Li^+ + O_2 + 2H_2O \tag{5.4}$$

$$4Fe^{2+} + O_2 + 4H^+ \xrightarrow{\textit{A. ferroxidans}} 4Fe^{3+} + 2H_2O \tag{5.5}$$

$$Fe^{3+} + 3H_2O \rightarrow Fe(OH)_3 + 3H^+ \tag{5.6}$$

$$3Fe^{3+} + K^+ + 2So_4^{2-} + 6H_2O \rightarrow KFe_3(SO_4)_2(OH)_6 + 6H^+ \tag{5.7}$$

Ferric iron is hydrolyzed into ferric hydroxide and jarosite, resulting in a decrease in pH, consequently inhibiting the leaching of $LiCoO_2$. During the first 5 days, H^+ consumption was much higher than H^+ generation, resulting in an increase in the pH value.

Ag^+ lowers the redox potential of the solution, and the minimum was observed at an Ag^+ concentration of 0.02 g/L, which further indicates an acceleration of the reduction rate of ferric ions. Leaching of cobalt increased from only 43.1% to 98.4% in a period of 7 days in the presence of Ag^+(0.02 g/L), as shown in Figure 5.3.

When compared with previous results (Zeng et al. 2012) using Cu^{2+} as a catalyst, the use of Ag^{2+} attained a higher leaching ratio and a much better leaching efficiency in terms of equal mass of copper and silver within a period of 5 days.

Based on the literature and results of SEM, XRD, and EDX investigations, the authors proposed a mechanism for bioleaching of $LiCoO_2$ as follows (Zeng et al. 2013):

$$Ag^+ + LiCoO_2 \rightarrow AgCoO_2 + Li^+ \tag{5.8}$$

$$AgCoO_2 + 3Fe^{3+} \rightarrow 3Fe^{2+} + Ag^+ + 2O_2 + Co^{2+} \tag{5.9}$$

$$4Fe^{2+} + O_2 + 4H^+ \xrightarrow{A.\,ferroxidans} 4Fe^{3+} + 2H_2O \tag{5.5}$$

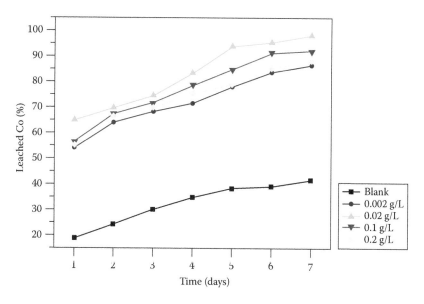

FIGURE 5.3 Leaching of Co catalyzed by different concentrations of Ag^+. (Reprinted from *Miner. Eng.*, 49, Zeng, G., Luo, S., Deng, X., Li, L., and Au, C., Influence of silver ions on bioleaching of cobalt from spent lithium batteries, 40–44, Copyright 2013, with permission from Elsevier.)

It was deduced that Ag^+ reacts with cobalt and facilitates the formation of $AgCoO_2$ as an intermediate (Equation 5.8). The consumed Ag^+ is released for further catalytic action once $AgCoO_2$ is oxidized by ferric ions (Equation 5.9). Due to the activity of bacteria, the resulted ferrous ions are oxidized back to ferric ions (Equation 5.5). Thus, the authors suggest the use of catalysts such as Ag^{2+}, for extraction of cobalt from spent LIBs, to increase the efficiency of bioleaching (Zeng et al. 2013).

Li et al. (2013) reported the influence of solution pH and redox potential on bioleaching of $LiCoO_2$ from spent LIBs using *At. ferrooxidans*. Bioleaching at different initial pH and ferrous ion (Fe^{2+}) concentrations was carried out and the electrochemical behavior of $LiCoO_2$ dissolution was examined to study the effect of solution redox potential on the bioleaching process. The maximum Co dissolution was obtained at an initial pH of 1.5 and an initial Fe^{2+} concentration of 35 g/L, and cobalt dissolution showed only a minor relationship with the pH of the solution. However, the dissolution improved at higher redox potentials. Cyclic voltammograms showed an increase in the dissolution rate at solution redox potentials higher than 0.4 V and a decrease at 1.3 V. Anodic polarization curves indicated that the corrosion, primary passive, and passivation potentials were 0.420, 0.776, and 0.802 V, respectively.

Niu et al. (2014) explored the process controls for improving the bioleaching efficiency of Li and Co from spent LIBs using a mixed culture of *Alicyclobacillus* sp. as SOB and *Sulfobacillus* sp. as IOB.

Pulp density is a very important factor that determines the commercial application of bioleaching to some degree. An increase in pulp density from 1% to 2% (w/v) means a sharp reduction of 50% in both the quantity of the leaching medium and the size of the bioreactor, resulting in a great drop in bioleaching cost. Pulp density exerted a noteworthy influence on the bioleaching performance of both Co and Li from spent LIBs. With an increase in pulp density from 1% to 4%, the bioleaching efficiency decreased from 52% to 10% and from 80% to 37% for Li and Co, respectively, after 11 days. Also, the pH value increased from 1.97 to 4.46, and the ORP value decreased from 509 to 335 mV, as a result of extensive consumption of H^+ for mobilization of both Co and Li. There was a decline in the bioleaching efficiency of a mixed culture at higher pulp densities. An increased amount of a toxic electrolyte decreased the growth and activity of microorganisms. Lower generation of H_2SO_4 consequently increased the pH of the medium as a result of reduced SOB activity; simultaneously, lower amounts of Fe^{3+} decreased the ORP of the medium due to the reduced activity of IOB. Therefore, there was a great descent in the bioleaching efficiency of both Co and Li at high pulp densities, as Li recovery depended on the acid dissolution mechanism by SOB, whereas Co recovery relied on the combined action of both acid dissolution and Fe^{2+} reduction by both SOB and IOB, as per Xin et al. (2009). However, the maximum extraction efficiency of 89% for Li and 72% for Co was obtained at a pulp density of 2% by process controls.

The bioleaching mechanism of Co and Li from spent LIBs was described by the following equation, which was mentioned in a previous work (Xin et al. 2009):

$$10O_2 + S + 2FeS_2 + 2LiCoO_2 = Li_2SO_4 + 2CuSO_4 + Fe_2(SO_4)_3 \qquad (5.10)$$

Bioleaching of spent LIBs has a much greater potential to occur than traditional chemical leaching based on thermodynamics analysis. The product bioleaching behavior of Co and Li was best described by layer diffusion, due to the highest correlation coefficients.

Further, Xin et al. (2015) performed the recovery of valuable metals from spent electric vehicle (EV) LIBs. SOB (*At. thiooxidans*) and IOB (*Leptospirillum ferriphilum*) were used in this process. In this work, the release of valuable Li, Co, Mn, and Ni from three typical spent EV LIB cathodes, $LiFePO_4$, $LiMn_2O_4$, and $LiNi_xCo_yMn_{1-x-y}O_2$, by bioleaching at 1% pulp density was explored using three types of bioleaching systems, that is, sulfur–*At. thiooxidans* system, pyrite–*L. ferriphilum* system, and mixed energy source–mixed culture (MS–MC) system.

The results showed that the maximum extraction efficiency of Li occurred in the sulfur–*At. thiooxidans* system, indicating that Li release was due to acid dissolution by biogenic H_2SO_4, whereas the MS–MC system displayed the highest dissolution yield for Co, Ni, and Mn, owing to the combined action of Fe^{2+} reduction and acid dissolution. Moreover, a noncontact mechanism accounted for Li extraction, whereas a contact mechanism between the cathodes and cells was required for the mobilization of Co, Ni, and Mn. The sulfur–*At. thiooxidans* system harvested 98% of Li release from $LiFePO_4$ and the MS–MC system attained an extraction efficiency of 95% for Li and 96% for Mn from $LiMnO_2$. Adjustment of the pH enhanced bacterial growth and consequently improved the bioleaching performance. An average extraction efficiency of four valuable metals from the resistant $LiNi_xCo_yMn_{1xy}O_2$ reached more than 95%, especially from 43.5% to 96% and from 38.3% to 97% for Co and Ni, respectively. The continuous production of H_2SO_4 and contact mechanisms by cells enabled a better bioleaching performance compared with chemical leaching (Xin et al. 2015).

Horeh et al. (2016) evaluated the biohydrometallurgical route for the detoxification and recovery of various metals from spent lithium-ion mobile phone batteries using *Aspergillus niger* under various conditions (one-step, two-step, and spent-medium bioleaching). The high performance liquid chromatography (HPLC) results showed that the organic acids secreted by *A. niger* were mainly malic, gluconic, oxalic, and citric acid. Based on the results of growth characteristics, the third day of growth was chosen for the two-step experiments and the 14th day of growth for spent-medium bioleaching.

The results indicated that the total recovery in spent-medium bioleaching was higher than that in other methods. Nearly 100% Cu, 95% Li, 70% Mn, 65% Al, 45% Co, and 38% Ni were leached in spent-medium experiments, while in one-step bioleaching, 100% Li, 58% Al, 11% Cu, and 8% Mn were leached, with Ni and Co showing negligible leaching. In two-step bioleaching, 100% Li, 61% Al, 10% Mn, 6% Cu, and 1% Co were leached (the amount of leached Ni was negligible).

As indicated by the HPLC results, citric acid played a significant role in effective bioleaching of battery powder in spent medium using *A. niger* as compared with other acids (gluconic, oxalic, and malic acid) and resulted in higher metal recovery.

Along with the higher metal recovery, spent-medium bioleaching involved a shorter processing time and a more facile handling. Moreover, in this approach, optimizations for higher organic acid production and higher metal recovery can be done separately; also there is no question of metal toxicity toward fungi (Aung and Ting 2005). Therefore, among all methods of bioleaching of battery powder using *A. niger*, spent-medium bioleaching is the most propitious for leaching of heavy metals.

Chemical leaching was performed to compare the bioleaching efficiency of spent LIBs using commercial organic acids (citric, oxalic, malic, and gluconic acid) as leaching agents and investigate the role of organic acids secreted by fungi. The recovery of all metals was higher using the bioleaching method than the chemical one, except for Li and Mn. Maybe other fungal metabolites such as amino acids and undetermined organic acids might be involved in the bioleaching process.

The surface morphology of the battery powder before and after bioleaching was studied using field emission scanning electron microscopy (FE-SEM) (Figure 5.4).

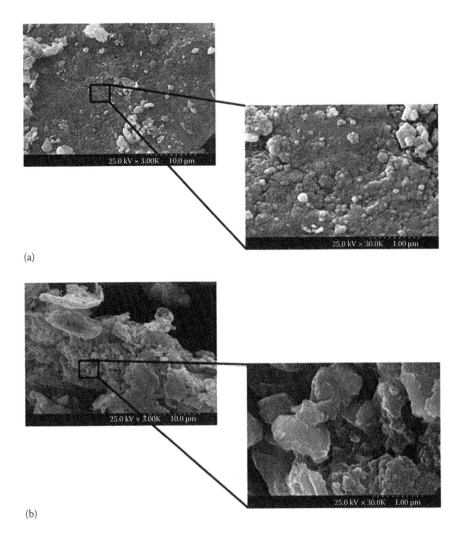

(a)

(b)

FIGURE 5.4 FE-SEM photomicrograph of (a) the original powder and (b) the bioleached residue. (Reprinted from *J. Power Sources*, 320, Horeh, N.B., Mousavi, S.M., and Shojaosadati, S.A., Bioleaching of valuable metals from spent lithium-ion mobile phone batteries using *Aspergillus niger*, 257–266, Copyright 2016, with permission from Elsevier.)

The bioleached residue of battery powder shows a rough surface with small holes. This micromorphological surface change suggests that fungal metabolites such as organic acids and amino acids slowly eroded battery powder particles through a chemical corrosive action and lead to metal mobilization into the solution phase (Qu et al. 2013). Further, the results of Fourier transform infrared spectroscopy (FTIR), XRD, and FE-SEM analyses of battery powder before and after bioleaching indicated the change in the appearance and structure of battery powder during bioleaching and confirmed the effectiveness of fungal activities. Thus, the authors present this fungal leaching as an eco-friendly and cost-effective way to recover heavy metals from spent batteries.

5.2.1.2   Ni–Cd and Ni–MH Batteries

In an innovating study, Cerruti et al. (1998) used *Thiobacillus ferrooxidans* for bioleaching of spent Ni–Cd batteries. Preliminary experiments of dissolution were carried out in order to evaluate the feasibility of a contact leaching process. Depending on the obtained results and the feasibility for commercial application, indirect bioleaching process was utilized. Two percolators were arranged in series, as shown in Figure 5.5, wherein in the first percolator, *T. ferrooxidans* was immobilized

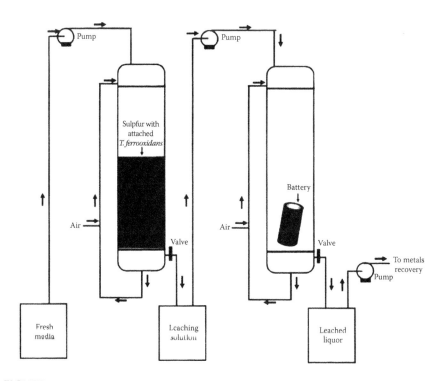

FIGURE 5.5   Schematic of the two-percolator system used for the dissolution of a spent nickel–cadmium battery. (Reprinted from *J. Biotechnol.*, 62, Cerruti, C., Curutchet, G., and Donati, E., Bio-dissolution of spent nickel–cadmium batteries using *Thiobacillus ferrooxidans*, 209–219, Copyright 1998, with permission from Elsevier.)

on elemental sulfur and produced an acidic growth medium, which was used in the second percolator for the recovery of nickel and cadmium from a spent battery.

T. ferrooxidans produced sulfuric acid using sulfur as an energy source. The maximum sulfuric acid productivity of 80 mmol/kg/day was attained. Sulfuric acid acts as a strong acid and a moderately oxidizing agent capable of oxidizing metals such as nickel, cadmium, and iron. Moreover, the authors suggest that the intermediates generated during the bio-oxidation of sulfur could contribute to the dissolution of nickel oxyhydroxide by prior reduction to nickel(II) (more soluble than nickel oxyhydroxide in an acidic solution). Nearly 100% cadmium, 96.5% nickel, and 95.0% iron were recovered from spent Ni–Cd batteries within a time span of 93 days. Also, a more than 90.0% recovery was obtained by the direct addition of anodic and cathodic materials to *T. ferrooxidans* cultures with sulfur as the sole energy source. Thus, the authors recommend this method be considered as the first effective step in recycling spent and discarded batteries from an economic and environmental standpoint (Cerruti et al. 1998).

Zhu et al. (2003) introduced a novel continuous-flow two-step bioleaching process for recycling of spent Ni–Cd batteries using indigenous thiobacilli from sludge of wastewater treatment plants.

A bioreactor system was constructed using two reactors, namely a bioreactor and a leaching reactor. The bioreactor was made of two parts, a reactor for growing bacteria and a settling tank. In the bioreactor, indigenous thiobacilli produced sulfuric acid utilizing sewage sludge, which provided nutritive elements, and externally supplied elemental sulfur was used as substrates to produce sulfuric acid. The overflow from this bioreactor was transferred to the settling tank, where it was used for metal recovery from the anode and cathode of Ni–Cd batteries.

The thiobacilli produced sulfuric acid, reducing power compounds using sewage sludge and elemental sulfur. The authors also noted the significant influence of residence time of sludge in the bioreactor (RTB) on the process, which affected the pH, metal concentration of the overflow, and population of acidophilic thiobacilli in the sludge. The pH value increased from 4 to 1 day RTB, but metal concentration and thiobacilli population decreased. However, the data were very similar when the RTB was above 4 days. In the case of 4–6 days' RTB, the pH value was 1.81–1.88, and at 1 day RTB, the pH value was 2.93.

Metal concentration increased with longer RTB (metal concentration at 5 days' RTB was more than three times compared with that at 3 days' RTB). However, due to the different total amount of solution used in each case, the metal leaching efficiency did not change based on the RTB only. During the 50 days' leaching experiment, all the Cd was recovered, but nickel recovery was only 66.1%, 75.6%, and 40.8% for an RTB of 5, 4, and 3 days, respectively. Thus, the release of nickel was slower than that of cadmium; also the solubility of cadmium hydroxide was higher than that of nickel hydroxide. Therefore, an RTB of 4 days was effective for metal leaching. Moreover, the sludge drained from the settling tank meets the requirements of environmental protection agencies and can be used in agricultural application after neutralization by lime (Zhu et al. 2003).

In another study, Zhao et al. (2007) employed indigenous thiobacilli from sewage sludge for biohydrometallurgical recovery of metals from spent Ni–Cd batteries using a continuous-flow two-step system. The acid supernatant produced by thiobacilli was transferred to the leaching reactor to dissolve electrode materials. The batch treatment of batteries showed that the complete dissolution of two AA-sized Ni–Cd batteries with 0.6 L/day acid supernatant took about 30, 20, and 35 days for Ni, Cd, and Co, respectively.

But the dissolution ability of the three metals differed from each other, where Cd and Co were recovered mostly at a pH below 4.0, while the complete dissolution of Ni was achieved at a pH below 2.5. Further, the bacterial strain (named *Thiooxidans. WL*) that attributed to the reduction in pH was isolated and sequenced. It was identified to be 100% similar to *At. ferrooxidans* strain Tf-49 based on 16S rDNA sequence analysis. The construction of the relevant phylogenetic tree indicated that the strain belonged to the genus *At. ferrooxidans*.

Further, Zhao et al. (2008b) compared the effect of two different substrates, ferrous sulfate and elemental sulfur, on the recovery of metals. A continuous-flow two-step leaching system comprising an acidifying reactor used to culture indigenous thiobacilli in sewage sludge and a leaching reactor to leach metals from spent batteries, with two settling tanks following them, respectively, was utilized in the process, as shown in Figure 5.6.

Indigenous acidophilic thiobacilli (*T. ferrooxidans* or *T. thiooxidans*) in sewage sludge can grow and produce sulfuric acid or ferric ions using elemental sulfur (Equation 5.11) or ferrous ions (Equations 5.12 through 5.14) as the energy source, with oxygen as the terminal electron acceptor, and produce biosulfuric acid through the oxidation of elemental sulfur or ferrous sulfate as the energy source, with oxygen as the terminal electron acceptor and carbon dioxide as the sole carbon source. The use of different substrates resulted in different dominant populations, leading to different pH and ORP changes, and thus, the different media for metal solubilization. Chemically or microbiologically mediated oxidation of ferrous iron shifted the pH to acidic values and increased the ORP. The consumption and release of protons in the oxidation of ferrous iron and ferric iron precipitation consequently decreased the pH:

$$2S_0 + 3O_2 + 2H_2O \xrightarrow{\text{\textit{T.thiooxidans}}} 2H_2SO_4 \qquad (5.11)$$

$$2Fe^{2+} + 1/2O_2 + 2H^+ \xrightarrow{\text{\textit{T.ferroxidans}}} 2Fe^{3+} + 2H_2O \qquad (5.12)$$

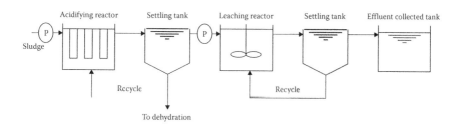

FIGURE 5.6 Diagram of the continuous-flow two-step system. (Reprinted from *J. Hazard. Mater.*, 160, Zhao, L., Yang, D., and Zhu, N.W., Bioleaching of spent Ni–Cd batteries by continuous flow system: Effect of hydraulic retention time and process load, 648–654, Copyright 2008a, with permission from Elsevier.)

$$Fe^{3+} + 3H_2O \leftrightarrow Fe(OH)_3 \downarrow + 3H^+ \qquad (5.13)$$

$$3Fe^{3+} + X^+ + 2HSO_4^- + 6H_2O \leftrightarrow XFe_3(SO_4)_2(OH)_6 + 8H^+ \qquad (5.14)$$

In the acidifying reactor, the final pH with sulfur was lower (~1.2) than with $FeSO_4 \cdot 7H_2O$ (~2.2), as $Fe(OH)_3$ precipitation cannot occur at a pH lower than 2.2. But when the supernatant was conducted to the leaching reactor, the pH was higher in the sulfur-oxidizing system, which leads to the slower dissolution of the metals. The ORP presented a reverse trend with the pH. pH was the major factor responsible for the dissolution of Ni, Cd, and Co. The pH and ORP of the acidifying reactor were stabilized around 2.3 and 334 mV for the iron-oxidizing system and 1.2 and 390 mV for the sulfur-oxidizing system, respectively. This was reverse in the acidifying reactor; the pH/ORP of the leaching reactor of the iron-oxidizing system was relatively lower/higher than that of the sulfur-oxidizing system in the first 17 days. Metal dissolution, in the first 12–16 days, was faster in the iron-oxidizing system as compared with the sulfur-oxidizing system because of the lower pH. In the iron-oxidizing system, the maximum solubilization of cadmium and cobalt reached at 2500 and 260 mg/L on day 6–8, respectively, and most of the metallic nickel was leached in the first 16 days. However, there was a lag period of 4–8 days in the sulfur-oxidizing system to reach the maximum solubilization of cadmium and cobalt. The maximum dissolution occurred at about day 12 and day 20 for nickel hydroxide (1400 mg/L) and metallic nickel (2300 mg/L), respectively.

The results show the iron-oxidizing system as more efficient and suitable for plants with a high loading rate owing to its shorter bioleaching time. However, the stronger acid generation capacity of the sulfur-oxidizing system makes it more suitable for a long leaching time. Cadmium hydroxide, elemental cadmium, and cobalt were more easily leached than nickel hydroxide and metallic nickel in both systems due to the higher solubility of cadmium hydroxide compared with that of nickel hydroxide and the higher reducing power of cadmium compared with that of nickel.

In continuation of this study, Zhao et al. (2008a) studied the effect of hydraulic retention time (HRT) and process load. HRT of biosulfuric acid in the leaching reactor is an important factor that represents the time for which the medium is in contact with the battery material. A very short HRT results in insufficient metal recovery from electrode materials, whereas at longer HRT, a high volume of the leaching reactor is required. The observations inferred that more time was required for complete leaching of Ni, Cd, and Co with longer HRT (1, 3, 6, 9, and 15 days). pH was an essential factor that affected metal recovery. The maximum recovery of cadmium and cobalt was achieved at higher pH values (3.0–4.5), while leaching of nickel hydroxide and nickel in metallic form (Ni_0) was achieved separately at different acidities (pH 2.5–3.5). Nearly 25, 30, and more than 40 days were required to remove all of the three heavy metals, with a processing load of two, four, and eight Ni–Cd batteries, under the following conditions: ingoing biosulfuric acid, 1 L/day; HRT, 3 days.

Velgosová et al. (2010) studied various factors influencing the bioleaching of metals from spent Ni–Cd batteries using a pure culture of *At. ferrooxidans* and a mixed culture of *At. ferrooxidans* and *At. thiooxidans*. Results demonstrated that the selection of suitable parameters increases leaching productivity.

Velgosová et al. (2012) examined the extraction of Ni and Cd from the electrode material of spent Ni–Cd batteries using the bacterium *At. ferrooxidans*. During the bioleaching process, no significant change was observed in pH. The pH value increased from 1.5 to 2.8 in up to 21 days and remained constant thereafter. The bioleaching efficiency of Ni and Cd reached 5.5% and 98% and 45% and 100% for anode powder and cathode powder, respectively.

The efficiency of Cd bioleaching from anode and cathode powders reached almost 100%, as cadmium is a less noble element which is oxidized more easily than metallic nickel and because cadmium hydroxides also dissolve faster than nickel hydroxides. The leaching of metallic Ni was found to be negligible. Further x-ray analysis confirmed jarosite formation and the removal of hydroxides from the solution.

In continuation of this, Velgosová et al. (2013) investigated the influence of sulfuric acid and ferric ions on the recovery of cadmium from spent Ni–Cd batteries. Leaching of Cd from an electrode material in media supplemented with sulfuric acid and ferric iron, respectively, was studied using *At. ferrooxidans*, and the influence and effect of different Fe(III) concentrations were evaluated. This study was useful in finding the major factor responsible for Cd recovery, as the bioleaching mechanism of nonsulfide ores and wastes had yet not been explained adequately.

Fe ions had an influence on Cd leaching. The best leaching efficiency of Cd (100%) was obtained by bioleaching and also by leaching in an $Fe_2(SO_4)_3$ solution. XRD results confirmed the complete dissolution of Cd by bioleaching as well as sole ferric iron.

The authors proposed that during the bioleaching process, the bacteria use Fe(II) ions as an energy source. Fe(II) ions get oxidized into Fe(III) ions, which act as strong oxidizing agents, leading to the solubilization of Cd (5.15), (5.16) (Velgosová et al. 2013):

$$Fe^{2+} + 1/4O_2 + H^+ \xrightarrow{\ A.\ ferrooxidans\ } Fe^{3+} + 1/2H_2O \qquad (5.15)$$

$$Cd^0 + 2Fe^{3+} \rightarrow 2Fe^{2+} + Cd^{2+} \qquad (5.16)$$

The bacteria produce sulfuric acid using sulfur as an energy source, which plays the major role in dissolving Cd from spent Ni–Cd batteries:

$$Cd(OH)_2 + H_2SO_4 \rightarrow Cd(SO)_4 + 2H_2O \qquad (5.17)$$

The presence of hydroxides in electrode materials enabled neutralization of the leaching solution and thus inhibited Cd leaching in the H_2SO_4 solution. Thus, ferric iron not only acts as a strong oxidizing agent but also as a factor that can maintain the low pH, the environment needed for the Cd release, due to its hydrolysis. The Cd bioleaching process rate was higher in contrast with the chemical leaching with solely ferric iron at the same concentration as that in 9 K medium (9 g/L) since almost 100% Cd was released in the first 7 days.

In the absence of the bacteria, in the acidic medium with conditions similar to bioleaching, the pH increased rapidly due to acid consumption, resulting in the inhibition of Cd leaching. The addition of sulfuric acid is required to maintain the low pH, which is ineffective from an economical point of view. Velgosová et al. (2010, 2012, 2013) suggested that optimum metal leaching efficiency can be achieved if separate leaching systems are used for the recovery of Cd and Ni.

Ijadi Bajestani et al. (2014) used the Box–Behnken design of response surface methodology (RSM) to study the bioleaching of metals from spent Ni–Cd and Ni–MH batteries using *At. ferrooxidans*. The microorganisms were adapted to the the solid-to-liquid ratio of 10 g/L (battery powder weight/volume of medium). The effects of initial pH, powder size, and initial Fe^{3+} concentration on the percentage of metal recovery were studied. Ni–MH battery waste was mixed with Ni–Cd battery waste and the bioleaching process for the mixed anode and cathode materials was studied, which would be helpful in developing an industrial process.

The proposed statistical method was used to accurately evaluate the interactions of the factors and their effects on the recovery efficiency of nickel, cadmium, and cobalt during bioleaching. The Box–Behnken design was able to determine the interactions and model the bioleaching process at R2 > 0.9, which indicates a high correlation of experimental responses with model predictions. An increase in particle size and initial pH and a decrease in ferric ion concentration resulted in the maximum recovery efficiency of nearly 99% for Cd, whereas for Ni and cobalt hydroxides, the minimum particle size and initial pH and the maximum initial ferric ion concentration led to the highest biorecovery efficiency. The highest simultaneous extraction for the three metals can be attained at the optimal initial pH value of 1, a particle size of 62 μm, and an initial ferric concentration of 9.7 g/L. Under these conditions, the recoveries predicted by the software (Design-Expert version 7.1.4) were 85.6% for Ni, 66.1% for Cd, and 90.6% for Co.

These results were confirmed using an experiment under optimum conditions, which calculated a recovery of 87%, 67%, and 93.7% for Ni, Cd, and Co, respectively. During the simultaneous recovery of metals under optimum conditions, the recovery of Cd decreased due to the different chemical characteristics of Cd from Ni and Co, whereas it was higher when Cd was the sole target for extraction.

An increase in particle size and initial pH elevated the Cd leaching efficiency, while it was reduced with a decrease in ferric ion concentration. For nickel and cobalt hydroxides, however, the minimum particle size and initial pH and the maximum initial ferric ion concentration led to the highest bioleaching efficiency.

In another study, Kim et al. (2016) explored the use of six *Aspergillus* species because of their ability to recover metals from spent ZMBs or Ni–Cd batteries using malt extract (ME) and sucrose as a nutrient source. As organic acids are known to chelate metal ions, which facilitate the dissolution of metals, organic acids produced by six *Aspergillus* species under different nutrient sources were evaluated and compared for their efficiency in bioleaching of spent ZMBs or Ni–Cd batteries.

Six *Aspergillus* species were grown on two different nutrients, ME and sucrose, to produce different types of organic acids. Oxalic acid and citric acid were found to be the dominant organic acid in the ME and sucrose media, respectively.

Citric acid was the major organic acid produced in the sucrose medium, and oxalic acid in the ME medium. However, *A. fumigatus* KUC1520 and *A. flavipes* KUC5033 did not produce both oxalic acid and citric acid. And *A. versicolor* KUC5201 showed the greatest organic acid production ability and had the lowest pH (2.08) in both the sucrose and ME media. The amount of oxalic acid and citric acid produced was different in both media, suggesting a change in the metabolic pathway of *Aspergillus* depending on the availability of nutrients. Biosynthesis of oxalic acid from glucose occurs when oxaloacetate is hydrolyzed to oxalate and acetate catalyzed by cytosolic oxaloacetate, whereas citric acid is produced as an intermediate in the tricarboxylic acid cycle involving a polysaccharide such as sucrose (Gadd 1999).

Higher metal recovery in the sucrose medium, as compared with the ME medium, suggests citric acid as the most suitable agent for metal retrieval from both spent ZMBs and Ni–Cd batteries, since oxalic acid tends to form insoluble metal oxalates. All species, except *A. niger* KUC5254, showed more than 90% removal of metals from ZMBs. For Ni–Cd batteries, more than 95% of metals was extracted by *A. niger* KUC5254 and *A. tubingensis* KUC5037. Thus, *A. tubingensis* KUC5037, a nonochratoxigenic fungus, has a great potential in industrial bioleaching and in improving the safety and efficiency of bioleaching. Further, the authors suggest the use of immobilization and biofilms and optimal treatment conditions to improve the efficiency of the process and overcome the limitations of fungal bioleaching (Kim et al. 2016).

5.2.1.3 Zn–Mn Batteries

Xin et al. (2012a) extracted valuable Zn and Mn from spent ZMBs by bioleaching. A mixed culture of *Alicyclobacillus* sp. (SOB) and *Sulfobacillus* sp. (IOB) was used in this process. A comparison of recovery of both Zn and Mn by different bioleaching bacteria under various energy source types showed that 96% of Zn extraction was achieved within 24 h regardless of energy source types and bioleaching bacteria species, whereas Mn bioleaching was a relatively slow-release process. Moreover, different bacteria exhibited different rates of Mn leaching, wherein the mixed culture showed 97% extraction, while the SOB only 56%.

Energy source types also had no effect on the extraction of Zn. However, initial pH had a remarkable influence on Zn release; the extraction dose sharply decreased from 2200 to 500 mg/L when the initial pH value was increased from 1.5 to 3.0 or higher. In contrast to Zn case, energy source types had a great influence on the extraction efficiency of Mn; the maximum released dose of 3020 mg/L was obtained under optimum conditions. Acidic dissolution by biogenic H_2SO_4 by a noncontact mechanism was responsible for Zn extraction, while Mn extraction was owed to both contact/biological and noncontact mechanisms.

Although acid dissolution by biogenetic H_2SO_4 could also extract Mn^{2+} from Mn_2O_3 and Mn_3O_4, Mn^{4+} in the form of residual MnO_2 was resistant and insoluble (Sayilgan et al. 2009). In the presence of IOB, FeS_2 could be biochemically oxidized to generate H_2SO_4, accompanied with the release of biogenic Fe^{2+}; on the other hand, biogenic Fe^{3+} from bio-oxidization of Fe^{2+} triggered a series of chemical reactions to produce chemical-origin Fe^{2+} (Xin et al. 2009). Biogenic

Fe^{2+} along with chemical-origin Fe^{2+} reduced insoluble Mn^{4+} to soluble Mn^{2+} in the presence of H^+ as follows:

$$MnO_2 + 2Fe^{2+} + 4H^+ = Mn^{2+} + 2Fe^{3+} + H_2O \quad \left(\Delta G_f^0 = -18.17\right) \quad (5.18)$$

As a result, both the IOB culture system and the mixed-culture system exhibited higher extraction of Mn than the SOB culture system due to the reduction of insoluble Mn^{4+} to soluble Mn^{2+} by Fe^{2+}, reflected by the strong fluctuation in ORP during bioleaching. Furthermore, the lower pH value in the mixed-culture system promoted a stronger generation of Fe^{3+} than the IOB culture system (Xin et al. 2009). As a result, more Fe^{3+} led to a greater production of Fe^{2+}, which strongly accelerated the reduction of Mn^{4+} and extraction of resulting Mn^{2+}. Therefore, the mixed-culture system resulted in the highest Mn extraction. As the extraction of Mn was a result of both acidic dissolution and reduction of Fe^{2+}, the release of Mn by bioleaching was slower than that of Zn with only acidic dissolution.

The combined action of acidic dissolution of soluble Mn^{2+} by biogenic H_2SO_4 and reductive dissolution of insoluble Mn^{4+} by Fe^{2+} resulted in 60% of Mn extraction, while contact of microbial cells with the spent battery material and incubation for more than 7 days was required to achieve the maximum extraction of Mn (Xin et al. 2012a).

Pulp density is an important parameter that greatly affects the operational cost of bioleaching. If pulp density is increased from 1% to 2%, the bioleaching medium is decreased by 50%, which results in a 100% rise in metal recovery and, consequently, a sharp decline in the operational cost. However, as a result of a very high metal content and alkaline matter in spent batteries, which is toxic to bioleaching bacteria, the pulp density studied so far is 1% or lower (Cerruti et al. 1998, Mishra et al. 2008, Xin et al. 2009).

From this viewpoint, Xin et al. (2012b) scrutinized the cause for the decreased bioleaching efficiency for Zn and Mn at high pulp densities of spent battery powder and also explored the measures for improving the same. The main objectives were to compare the bioleaching dynamics of Zn and Mn from spent batteries at different pulp densities, ranging from 1% to 8%; to examine the reasons for the decrease in the bioleaching efficiency at high pulp densities; and to propose measures for improving the bioleaching efficiency at a high pulp density of 4%.

A mixed culture of *Alicyclobacillus* sp. (SOB) and *Sulfobacillus* sp. (IOB) was employed for bioleaching of Zn and Mn from spent batteries. The authors observed that the recovery efficiency of Zn and Mn dropped from 100% and 94% at 1% pulp density to only 29.9% and 2.5% at 8% pulp density, respectively. Thus, the increase is pulp density exerted an adverse effect on the recovery of Zn and, markedly, Mn. The experimental results suggested that the toxicity of released metals played no role in the decreased bioleaching efficiency. However, it was revealed that the deteriorated environmental conditions such as pH and ORP at high pulp densities resulted in a decrease in the cell number of mixed cultures, which attributed to the decrease in the bioleaching efficiency. The bioleaching efficiency of Zn declined due to a linear reduction in the activity of SOB with an increase in pulp density, whereas a complete inactivation of IOB at 2% pulp density or higher resulted in decreased bioleaching of Mn.

Reducing the initial pH value of the leaching medium, increasing the concentration of energy sources and periodic exogenous acid adjustment of pH during bioleaching resulted in the maximum extraction efficiency of almost 100% for Zn and 89% for Mn at 4% pulp density.

The bioleaching yield of Zn and Mn from spent ZMBs was improved by applying metal ion catalysis at a high pulp density of 10% (Niu et al. 2015). In biohydrometallurgy/bioleaching, pulp density is a vital parameter that greatly affects the design of the bioreactor and consumption of the medium (Cerruti et al. 1998, Xin et al. 2012b). Spent batteries are composed of high concentrations of alkaline or toxic compounds such as oxides or hydroxides, which decrease the growth and activity of leaching cells, owing to which pulp density is often only 1% or lower in bioleaching of spent batteries (Zhao et al. 2008a, Xin et al. 2009, Zeng et al. 2013, Ijadi Bajestani et al. 2014), meaning that a significant quantity of the medium and a great volume of the bioreactor were required. To accelerate the electron transfer involved in the bioleaching process and the slow kinetics of metal dissolution so as to improve metal extraction, accession of catalytic metal ions such as Cu^{2+}, Bi^{3+}, Hg^{2+}, Ag^+, and Co^{2+} is being applied as an effective way (Byerley et al. 1979, Ballester et al. 1990, Hu et al. 2002, Chen et al. 2008).

Efficient extraction of target metals at high pulp densities is very important for practical application of biohydrometallurgy/bioleaching. The catalytic bioleaching performance of Zn and Mn from spent batteries at 10% pulp density using various metal ions was studied. All the tested metal ions exerted different catalytic activities toward the extraction of Zn and Mn. Of the four metal ions, Co^{2+} and Ni^{2+} had no increasing impact on the release of Zn and Mn, whereas Ag^+ had an adverse effect on the dissolution of Zn and Mn by bioleaching. However, Cu^{2+} showed improved mobilization of Zn and Mn. Further studies on the catalytic bioleaching performance using Cu^{2+} under varied conditions revealed that an increase in Cu^{2+} content from 0 to 0.8 g/L increased the leaching efficiency from 47.7% to 62.5% and from 30.9% to 62.4% for Zn and Mn, respectively. Various process parameters such as ORP, SO_4^{2-} content, cell number, Fe^{2+} concentration, and Fe^{3+} concentration in Cu^{2+} catalytic bioleaching were monitored to explore the bioleaching mechanism. Cu^{2+} catalysis accelerated/boosted bioleaching of resistant hetaerolite by forming a possible intermediate, $CuMn_2O_4$, which was subject to attack by Fe^{3+} and H^+ based on a cycle of Fe^{3+}/Fe^{2+} (Equations 5.19 and 5.20):

$$Cu^{2+} + ZnMn_2O_4 \rightarrow CuMn_2O_4 + Zn^{2+} \tag{5.19}$$

$$CuMn_2O_4 + 2Fe^{3+} + 4H^+ \rightarrow 3Fe^{2+} + 2Mn^{2+} + O_2 + Cu^{2+} + 2H_2O \tag{5.20}$$

However, poor growth of cells, the formation of $KFe_3(SO_4)_2(OH)_6$, and its possible blockage between cells and energy sources destroyed the cycle of Fe^{3+}/Fe^{2+}, stopping the bioleaching of hetaerolite (Equation 5.21):

$$K^+ + 3Fe^{3+} + 6H_2O + 2SO_4^{2-} \rightarrow KFe_3(SO_4)_2(OH)_6 \downarrow + 6H^+ \tag{5.21}$$

A chemical reaction–controlled model fitted best for describing Cu^{2+} catalytic bioleaching of spent ZMBs.

In another study, Niu et al. (2016) carried out the optimization of different parameters for bioleaching of metals from spent ZMBs using RSM. In order to make a process commercially successful, the values for effective parameters must be optimized, especially to face competition with counterparts. The authors used central composite design (CCD) with RSM to obtain information about the interactions among multiple parameters involved in the bioleaching process. They optimized the bioleaching process of spent ZMBs by using a mixed culture at a higher pulp density of 8%–12% to compete with physical–chemical processes. Four important operating parameters, including energy sources concentration, pH control of bioleaching media, incubating temperature, and pulp density, were assessed to determine their role and interactions on extraction efficiency of Zn and Mn. Also the relationship between multiple parameters and the release of target metals were studied and confirmed using a suitable mathematical model.

The released dose of Zn and Mn was highly dependent on many parameters; thus, a clear understanding of the affecting parameters was useful to optimize the process. In particular, the exogenous acid adjustment of pH was found to be an efficient and economical way to improve the dissolution efficiency of spent ZMBs at high pulp densities. Nonlinear equations were formulated to describe the relationship between the bioleaching efficiency of Zn and Mn and the four important parameters. Results revealed the optimum parameter values as dose of mixed energy sources (28 and 29 g/L), exogenous acid adjustment of pH (1.9 and 1.8), incubating temperature ($33°C$ and $36.7°C$), and pulp density (9.7% and 8%), for Zn and Mn release, respectively. The maximum efficiency of extraction was found to be 52.5% for Zn and 52.4% for Mn under optimum conditions after 9 days of bioleaching. Moreover, the results of XRD analyses suggested that Zn and Mn existed mainly as hetaerolite ($ZnMn_2O_4$) in spent ZMBs, which gradually disappeared and released Zn^{2+} and Mn^{2+} into the solution during the bioleaching process.

5.2.2 Button Cells

Button cell batteries are abundantly used in various electronic equipment owing to their small size, long operating life, and high capacity per unit mass (Sathaiyan et al. 2006). Although various pyro- and hydrometallurgical methods are used for recycling them, many disadvantages of these processes make it inevitable to develop a new sustainable process.

Jadhav and Hocheng (2013) demonstrated such a process for the extraction of silver from spent silver oxide zinc button cell battery using *At. ferrooxidans* culture supernatant. Preliminary experiments were carried out to assess the practicability of the indirect leaching (two-step) process. Depending on a review of the literature and comparative results, indirect leaching by spent culture was found to be efficient.

The culture of *At. ferrooxidans* was supplemented with 1% sulfuric acid and $FeSO_4$ as an energy source. Indirect bioleaching was employed, where the battery powder was added to the bacterial culture supernatant. Results revealed the importance of reaction time, as silver extraction increased from 45.8% to 98% when the reaction time was increased from 1 to 3 h. Silver solubilization in the culture supernatant was attributed to the Fe^{3+} oxidant, which was most likely regenerated by *At. ferrooxidans*. A comparison of ferrous/ferric ion concentrations showed that after 3 h of bioleaching, 80% Fe^{3+} was converted into Fe^{2+}, thus confirming the role of Fe^{3+}. The ferrous–ferric couple serves as a redox mediator and catalyzes the reaction.

The use of microbial cells for silver recovery took 30 h for 98.5% silver solubilization which was may be due to the slower availability of Fe^{3+}. A comparable leaching efficiency was achieved by simply using culture supernatant. During the recovery process, biologically produced ferric ions react with silver, dissolve it, and themselves get converted into ferrous ions, which could be further used for bacterial growth. Thus, a cleaner metal recovery process can be developed.

Silver extraction increased with an increase in the volume of culture supernatant and initial $FeSO_4$ concentration. The use of 40 and 60 g/L $FeSO_4$ culture supernatant decreased the time of silver dissolution to 2 and 1 h, respectively. Other parameters such as pH (2.5) and shaking speed (150 rpm) were found to be optimum; however, the change in incubation temperature did not have any effect on silver recovery.

Furthermore, a comparison of these bioleaching results with the previously reported chemical leaching using nitric acid (Sathaiyan et al. 2006, Aktas 2010) suggested that silver leaching using biologically produced ferric ions was as efficient as nitric acid leaching. About 98% silver was dissolved in 60 min. Therefore, the use of biologically produced ferric sulfate is advantageous over chemical leaching due to its efficiency and minimal impact on the environment.

Hence, the two-step process is more suitable for industrial application and for increasing the efficiency of leaching. In general, the advantages of such a process are that independent lixiviant generation removes the link between the bioprocess and the chemical process and thus makes it possible to optimize each process independently to maximize productivity. This strategy can be used when the ore/waste does not contain the necessary mineral components in sufficient quantities to sustain a viable bacterial population. Furthermore, higher waste concentrations can be treated, compared with the one-step process, which results in increased metal yields.

5.3   CONCLUSION

Recovery of metals from energy storage wastes such as batteries, button cells, etc. is financially considerable due to the presence of high concentrations of valuable metals. The higher solid/liquid ratio of waste batteries in media adversely affects bioleaching owing to its toxicity toward microbial cells. The use of indirect bioleaching is an effective step to overcome this situation. Moreover, the two-step process is more suitable for increasing the leaching efficiency and is also applicable at industrial level. The use of mixed bacterial cultures enhances the bioleaching efficiency, as these microorganisms can survive the high metal concentration. However, the high pulp density of spent batteries greatly affects the operational cost of bioleaching as a result of decreased metal recovery due to the high metal content and alkalinity, which are toxic to bioleaching bacteria. It has been observed that the deteriorated environmental conditions due to high pulp densities affect the bacteria, thereby decreasing the bioleaching efficiency. This problem can be dealt with by reducing the initial pH of the leaching medium, increasing the concentration of energy sources and intermittent exogenous acid adjustment of pH during bioleaching. Many reports state the use of catalytic ions in the bioleaching solution to improve the process. Also, the use of immobilized cells and biofilms and optimal process parameters can improve the efficiency of the process and overcome the hurdles involved in bioleaching.

REFERENCES

Aktas, S. 2010. Silver recovery from spent silver oxide button cells. *Hydrometallurgy* 104(1): 106–111. doi:10.1016/j.hydromet.2010.05.004.

Aung, K. M. M. and Y. P. Ting. 2005. Bioleaching of spent fluid catalytic cracking catalyst using *Aspergillus niger*. *Journal of Biotechnology* 116(2): 159–170. doi:10.1016/j.jbiotec.2004.10.008.

Ballester, A., F. González, M. L. Blázquez, C. Gómez, and J. L. Mier. 1992. The use of catalytic ions in bioleaching. *Hydrometallurgy* 29(1): 145–160. doi:10.1016/0304-386X(92)90010-W.

Ballester, A., F. González, M. L. Blázquez, and J. L. Mier. 1990. The influence of various ions in the bioleaching of metal sulphides. *Hydrometallurgy* 23(2–3): 221–235. doi:10.1016/0304-386X(90)90006-N.

Byerley, J. J., G. L. Rempel, and G. F. Garrido. 1979. Copper catalysed leaching of magnetite in aqueous sulfur dioxide. *Hydrometallurgy* 4(4): 317–336. doi:10.1016/0304-386X(79)90031-8.

Cerruti, C., G. Curutchet, and E. Donati. 1998. Bio-dissolution of spent nickel–cadmium batteries using *Thiobacillus ferrooxidans*. *Journal of Biotechnology* 62(3): 209–219. doi:10.1016/S0168-1656(98)00065-0.

Chen, S., W. Q. Qin, and G. Z. Qiu. 2008. Effect of Cu^{2+} ions on bioleaching of marmatite. *Transactions on Nonferrous Metals Society of China* 18: 1518–1522.

Gadd, G. M. 1999. Fungal production of citric and oxalic acid: Importance in metal speciation, physiology and biogeochemical processes. *Advances in Microbial Physiology* 41: 47–92. doi:10.1016/S0065-2911(08)60165-4.

Horeh, N. B., S. M. Mousavi, and S. A. Shojaosadati. 2016. Bioleaching of valuable metals from spent lithium–ion mobile phone batteries using *Aspergillus niger*. *Journal of Power Sources* 320: 257–266. doi:10.1016/j.jpowsour.2016.04.104.

Hu, Y. H., G. Z. Qiu, and J. Wang. 2002. The effect of silver-bearing catalysts on bioleaching of chalcopyrite. *Hydrometallurgy* 64: 81–88.

Huang, K., J. Li, and Z. Xu. 2010. Characterization and recycling of cadmium from waste nickel–cadmium batteries. *Waste Management* 30(11): 2292–2298. doi:10.1016/j.wasman.2010.05.010.

Ijadi Bajestani, M., S. M. Mousavi, and S. A. Shojaosadati. 2014. Bioleaching of heavy metals from spent household batteries using *Acidithiobacillus ferrooxidans*: Statistical evaluation and optimization. *Separation and Purification Technology* 132: 309–316. doi:10.1016/j.seppur.2014.05.023.

Jadhav, U. and H. Hocheng. 2013. Extraction of silver from spent silver oxide–zinc button cells by using *Acidithiobacillus ferrooxidans* culture supernatant. *Journal of Cleaner Production* 44: 39–44. doi:10.1016/j.jclepro.2012.11.035.

Kim, M. J., J. Y. Seo, Y. S. Choi, and G. H. Kim. 2016. Bioleaching of spent Zn–Mn or Ni–Cd batteries by *Aspergillus* species. *Waste Management* 51: 168–173. doi:10.1016/j.wasman.2015.11.001.

Li, J., P. Shi, Z. Wang, Y. Chen, and C. C. Chang. 2009. A combined recovery process of metals in spent lithium-ion batteries. *Chemosphere* 77(8): 1132–1136. doi:10.1016/j.chemosphere.2009.08.040.

Li, L., G. S. Zeng, S. L. Luo, X. R. Deng, and Q. J. Xie. 2013. Influences of solution pH and redox potential on the bioleaching of LiCoO$_2$ from spent lithium–ion batteries. *Journal of the Korean Society for Applied Biological Chemistry* 56(2): 187–192. doi:10.1007/s13765-013-3016-x.

Li, Y. and G. Xi. 2005. The Dissolution mechanism of cathodic active materials of spent Zn–Mn batteries in HCl. *Journal of Hazardous Materials* 127: 244–248. doi:10.1016/j.jhazmat.2005.07.024.

Mishra, D., D. J. Kim, D. E. Ralph, J. G. Ahn, and Y. H. Rhee. 2008. Bioleaching of metals from spent lithium ion secondary batteries using *Acidithiobacillus ferrooxidans*. *Waste Management* 28(2): 333–338. doi:10.1016/j.wasman.2007.01.010.

Muñoz, J. A., D. B. Dreisinger, W. C. Cooper, and S. K. Young. 2007a. Silver-catalyzed bioleaching of low-grade copper ores. Part I: Shake flasks tests. *Hydrometallurgy* 88(1): 3–18. doi:10.1016/j.hydromet.2007.04.004.

Muñoz, J. A., D. B. Dreisinger, W. C. Cooper, and S. K. Young. 2007b. Silver-catalyzed bioleaching of low-grade copper ores. Part II: Stirred tank tests. *Hydrometallurgy* 88(1): 19–34. doi:10.1016/j.hydromet.2007.01.007.

Muñoz, J. A., D. B. Dreisinger, W. C. Cooper, and S. K. Young. 2007c. Silver catalyzed bioleaching of low-grade copper ores. Part III: Column reactors. *Hydrometallurgy* 88(1): 35–51. doi:10.1016/j.hydromet.2007.04.003.

Niu, Z., Q. Huang, J. Wang, Y. Yang, B. Xin, and S. Chen. 2015. Metallic ions catalysis for improving bioleaching yield of Zn and Mn from spent Zn–Mn batteries at high pulp density of 10%. *Journal of Hazardous Materials* 298: 170–177. doi:10.1016/j.jhazmat.2015.05.038.

Niu, Z., Q. Huang, B. Xin, C. Qi, J. Hu, S. Chen, and Y. Li. 2016. Optimization of bioleaching conditions for metal removal from spent zinc–manganese batteries using response surface methodology. *Journal of Chemical Technology and Biotechnology* 91(3): 608–617. doi:10.1002/jctb.4611.

Niu, Z., Y. Zou, B. Xin, S. Chen, C. Liu, and Y. Li. 2014. Process controls for improving bioleaching performance of both Li and Co from spent lithium ion batteries at high pulp density and its thermodynamics and kinetics exploration. *Chemosphere* 109: 92–98. doi:10.1016/j.chemosphere.2014.02.059.

Qu, Y., B. Lian, B. Mo, and C. Liu. 2013. Bioleaching of heavy metals from red mud using *Aspergillus niger*. *Hydrometallurgy* 136: 71–77. doi:10.1016/j.hydromet.2013.03.006.

Sathaiyan, N., V. Nandakumar, and P. Ramachandran. 2006. Hydrometallurgical recovery of silver from waste silver oxide button cells. *Journal of Power Sources* 161: 1463–1468. doi:10.1016/j.jpowsour.2006.06.011.

Sayilgan, E., T. Kukrer, G. Civelekoglu, F. Ferella, A. Akcil, F. Veglio, and M. Kitis. 2009. A review of technologies for the recovery of metals from spent alkaline and zinc–carbon batteries. *Hydrometallurgy* 97(3): 158–166. doi:10.1016/j.hydromet.2009.02.008.

Velgosová, O., J. Kaduková, and R. Marcinčáková. 2012. Study of Ni and Cd bioleaching from spent Ni–Cd batteries. *Nova Biotechnologica et Chimica* 11(2): 117–123. doi:10.2478/v10296-012-0013-0.

Velgosová, O., J. Kaduková, R. Marcinčáková, P. Palfy, and J. Trpčevská. 2013. Influence of H$_2$SO$_4$ and ferric iron on Cd bioleaching from spent Ni–Cd batteries. *Waste Management* 33(2): 456–461. doi:10.1016/j.wasman.2012.10.007

Velgosová, O., J. Kaduková, A. Mrazikova, A. Blaskova, M. Petoczova, H. Horvathova, and M. Stofko. 2010. Influence of selected parameters on nickel bioleaching from spent Ni–Cd batteries. *Mineralia Slovaca* 42: 365.

Xin, B., W. Jiang, H. Aslam, K. Zhang, C. Liu, R. Wang, and Y. Wang. 2012a. Bioleaching of zinc and manganese from spent Zn–Mn batteries and mechanism exploration. *Bioresource Technology* 106: 147–153. doi:10.1016/j.biortech.2011.12.013.

Xin, B., W. Jiang, X. Li, K. Zhang, C. Liu, R. Wang, and Y. Wang. 2012b. Analysis of reasons for decline of bioleaching efficiency of spent Zn–Mn batteries at high pulp densities and exploration measure for improving performance. *Bioresource Technology* 112: 186–192. doi:10.1016/j.biortech.2012.02.133.

Xin, B., D. Zhang, X. Zhang, Y. Xia, F. Wu, S. Chen, and L. Li. 2009. Bioleaching mechanism of Co and Li from spent lithium–ion battery by the mixed culture of acidophilic sulfur-oxidizing and iron-oxidizing bacteria. *Bioresource Technology* 100(24): 6163–6169. doi:10.1016/j.biortech.2009.06.086.

Xin, Y., X. Guo, S. Chen, J. Wang, F. Wu, and B. Xin. 2015. Bioleaching of valuable metals Li, Co, Ni and Mn from spent electric vehicle Li–Ion batteries for the purpose of recovery. *Journal of Cleaner Production* 116: 249–258. doi:10.1016/j.jclepro.2016.01.001.

Yuehua, H., Q. Guanzhou, W. Jun, and W. Dianzuo. 2002. The effect of silver-bearing catalysts on bioleaching of chalcopyrite. *Hydrometallurgy* 64(2): 81–88. doi:10.1016/S0304-386X(02)00015-4.

Zeng, G., X. Deng, S. Luo, X. Luo, and J. Zou. 2012. A copper-catalyzed bioleaching process for enhancement of cobalt dissolution from spent lithium–ion batteries. *Journal of Hazardous Materials* 199–200: 164–169. doi:10.1016/j.jhazmat.2011.10.063.

Zeng, G., S. Luo, X. Deng, L. Li, and C. Au. 2013. Influence of silver ions on bioleaching of cobalt from spent lithium batteries. *Minerals Engineering* 49: 40–44. doi:10.1016/j.mineng.2013.04.021.

Zhao, L., W. Liang, Y. Dong, and Z. Nanwen. 2007. Bioleaching of spent Ni–Cd batteries and phylogenetic analysis of an acidophilic strain in acidified sludge. *Frontiers in Environmental Science and Engineering* 1: 459–465.

Zhao, L., D. Yang, and N. W. Zhu. 2008a. Bioleaching of spent Ni–Cd batteries by continuous flow system: Effect of hydraulic retention time and process load. *Journal of Hazardous Materials* 160(2–3): 648–654. doi:10.1016/j.jhazmat.2008.03.048.

Zhao, L., N. W. Zhu, and X. H. Wang. 2008b. Comparison of bio-dissolution of spent Ni–Cd batteries by sewage sludge using ferrous ions and elemental sulfur as substrate. *Chemosphere* 70(6): 974–981. doi:10.1016/j.chemosphere.2007.08.011.

Zhu, N., L. Zhang, C. Li, and C. Cai. 2003. Recycling of spent nickel–cadmium batteries based on bioleaching process. *Waste Management* 23(8): 703–708. doi:10.1016/S0956-053X(03)00068-0.

6 Recycling of Metal Production Wastes

6.1 INTRODUCTION

Solid wastes are harmful to the environment but can be turned into a valuable resource if the metals can be recovered. One such waste is industrial waste slag, which is produced as by-products in industrial metallurgical processes or as residues in incineration processes. According to the origins and the characteristics, the main slags can be classified into three categories, namely, ferrous slag, nonferrous slag, and incineration slag. First, rather than an end-waste, slags are actually secondary resource materials. Second, they are comparatively better than their counterparts in some applications. Third, they contain a remarkable amount of hazardous heavy metals (Shen and Forssberg 2003).

Ferrous slag mainly includes iron slag (blast furnace slag), steel slag, alloy steel slag, and ferroalloy slag. Blast furnace slag (BF slag) and steel slag make up the major part of ferrous slag. BF slag and steel slag have found wide application, such as cement production, road construction, civil engineering work, fertilizer production, landfill daily cover, and soil reclamation. Alloy steel slag and ferroalloy slag usually contain high amounts of alloy elements, such as Cr, Ni, Mn, Ti, V, and Mo. Nonferrous slags include copper slag, salt slag, pyrite slag, tin slag, etc.

Industrial dust also contains many such heavy metals. They are recycled back to the industry or used externally as raw materials in other industries. Recycling them back in the same industrial process reduces their efficiency, increases the energy input, and damages the furnace structure. Concentrations of heavy metals in these wastes vary significantly depending on the industrial process from which they are generated. This chapter summarizes the bioleaching studies of industrial waste slags and dust.

6.2 BIOLEACHING OF SLAGS

6.2.1 Cu Smelter Slags

Sukla et al. (1995) studied bioleaching of copper converter slag using *Aspergillus niger* isolated from lateritic nickel ore. Leaching studies were carried out in potato dextrose broth with different solid to liquid ratios. The results were compared with chemical leaching using oxalic, acetic, and succinic acids. 47%, 50%, and 23% of Cu, Co, and Ni were recovered, respectively, at 2% solid–liquid ratio. Of the three organic acids tested, succinic acid was found to be relatively better for metal leaching.

Bench scale bioleaching of the convertor slag was studied by Mehta et al. (1999) using *Thiobacillus ferrooxidans*. Various process parameters like particle size, pH, pulp density, Fe(II)/Fe(III) ratio, and the amount of Fe(II), additives, etc., were optimized. In the initial stages, dissolution of copper, nickel, and cobalt from the slag was governed by chemical leaching. *Thiobacillus ferrooxidans* oxidized the ferrous ions to ferric form, which enhanced the recovery of metals from the slag, suggesting the indirect mechanism of bioleaching. Further, low concentrations of Fe(II) and Fe(II)/ Fe(III) ratio in bioassisted leaching as compared to the chemical one confirms the role of microorganisms in metal dissolution.

It was observed that 80% and 99% of Cu dissolved in 1920 h, 18% and 22% of Ni in 1680 h, and 16% and 30% Co in 1200 h in the absence and presence of microorganisms, respectively. The low recovery of nickel and cobalt may be attributed to the association of these metals with magnetite/fayalite matrix, which allows an only partial attack on these metals during bioleaching.

At higher pH of 2.5–3.5, metal recovery decreased due to jarosite formation as identified by XRD analysis.

Optimum parameters for metal dissolution were observed as pH 2.0, pulp density (S/L) 1:20, and particle size of −75 μm fraction for Ni and Co, giving maximum recovery of 50% and 64%, respectively, in 1920 h. However, the optimum particle size for copper dissolution was found to be 100 + 75 μm fraction. Further, the addition of NaCl and $(NH_4)_2SO_4$ had a marginal effect on Cu recovery. The addition of NaCl increased the recovery of Co and Ni as compared to the addition of $(NH_4)_2SO_4$, which was due presumably to the effect of anions present in the solution.

Kaksonen et al. (2011) evaluated the bioleaching of metals from final smelter slags and precipitated the metals from the leach liquors using biogenic sulfide. A continuous stirred tank reactor (CSTR) was operated at 20°C–25°C with 5% pulp density (particle size 75% < 47 μm).

Metals in the solution after bioleaching can be recovered by applying different processes. Kaksonen and Puhakka (2007) suggested the use of biogenic H_2S for precipitation of metals as sulfides instead of chemical hydroxide precipitation as the former can recover metals from lower concentrations of effluent and better thickening characteristics of the metal sludge, and there is the possibility of recovering valuable metals. Further, they describe the process as being based on biological H_2S and alkalinity $2HCO_3^-$ production by sulfate-reducing bacteria (SRB) (reaction 6.1), metal sulfide precipitation with biogenic H_2S (reaction 6.2), and neutralization of the acidity of the water with alkalinity produced by the SRB in the oxidation of provided electron donors (reaction 6.3).

$$SO_4^{2-} + 2CH_2O \rightarrow H_2S + 2HCO_3^- \quad \text{where } CH_2O = \text{electron donor} \quad (6.1)$$

$$H_2S + M^{2+} \rightarrow MS(s) + 2H^+ \quad \text{where } M^{2+} = \text{metal such as } Zn^{2+} \quad (6.2)$$

$$HCO_3^- + H^+ \rightarrow CO_2(g) + H_2O \quad (6.3)$$

A 125 L CSTR with an 85 L working volume was employed for the bioleaching of slag using the mixed bacterial culture enriched from acidic waters of the Talvivaara mine, Finland. The culture was supplemented with elemental sulfur and 5% final slag. The yields of metal solubilization after 29 days of contact were 41% Fe, 62%

Cu, 35% Zn, and 44% Ni. However, Fe and Cu solubilization increased to 58% and 65%, respectively, by day 39, whereas the yields of Zn and Ni decreased, suggesting precipitation.

Further, the metals from the bioleach liquors were precipitated using two sulfidogenic fluidized-bed reactors (FBRs A and B) with a working volume of 0.55 L and with 0.32 L of silicate mineral as the biomass carrier at 25°C as shown in Figure 6.1.

The recovery of metals from the bioleach liquor obtained from final slag was studied in a CSTR by titrating the bioleach solution with sulfide-containing alkaline effluent from FBR A to the desired pH values. Over 98% of the Cu precipitated above pH 2.8 and 99% of the Zn precipitated above pH 3.9. The precipitation of Ni and Fe required higher pH values and was less efficient. Ni (43% and 63%) and Fe (14% and 28%) were precipitated at pH 6.1 and 6.5, respectively.

Due to the limited acetate oxidation and alkalinity production in the FBR, the volume of the FBR effluent needed to adjust the pH of the bioleach solution was relatively large. The effluent/bioleach liquor ratio could be decreased by leaching the metals at a higher pH and increasing the alkalinity production in the FBR by enriching for acetate oxidizers. Bulk precipitation of metals from the bioleach solution in the sulfidogenic FBR required at least tenfold dilution of the leach solution and preadjustment of the pH from 0.6 to approximately 4.

In another study, Kaksonen et al. (2016) compared the dissolution of Si, Fe, Cu, and Zn from a smelter slag sample using chemical and bacterial leaching processes.

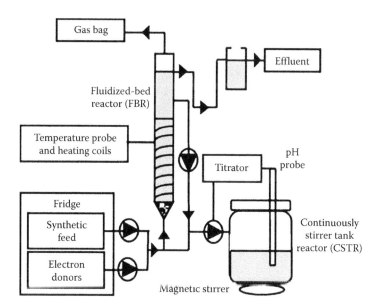

FIGURE 6.1 Design of the sulfate-reducing FBR and CSTR. (Reprinted from *Miner. Eng.*, 24(11), Kaksonen, A.H., Lavonen, L., Kuusenaho, M., Kolli, A., Närhi, H., Vestola, E., Puhakka, J.A., and Tuovinen, O.H., Bioleaching and recovery of metals from final slag waste of the copper smelting industry, 1113–1121, doi:10.1016/j.mineng.2011.02.011, Copyright 2011, with permission Elsevier.)

The enrichment of eleven samples from the lagoon area of a smelter site yielded two iron- and sulfur-oxidizing enrichment cultures, namely, HB1 and HB2. The bacterial community of culture HB1 was further characterized by denaturing gradient gel electrophoresis (DGGE) of PCR-amplified partial 16S rRNA gene sequences followed by sequencing. Before chemical and bacterial leaching, slag was preleached for 24 h with acidified water because of the initial acid demand. Bioleaching experiments were carried out in shake flasks and stirred tank reactors at 10% pulp density.

HB1 culture contained at least *Acidithiobacillus ferrivorans* and *Alicyclobacillus cycloheptanicus*, with sequences of three DGGE bands matching distantly with *Alicyclobacillus tolerans* and *Alicyclobacillus herbarium* in the database. *Alicyclobacillus* spp. has not been previously associated with slag lagoons or slag bioleaching.

The acid consumption of slag varied between 412 and 886 g H_2SO_4 at pH 2.0 and 1.0, respectively, during preleaching treatment. Approximately 80% Cu and 25% Zn were dissolved from the slag (10% pulp) in shake flasks when S^0 was provided for the bacteria to produce H_2SO_4. The addition of elemental sulfur for H_2SO_4 production increased the Zn dissolution but no such effect was observed in the case of Cu.

Bioleaching in stirred tanks was conducted at controlled pH values and was practiced at pH levels promoting metal dissolution and suppressing iron and silicate solubilization from fayalite and Na-silicate. Moreover, it was sensitive to pH changes and increase in pH decreased the solubility of both Cu and Zn. The alkali demand was 180–230 g NaOH/kg slag in stirred tank bioleaching owing to the oxidation of S_0 to H_2SO_4 and proton-yielding hydrolysis and precipitation of ferric iron as Fe(III)-hydroxy sulfate-type solid phases in sulfate-rich solutions. Acid and alkali demand could be optimized if the net acid or alkali-forming potential of the bioleaching reactions is elucidated.

Chemical leaching at pH 2.3–4.0 did not yield substantial dissolution of valuable metals and thus was insignificant when compared to bioleaching.

Muravyov et al. (2012) explored the use of sulfuric acid and biologically generated ferric iron for leaching of copper and zinc from copper converter slag flotation tailings. This process of two-stage bacterial–chemical leaching was employed by many researchers for the treatment of sulfide raw materials and concentrates (Romero et al. 2003, Kinnunen et al. 2006, Muravyov et al. 2011). In the first stage, the metals are leached by microbially produced ferric sulfate. As compared to conventional biohydrometallurgical technology, higher pulp density and temperature up to 80°C can by employed in this process, which significantly increases the leaching rate. During the second stage, ferric iron is regenerated, elemental sulfur is oxidized, and sulfides are terminally oxidized. Depending on the temperature conditions at the first stage, regeneration is carried out by different communities of acidophilic chemolithotrophic microorganisms. Ferric iron can be regenerated at a low cost by biological oxidation. Moreover, separate stages of chemical and biological processes aid in creating favorable conditions for the chemical processes of mineral oxidation and for microbial activity. This technique is flexible and can be applied to any material containing nonferrous metals as sulfide minerals.

The process was carried out as follows: (1) the chemical leaching by a biogenic ferric sulfate solution in sulfuric acid medium; (2) removal of Fe^{3+} ions after the chemical leaching of slag by addition of CaO or $CaCO_3$ into the pulp with subsequent solid–liquid separation; (3) copper cementation from the liquid phase; (4) removal of zinc (as ZnS) by addition of NaHS (Khalezov et al. 2005) with subsequent solid–liquid separation; and (5) separation of obtained liquor (after nonferrous metal removal) into two parts, the first part of the flow may be used as a coagulator, the second part (after dilution with an aqua) is sent to bioregeneration and further to step (1).

Flotation tailings contained 0.56% Cu and 4.74% Zn. Copper was present mostly as digenite, bornite, and free metal. Zinc was present as a ferrite (franklinite) $ZnFe_2O_4$ and, to a lesser degree, as silicate. Biooxidation of ferrous iron ($FeSO_4 \cdot 7H_2O$) was carried out in a 5.0 L vessel with 4.0 L of the liquid at 40°C and aeration rate 4/min using a consortium of moderately thermophilic acidophilic chemolithotrophic microorganisms, including *Sulfobacillus* species. The obtained solution was used as lixiviant in the chemical leaching process.

Depending upon the mineralogical composition of the converter slag flotation tailings, authors suggest that chemical leaching of nonferrous metals with ferric iron containing sulfuric acid solutions should be based mainly on the following reactions of oxidation or dissolution of metallic copper, digenite, bornite, magnetite, and copper, and franklinite:

$$Cu^0 + 2Fe^{3+} = Cu^{2+} + 2Fe^{2+} \tag{6.4}$$

$$Cu_9S_5 + 8Fe^{3+} = 5CuS + 4Cu^{2+} + 8Fe^{2+} \tag{6.5}$$

$$Cu_5FeS_4 + 4Fe^{3+} = 2CuS + CuFeS_2 + 2Cu^{2+} + 4Fe^{2+} \tag{6.6}$$

$$CuS + 2Fe^{3+} = Cu^{2+} + 2Fe^{2+} + S^0 \tag{6.7}$$

$$CuO \cdot Fe_2O_3 + 8H^+ = Cu^{2+} + 2Fe^{3+} + 4H_2O \tag{6.8}$$

$$Fe_3O_4 + 8H^+ = Fe^{2+} + 2Fe^{3+} + 4H_2O \tag{6.9}$$

$$ZnFe_2O_4 + 8H^+ = Zn^{2+} + 2Fe^{3+} + 4H_2O \tag{6.10}$$

The higher the temperature and pH of the pulp, the higher the rate of the reaction with jarosite formation (Dutrizac 1980):

$$3Fe_2(SO_4)_3 + 12H_2O + M_2SO_4 = 2M[Fe_3(SO_4)_2(OH)_6] \downarrow + 6H_2SO_4 \tag{6.11}$$

where $M = K^+, Na^+, NH_4^+, H_3O^+$.

The effects of pH, pulp density, temperature, initial Fe^{3+} concentration, and the presence of oxygen in the leaching solution on the leaching dynamics of copper, zinc, and iron were investigated. Higher solubilization of copper and zinc was obtained at pH 1.5 along with a low increase in fayalite solubility and less intense formation of the jarosite precipitate. Variations in pulp density from 10% to 30% had no effect on zinc recovery, whereas copper recovery decreased by only 1.25-fold after 6 h with increasing pulp density. However, 30% pulp density was considered to be optimum as the economic efficiency of the process increases at higher pulp densities and the decrease in copper recovery may be compensated by increased Fe^{3+} concentration in the leaching solution.

High Fe^{3+} concentrations had an inhibitory effect on leaching of zinc and iron, but copper remained unaffected. Recovery of copper increased from 21% to 42.6% in the presence of 5.3 g/L Fe^{3+} after 30 min, while zinc recovery decreased from 25.4% to 13.6%, respectively. Also, iron recovery decreased from 25.1% and 9.3%, respectively, after 30 min in the presence of ferric iron. Chemical leaching at the initial concentration of ferric iron of 15.7 g/L was shown to be the best condition, characterized by higher recovery of the more valuable component (copper) from converter slag flotation tailings, lower recovery of a less valuable component (zinc), and a twofold decrease in iron recovery during 6 h of the process.

The temperature showed the significant influence on the kinetics of copper and zinc leaching. Kinetic analysis of copper and zinc leaching revealed the activation energies of 51.3 and 53.8 kJ/mol, respectively, indicating that these processes are controlled by a chemical reaction on the particle surface. 81.6% copper, 37.7% zinc, and 26.2% iron were recovered from the converter slag flotation tailings at a pulp density of 30% (w/v), pH 1.5, a temperature of 70°C, an initial Fe^{3+} concentration of 15.7 g/L, no aeration, and leaching for 1.5 h. The leach residue contained 0.13% Cu and 3.69% Zn.

Vestola et al. (2012) applied the bioleaching process for metal recovery from waste slags of copper and steel industries employing iron and sulfur-oxidizing acidophiles. Two different metal-containing waste samples were used: slag from copper smelting and converter sludge from the steel industry. Mixed culture of *Acidithiobacillus* spp. and *Leptospirillum* spp. was supplemented with elemental sulfur and ferrous sulfate in the growth medium.

Biological oxidation of sulfur decreased the pH of media containing both wastes. pH values decreased when the rate of biological sulfur oxidation was higher than the acid consumption by materials. The higher metal recoveries were obtained in the culture supplemented with S_0 at pH 1.0 and 0.5, suggesting that the acid attack was responsible for metal solubilization. Complete solubilization of copper and nickel was obtained in slag samples, whereas it was lower with the converter sludge sample. The authors suggested that the final slag sample did not contain inhibitory substances and the metal content was higher as compared to the converter sludge.

Comparable results were obtained with the chemical leaching process but eventually it had higher acid consumption as compared to bioleaching where sulfuric acid was generated from supplemented sulfur. The addition of ferrous iron and chloride ions had an insignificant effect on metal solubilization, as metals in these wastes are present mainly in oxide form, whose solubilisation does not involve a redox reaction. Supplemental Fe^{2+} only serves as an electron donor for the bacteria in the leaching process. Separation inhibition tests revealed that the Fe^{2+} oxidation rate was faster in the presence of both wastes as compared to the control. However, none of the wastes inhibited the biological sulfur oxidation.

Panda et al. (2015) studied in detail mineralogical characterization, physical beneficiation, and bioleaching studies of copper from copper slag. Sample characterization studies showed the presence of fayalite and magnetite along with metallic copper disseminated within the iron and silicate phases. Physical beneficiation of the slag sample resulted in poor copper recovery as compared with a mixed meso-acidophilic bacterial leaching test. Physical beneficiation of the slag was carried out in a 2 L working volume flotation cell using sodium isopropyl xanthate and resulted in 2%–4% Cu beneficiation, and final recovery was only 42%–46%. However, the use of mixed mesoacidophilic bacterial consortium comprised a group of iron and/or sulfur-oxidizing bacteria showed the enhanced recovery of Cu (92%–96%) in a 12-day leaching period with optimized conditions at pH of 1.5, 15% (v/v) bacterial inoculum in media, 3 g/L initial Fe (II) concentration, and 7% (w/v) pulp density. After leaching, SEM analyses of slag samples revealed the loss of structure showing massive coagulated texture with severely weathered structures. However, some cubic grains of magnetite, which is hard to leach because of its spinel and cubic structure, were also observed as shown in Figure 6.2.

These observations confirmed the role of the mixed mesoacidophilic bacterial consortium in Cu recovery. Furthermore, mineralogical analysis by means of FE-SEM elemental mapping and EDAX studies provided better insights into the copper recovery patterns from the bioleached slag residue, indicating that the bioleached residues were devoid of copper, which clearly confirmed the efficacy of the microbial leaching over physical beneficiation.

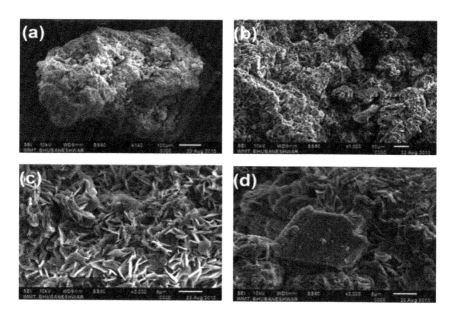

FIGURE 6.2 (a) Massive and patchy nature of the slag residue after bioleaching, (b) Fibrous, needle, and petal shapes of granules after bioleaching, (c) Petal-shaped granules in the bioleached residue, (d) Part of cubic crystal remaining in the bioleached residue. (Reprinted with kind permission from Springer Science+Business Media: *Korean J. Chem. Eng.*, Extraction of copper from copper slag: Mineralogical insights, physical beneficiation and bioleaching studies, 32(4), 2015, 667–676, doi:10.1007/s11814-014-0298-6, Panda, S., Mishra, S., Rao, D.S., Pradhan, N., Mohapatra, U., Angadi, S., and Mishra, B.K.)

Potysz et al. (2016) compared the feasibility of metal bioleaching from Cu-metallurgical slags using heterotrophic bacteria, *Pseudomonas fluorescens* (DSM 50091), known to produce siderophores and lithoautotroph, and *Acidithiobacillus thiooxidans*, known to produce sulfuric acid from reduced sulfuric compounds. The effect of particle size and pulp density of slag on the bioleaching of metals by these organisms was evaluated. Two types of slags, granulated amorphous and massive crystalline with particle sizes of <0.3 mm and 1–2 mm, were chosen for the experiments. pH profile studies suggested that the acidic pH (pH 2.5) conditions of the *A. thiooxidans* experiments and that of *P. fluorescens* were at circumneutral pH (7.0).

A. thiooxidans was found to be efficient in leaching of both slags. Up to 79% Cu, 76% Zn, and 45% Fe were extracted from crystalline slag at 1 wt% pulp density and particle size <0.3 mm after 21 days of incubation. In contrast, with amorphous slag, 81% Cu, 79% Zn, and 22% Fe were extracted at 1% pulp density and 1–2 mm particle size. For *P. fluorescens*, low leaching efficiencies were observed with the extraction of 10% Cu, 4% Zn, and 0.3% Fe for crystalline slag at 1% pulp density and particle size <0.3 mm, and only 4% Cu, 3% Zn, and 0.7% Fe for amorphous slag. The higher pH values during the *P. fluorescens* incubations were speculated to be the reason for poor leaching. The author suggests the further optimization of parameters like pulp, density, particle size, and pH to improve the leaching efficiency.

6.2.2 Pb/Zn Slags

Various physical, chemical, and biological factors in the system, like the nature of the contaminated material, pulp density, temperature, pH, concentrations of dissolved oxygen and carbon dioxide, oxidation–reduction potential (ORP), composition of the medium, bacterial strain, and cell concentration, have a considerable effect on bioleaching efficiency (Ojumu et al. 2006, Halinen et al. 2009a,b). Solution pH plays a key role in metal solubilization during bioleaching and is rarely controlled at a set value in the process (Yang et al. 2003). The buffering capacity of pulp density can bring the change in pH and can also affect the gas–liquid mass transfer rate in the biological system, which may result in inhibition of the growth of the microorganisms and greatly affect the metal extraction rate (Mousavi et al. 2005). Moreover, the particle size of the materials is also an important factor in evaluating both preprocess cost and metal recovery as it can affect the activity of microbes by affecting physical attrition, availability of materials for leaching, and mass transfer (Deveci 2004).

In a preliminary study, Guo et al. (2008) used an orthogonal experiment to study the metal leaching efficiency based on pH, pulp density, and temperature. It was found that the maximum leaching rate in the biological system can be reached when pH ranges from 1.0 to 2.0 and pulp density ranges from 5% to 15% at a temperature of 65°C. The effect of pH, pulp density, and particle size on solubilization of heavy metals, such as As, Fe, Cu, Zn, Pb, and Mn, from a Pb/Zn smelting slag using indigenous moderate thermophilic bacteria was evaluated in the study by Guo et al. (2010).

The indigenous moderate thermophilic bacteria employed in these experiments included *Bacillus* spp., *Sporosarcina* spp., and *Pseudomonas* spp. Bioleaching experiments showed that the pH of the solution had a significant influence on the solubilization of As, Fe, Cu, Pb, and Zn in slag when the pH value ranged from 1.0 to

2.0; however, it had only a slight influence on Mn. The optimized pH of the solution for bioleaching of As, Cu, Fe, and Zn from the Pb/Zn smelting slag fell in the range of 1.25–1.5. Consistent with the pH, pulp density also had a significant effect on metal solubilization. The lower the pulp density, the higher was the extraction of As, Fe, Cu, Pb, and Zn, but not Mn. The pulp density of 10% was found to be appropriate for the leaching process. Further, the particle size of the slag had a slight influence on the metal solubilization when the particle size was less than 0.83 mm. With the optimum condition of pH 1.5, 10% pulp density, and <0.83 mm particle size, about 86%–91% of As, 90%–93% of Cu, 90%–94% of Mn, and 81%–87% of Zn were bioleached from the slag after 6 days.

The surface morphological characteristics of the residues after 6 days of bioleaching were analyzed by SEM. The results showed the formation of small holes and pits located around bacteria. It was assumed that the bacteria adhering tightly to the surface of the slag either directly eroded the minerals to form numerous holes and pits, or excreted some bioactive metabolites such as extracellular polymeric substances, which directly/indirectly erode the minerals and mobilize the metals from the Pb/Zn smelting slag. Thus, SEM results reveal that although the metal solubilization can be improved by acid leaching, the activities of the ingenious bacteria significantly contributed to the metal extraction.

Cheng et al. (2009) studied the metal recovery from Pb/Zn smelting slag and applied a culture-independent RFLP separation of 16S rRNA genes approach to determine the community characteristics and the dominant bacteria in the bioleaching system. The bacteria employed for the experiments were collected from an old Pb/Zn smelting slag dumping site in Hunan Province of China. 82%–84% of Al, 85%–91% of As, 86%–88% of Cu, 85%–95% of Mn, 85%–90% of Fe, 95%–97% of Zn, and only 5% of Pb were leached from the Pb/Zn smelting slag at 65°C, pH 1.5, and pulp density 5%.

Phylogenetic analysis revealed that the bacteria in the bioleaching system mainly fell among *Firmicutes* (54.32%), *Gammaproteobacteria* (35.80%), and *Betaproteobacteria* (9.88%), and the dominant bacteria are affiliated with *Bacillus* spp. (33.33%), *Sporosarcina* spp. (20.99%), and *Pseudomonas* spp. (16.05%), which account for 70.37% of the detected microbial population.

Autotrophic bacteria are not well known for the bioleaching of metals like lead and arsenic, and hence mostly heterotrophic bacteria are used by supplementing expensive energy sources like yeast extract, thus increasing the bioleaching cost. To overcome this shortcoming, Wang et al. (2015) proposed the use of cheap sulfur and/or pyrite as an energy source for autotrophic bacteria during the bioleaching of Pb/Zn smelting slag. Also, its mechanism was studied and explained.

Lead/zinc smelting slag contains not only high doses of lead and zinc, but also many minor and trace heavy metals, such as Cd, As, Cr, Hg, In, Ag, Cu, Ni, and Co (Liu et al. 2008, Cheng et al. 2009, Guo et al. 2010). The acidophilic sulfur-oxidizing bacteria (SOB) *Acidithiobacillus thiooxidans* and iron-oxidizing bacteria (IOB) *Leptospirillum ferriphilum* were used as pure cultures or mixed culture for the experiments. Sulfur and pyrite were supplemented as respective energy sources alone or in combination for pure and mixed cultures referred to as S-SOB, P-IOB, and MS-MC. Noncontact bioleaching was performed by using the slag encapsulated with a dialysis bag.

Maximum dissolution efficiency of 90% was obtained for Zn independent of the bioleaching system, whereas it was only 12% for Pb. Seventy-one percent of In was extracted by the MS-MC system and 25% of As by the S-SOB system. Around 86% of Cd was extracted across three systems; however, none of the systems were found to dissolve Ag. The extraction efficiencies of the target metals followed the order Zn > Cd > In > As > Pb > Ag. Zn, Cd, and Pb rapidly reached the maximum extraction efficiencies after 24 h of contact and the final efficiencies were independent of the leaching system used, suggesting that they might be bioleached by the same mechanism of acid dissolution.

Although the mechanism of bioleaching for As mobilization was an acid dissolution, the formation of insoluble $FeAsO_4$ reduced the extraction of As in MS-MC and P-IOB. MS-MC yielded the highest extraction efficiency of In because of the coexistence of acid dissolution and Fe^{3+} oxidation. The combined work of acid dissolution and Fe^{3+} oxidation in MS-MC was responsible for the highest extraction efficiency of In. The presence of Ag in residual form as $AgPb_4(AsO_4)_3$ and $AgFe_2S_3$, which were very refractory and immune from attack by H^+, Fe^{2+}, and Fe^{3+}, inhibited its bioleaching by any of the systems involved.

6.3 BIOLEACHING OF FOUNDRY SAND

Foundries use high-quality silica sand for molding and casting operations, which is reused and recycled many times. Sand is used because it can absorb and transmit heat while allowing gases generated during the thermal degradation of binders to pass through the grains. When it is not possible to reuse it further, it turns into waste foundry sand (WFS). The automotive industry and its parts suppliers are the major generators of foundry sand. According to the literature, foundries generate approximately 9–12 million metric tons of WFS around the world (Abichou et al. 2004, Guney et al. 2006). Some of these sands have a high metal concentration which can cause pollution (Winkler and Bolshakov 2000). Heavy metals in WFS include chromium, cadmium, mercury, lead, nickel, arsenic, and silver (Ji et al. 2001). Due to the lack of landfilling space and its increasing costs, utilization of waste material and by-products is the only alternative. The landfilling of WFS can pollute groundwater (Riediker et al. 2000, Siddique et al. 2009, Miguel et al. 2012). The majority of molding sands are not considered to be hazardous in nature (Deng and Tikalsky 2008, Siddique and Noumowe 2008, Siddiquea et al. 2010, Penkaitis and Sígolo, 2012). Nevertheless, recycling is desirable due to regulations on disposal and also the economic burden of landfilling. There remains a dearth in the availability of reports on the concentration of metals present in WFS, as well as their biological recycling.

Recently, Jadhav et al. (2017) applied the culture supernatant of *Acidithiobacillus thiooxidans* to recovery of metals from waste foundry sand. The two-step bioleaching process was employed as it avoids direct contact of microorganisms with the metal-containing waste, thereby protecting the microbial cells from metal toxicity and improving the bioleaching efficiency. *At. thiooxidans* supernatant was thus used for bioleaching of WFS. Also, the effect of various physicochemical parameters like temperature, shaking speed, and increasing pulp density on the bioleaching process was studied. The results revealed complete recovery of Mg, Si, P, and Cu and nearly 98% Zn recovery during the process in 24 and 72 h, respectively. Change in temperature had no influence on the metal leaching; however, the efficiency increased with increase in shaking speed up to 150 rpm.

Further, the authors studied the phytotoxic effects of this treated and untreated WFS using *V. radiata*. The untreated WFS completely inhibited the growth of *V. radiata*, whereas the use of bioleached WFS in combination with soil (50:50) showed growth similar to that of the control with only soil. This shows the usefulness of pretreatment of WFS with *At. thiooxidans* culture supernatant. Thus, the bioleaching of WFS with *At. thiooxidans* was found to be effective and the treated WFS can be used for agricultural purposes (Jadhav et al. 2017).

6.4 BIOLEACHING OF INDUSTRIAL WASTE DUST

6.4.1 SMELTER DUST

Oliazadeh et al. (2006) explored the use of mixed culture of *Acidithiobacillus ferrooxidans* and *Acidithiobacillus thiooxidans* for the leaching of the copper dust emanating from smelting furnaces of Sarcheshmeh Copper Mine. The effects of culture medium and pulp density on copper leaching were studied in shake flasks. 9K medium and 1% pulp density were found to be optimum. Eighty-seven and 38% recovery of copper were obtained with 5% pulp density in inoculated and control flasks, respectively, after 22 days of incubation.

In another study, Massinaie et al. (2006) investigated the bioleaching of copper from melting furnace dust of Sarcheshmeh Copper Mine. A sulfuric acid leaching process was carried out before bioleaching to know the amount of acid leachable copper. Before bioreactor studies, shake flask experiments were carried out using a consortium of *Acidithiobacilli* and the effect of various parameters like culture medium, pulp density, and bacterial inoculation rate was studied. The increase in the pulp density resulted in higher toxicity and shear stress and decreased the mass transfer, inevitably reducing the process rate and copper recovery, and thus required more microorganisms and richer nutrient medium. In the agitated bioreactors, appropriate conditions like continuous aeration, effective mixing, and high mass transfer of gases, and the possibility of continuous control of operating variables, caused an evident increase in metal extraction as compared to the shake flask. Eighty-seven and 38% of Cu were recovered by bacterial leaching in shaking flasks and chemical leaching after 22 days, respectively, whereas in bioreactors, under the same conditions, 91% Cu was recovered within 6.5 days.

Bioleaching of copper from the flue dust of the Sarcheshmeh copper smelter was investigated in a two-stage series of airlift bioreactors (Bakhtiari et al. 2008a). Airlift reactors are employed/selected in many processes due to their simple design and construction, low power input, low shear, good mixing characteristics, and high heat and mass transfer (Chisti and Moo-Young 1987, Chisti, 1989). In airlift reactors, the different volumes of gas retained in the riser and the downcomer create a pressure gradient that forces the fluid from the bottom of the downcomer toward the riser, creating liquid velocity and mixing.

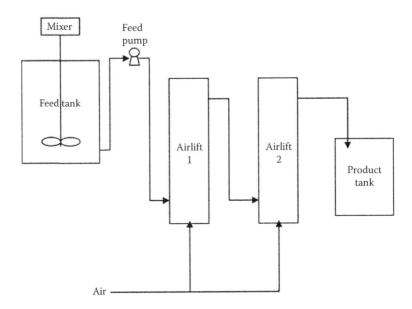

FIGURE 6.3 The design of airlift reactor. (Reprinted from *Hydrometallurgy*, 90(1), Bakhtiari, F., Zivdar, M., Atashi, H., and Seyed Bagheri, S.A., Bioleaching of copper from smelter dust in a series of airlift bioreactors, 40–45, doi:10.1016/j.hydromet.2007.09.010, Copyright 2008, with permission from Elsevier.)

The mixed culture of mesophilic, iron-sulfur-oxidizing bacteria (*A. ferrooxidans*, *A. thiooxidans* and *L. ferrooxidans*) isolated from acid mine drainage liquor of natural sulfide ore and adapted to high pulp density and copper content was employed in this process. A two-stage continuously operated airlift reactor (6.2 L) with an internal loop (as shown in Figure 6.3) comprised of a column of slurry divided into a gas-sparged riser and a downcomer, which was not aerated. The reactors were connected in series such that the exit stream of the first reactor was fed to the second reactor by means of gravity overflow. Also, they were fitted with condensers to minimize evaporation. The tests were started in batch mode in the first reactor, followed by the second reactor on the 12th day. The system was operated in batch mode until the ORP in both reactors reached its maximum level, after which the system was shifted to continuous mode operation.

During the 150th day of operation, the effects of different parameters such as pulp density, retention time, and temperature on the mesophile bioleach performance of the copper sulfide-rich dust were evaluated after preleaching with dilute acid. Pulp densities of 2% and 4% gave the same ORP in both reactors. However, increasing the average pulp density to 7% generated an unstable oxidation-reduction potential in the first bioreactor at 34°C. An increase in bioleaching intensity (i.e., a shortened retention time or increased feed pulp density) was detrimental to copper recovery. The ORP in the reactors decreased with increasing feed pulp densities and increased with increasing temperature. Copper recovery of 90%, 89%, and 86% were obtained at 2%, 4%, and 7% pulp densities with mean retention times of 2.7, 4 and 5, days at 32°C, respectively. At higher temperature (38°C) and 7% pulp density copper extraction was increased to 91% after 5 days mean retention time (Bakhtiari et al. 2008a).

In another study, Bakhtiari et al. (2008b) employed aerated stirred tank reactors for recovery of copper from smelter dust in a continuous two-stage process using adapted mixed bacterial culture. The system consisted of a feed pulp tank and two aerated stirred tank bioreactors (2.1 L working volume), with a container at the end for collecting the products. The tests were started in batch mode in the first reactor, followed by the second reactor on the eighth day. After the ORP attained a maximum level, the system was switched to continuous mode. During the 180th day of operation, the effects of pulp densities and residence times on the final copper recovery and redox potential were determined. Once established in the pilot plant, employing standard bioleach conditions, the mesophilic bacterial culture exhibited relatively stable leaching performance over the period of pilot plant operation. Despite the acid produced during the oxidation of sulfide minerals (especially pyrite) in copper concentrates, the net process was found to be acid consuming. Lower pulp densities (2% w/v and 4% w/v) resulted in a stable redox potential in both reactors. However, unstable redox potential was created in the first reactor when pulp density was increased to 7%. For copper, 89.7%, 90.3%, and 86.8% were recovered with residence times of 2.7, 4, and 6 days at pulp densities of 2%, 4%, and 7% (w/v), respectively (Bakhtiari et al. 2008b).

Further, Bakhtiari et al. (2011) studied the kinetics of copper bioleaching from the Sarcheshmeh preleached copper smelter dust using a mixed native culture of *A. ferrooxidans*, *A. thiooxidans*, and *L. ferrooxidans* isolated from acid mine drainage. The effect of pulp density, nutrient medium, temperature, and the amount of pyrite added to the bioleaching media on the bioleaching were assessed using a one factor at a time procedure.

The ability of the mixed culture isolates to grow on different concentrations of copper dust was evaluated. The mixed culture was able to grow at low pulp densities (below 2%) of copper dust; however, the growth was hampered at 5% pulp density due to increase in waste concentration and copper content. To overcome this problem, the culture was exposed gradually to increased pulp density and it was able to tolerate 7% (w/v) preleached dust and 20 g/L of copper after a 2-month period of adaptation. For the pulp density of 2%, 3%, 4%, and 7%, maximum copper recovery for control and bioleaching were 42.2% and 90.1% in 23 days, 45.9% and 83.1% in 20 days, 45.2% and 89.2% in 29 days, and 43.4% and 81.9% in 35 days, respectively.

The increase in pulp density resulted in a prolonged lag phase of bacterial growth along with an increase in acid consumption, the toxicity of metal ions, copper concentration, and shear stress accordingly decreasing the ORP and copper recovery. Pyrite addition to higher pulp densities hampered the ORP due to the high shear stress, toxicity of metal ions, and precipitation of jarosite. Although Norris nutrient medium increased the bacterial lag phase, it was found to be the most suitable medium because of higher copper extraction and lower concentrations of basal salts, making it economical. Pyrite addition and higher temperatures did not have a significant effect on the ORP and copper recovery. Increasing pulp density increased acid consumption, but the addition of pyrite did not have a considerable effect on the amount of acid consumption and thus no role in the acid production during copper dust bioleaching.

A Michaelis–Menten-type equation was used to relate the pulp density to the bioleaching rate. The equation has been given as

$$V = \frac{V_m \cdot S}{\left(K_s + S\right)} \tag{6.12}$$

where
  V is the extraction rate of metals
  V_m is the maximum metal extraction rate
  S is the pulp density
  K_s is the Michaelis constant

This constant gives an idea about the affinity of the bacteria to the mineral substrate. These results confirmed that iron dissolved better than copper from the copper flue dust, because $V_{m,Fe}$ is 28 times higher than $V_{m,Cu}$. The higher value of $K_{s,Cu}$ than $K_{s,Fe}$ might indicate a preference of the microorganisms for the copper from the dust.

The bioleaching data showed that at pH 1.8 and a pulp density less than 7%, the dissolution of copper followed the shrinking core kinetic model and the process was limited by diffusion of lixiviant at a temperature of 32°C and up to 4% of solid concentration. With the pulp density of 7%, however, the process was shown to be reaction limited as it may interfere with the mass transfer of oxygen and carbon dioxide, resulting in the change of the kinetic model to a combination of diffusion- and reaction-controlled mechanisms.

The authors suggest that this process is promising to cope with the problem of dust accumulation in the plant. Also, the limitations of this process can be solved with an increase in aeration and agitation in a laboratory reactor (Bakhtiari et al. 2008a,b).

6.4.2 EAF Dust

Acidithiobacillus ferrooxidans was utilized for bioleaching of zinc and iron from solid wastes (electric arc furnace steelmaking dust) at the Isdemir iron and steel plant (Bayat et al. 2009). The non-adapted bacteria extracted 6% and 2% of Zn and Fe, respectively, after 21 days, whereas adapted bacteria extracted 25% Zn and 16% Fe and hence were used in subsequent experiments. The effects of particle size, solids concentration, and pH on the efficiency of the bioleaching process were investigated. In each test, several variables were determined to assess the efficiency of leaching, including slurry pH and redox potential, temperature, bacteria population, and concentrations of zinc and iron in solution. The results revealed that pulp solids concentration, slurry pH, and solids particle size were important parameters in the bacterial leaching process. Maximum recovery of Zn (35%) and Fe (37%) was observed at pH 1.3 and a solids concentration of 1% w/v (Bayat et al. 2009).

6.4.3 ESP Dust

Jena et al. (2012) explored the ability of *Aspergillus niger, Aspergillus fumigatus,* and *Aspergillus flavus* to bioleach heavy metals from electrostatic precipitator (ESP) dust of a coal-based sponge iron plant. Pulp density significantly affected the bio-leaching capacity of fungus, inhibiting the leaching at higher pulp density due to increased toxicity of metals in ESP dust. Irrespective of fungal species, treatments with 5% pulp density showed significantly lower pH compared to treatments with 10% pulp density. The employed fungal species were able to dissolve Zn, Cu, Pb, Mn, and Fe from the ESP dust at 5% and 10% pulp densities over a period of 28 days. Metal leaching was in the order of *A. niger > A. flavus > A. fumigatus.* The constant decrease in pH was observed during the incubation period. Zn, Cu, Pb, Mn, and Fe were recovered at 81%, 76%, 74%, 72%, and 52% recovered, respectively. While fungal leaching requires a longer period of operation than chemical/acid leaching, it achieves a higher extraction of heavy metals at a lower cost. The results revealed that *A. niger* may be used for dissolution of Cu, Fe, Zn, Mn, and Pb; *A. fumigatus* for dissolution of Zn and Cu; and *A. flavus* for dissolution of Zn, Cu, and Pb from ESP dust generated by sponge iron plants.

6.4.4 Industrial Filter Dust

Bioreactor studies (stirred tank reactor) on leaching of zinc from filter dust were carried out using *Penicillium simplicissium* (Burgstaller et al. 1992). Fungi are suit-able for leaching of filter dust as they can tolerate high metal concentrations and the leaching is carried out at pH 2–7, while dust generally raises the pH of the medium. Zinc solutions of 0.41 M were obtained after 9 days of leaching. Doubled zinc con-centration was recovered in the bioreactor as compared to shake-flask experiments. The proton originating from citric acid produced by fungi played an important role in the solubilization of zinc. Higher concentrations of zinc and citric acid at the end of the leaching period resulted in the precipitation of both. If the proper cheap substrate is obtained for the growth of fungus, this process can be successfully applied at the industrial scale.

In another study, Müller et al. (1995) studied the leaching of industrial filter dust using three different microorganisms. The authors investigated if the secreted amino acids play a role in leaching of Zn from the zinc oxide–containing indus-trial dust. *Pseudomonas putida, Corynebacterium glutamicum,* and *Penicillium simplicissium* were employed for the same using a one-step process. *P. putida* and *P. simplicissium* are known to dissolve metals, whereas *C. glutamicum* is known to produce amino acids. The results revealed that all the organisms solubilized the zinc from the dust; however, they excreted higher amounts of citric acid than amino acids. Hence, citric acid rather than amino acids was the lixiviant in metal solubilization. Also, it was observed that *P. putida* was more resistant to the heavy metal containing filter dust.

6.5   BIOLEACHING OF INDUSTRIAL WASTE SLUDGE

Solisio et al. (2002) investigated the bioleaching of Zn and Al from two different industrial waste sludges (sludge 1 and sludge 2) using *Thiobacillus ferrooxidans*. Sludge 1 was a dust from the iron-manganese alloy production in an electric furnace and sludge 2 was a sludge coming from a process treatment plant of aluminum anodic oxidation. Even though the initial pH was 8, the sulfuric acid produced by *T. ferrooxidans* maintained the strongly acidic conditions (pH 2.5–3.5), which also played a role in metal solubilization.

At the initial pulp density of 1%, 76% and 78% of zinc and aluminum were solubilized from sludge 1 and 2, respectively, albeit the lag phase was not similar in both cases, suggesting a different affinity of microorganisms for the solid phase. The increase in the initial slurry concentration resulted in approximately 72%–73% of Zn recovery; however, it decreased continuously for Al. At 3% initial sludge concentration only 15% of Al was extracted. These results were confirmed by the increase in the lag phase that resulted in 2–4 days for zinc solubilization and 12–14 days for aluminum removal. Two kinetic models were proposed to describe the course of the metal dissolution.

Further, the authors suggest that as organisms utilize a substrate containing reduced sulfur for the production of sulfuric acid, this process can be applied when the substrate contains a source of sulfur, or when the addition of ferrous sulfate obtained as a by-product is possible.

6.6   CONCLUSION

Various metal production industries generate metal-laden wastes in the form of slags, dust, fumes, etc. These metals can be recovered by various methods. Biohydrometallurgy is one such process which is both eco-friendly and effective. Diverse groups of microorganisms are employed for bioleaching of these metals. Optimization of various process parameters can increase the efficiency of bioleaching. Also, many authors found the two-step bioleaching process efficient for metal recovery from industrial wastes. However, often the recovery is affected by the jarosite formation, and further research is required in order to overcome it. Furthermore, the problem of the drastic effects of increasing pulp density remains persistent and must be studied in detail.

REFERENCES

Abichou, T., C. H. Benson, T. B. Edil, and K. Tawfiq. 2004. Hydraulic conductivity of foundry sands and their use as hydraulic barriers. In *Recycled Materials in Geotechnics, Geotechnical Special Publication*, A. H. Aydilek and J. Wartman, pp. 127. Baltimore, MD: ASCE.

Bakhtiari, F., H. Atashi, M. Zivdar, and S. A. S. Bagheri. 2008b. Continuous copper recovery from a smelter's dust in stirred tank reactors. *International Journal of Mineral Processing* 86(1–4): 50–57. doi:10.1016/j.minpro.2007.10.003.

Bakhtiari, F., H. Atashi, M. Zivdar, S. Seyedbagheri, and M. H. Fazaelipoor. 2011. Bioleaching kinetics of copper from copper smelters dust. *Journal of Industrial and Engineering Chemistry* 17(1): 29–35. doi:10.1016/j.jiec.2010.10.005.

Bakhtiari, F., M. Zivdar, H. Atashi, and S.A. Seyed Bagheri. 2008a. Bioleaching of copper from smelter dust in a series of airlift bioreactors. *Hydrometallurgy* 90(1): 40–45. doi:10.1016/j.hydromet.2007.09.010.

Bayat, O., E. Sever, B. Bayat, V. Arslan, and C. Poole. 2009. Bioleaching of zinc and iron from steel plant waste using *Acidithiobacillus ferrooxidans*. *Applied Biochemistry and Biotechnology* 152(1): 117–126. doi:10.1007/s12010-008-8257-5.

Burgstaller, W., H. Strasser, H. Woebking, and F. Schinner. 1992. Solubilization of zinc oxide from filter dust with *Penicillium Simplicissimum*: Bioreactor leaching and stoichiometry. *Environmental Science & Technology* 26(2): 340–346. doi:10.1021/es00026a015.

Cheng, Y., Z. Guo, X. Liu, H. Yin, G. Qiu, F. Pan, and H. Liu. 2009. The bioleaching feasibility for Pb/Zn smelting slag and community characteristics of indigenous moderate-thermophilic bacteria. *Bioresource Technology* 100(10): 2737–2740. doi:10.1016/j.biortech.2008.12.038.

Chisti, Y. 1989. *Airlift Bioreactors*. New York: Elsevier.

Chisti, Y. and M. Moo-Young. 1987. Airlift reactors: Characteristics applications and design considerations. *Chemical Engineering Communications* 60: 195–242.

Deng, A. and P. J. Tikalsky. 2008. Geotechnical and leaching properties of flowable fill incorporating waste foundry sand. *Waste Management* 28(11): 2161–2170. doi:10.1016/j.wasman.2007.09.018.

Deveci, H. 2004. Effect of particle size and shape of solids on the viability of acidophilic bacteria during mixing in stirred tank reactors. *Hydrometallurgy* 71(3): 385–396. doi:10.1016/S0304-386X(03)00112-9.

Dutrizac, J. E. 1980. The physical chemistry of iron precipitation in the zinc industry. In *Lead-Zinc-Tin '80*, J. M. Cigan, T. S. Mackey, and T. J. O'Keefe (Eds.), pp. 532–564. New York: AIME.

Guney, Y., A. H. Aydilek, and M. M. Demirkan. 2006. Geoenvironmental behavior of foundry sand amended mixtures for highway subbases. *Waste Management* 26(9): 932–945. doi:10.1016/j.wasman.2005.06.007.

Guo, Z. H., Y. Cheng, G. Z. Qiu, X. D. Liu, and F. K. Pan. 2008. Optimization on bioleaching of metal values from Pb/Zn smelting slag. *The Chinese Journal of Nonferrous Metals* 18: 923–928. (In Chinese.)

Guo, Z., L. Zhang, Y. Cheng, X. Xiao, F. Pan, and K. Jiang. 2010. Effects of pH, pulp density and particle size on solubilization of metals from a Pb/Zn smelting slag using indigenous moderate thermophilic bacteria. *Hydrometallurgy* 104(1): 25–31. doi:10.1016/j.hydromet.2010.04.006.

Halinen, A.-K., N. Rahunen, A. H. Kaksonen, and J. A. Puhakka. 2009a. Heap bioleaching of a complex sulfide ore: Part I: Effect of pH on metal extraction and microbial composition in pH controlled columns. *Hydrometallurgy* 98(1): 92–100. doi:10.1016/j.hydromet.2009.04.005.

Halinen, A.-K., N. Rahunen, A. H. Kaksonen, and J. A. Puhakka. 2009b. Heap bioleaching of a complex sulfide ore: Part II. Effect of temperature on base metal extraction and bacterial compositions. *Hydrometallurgy* 98(1): 101–107. doi:10.1016/j.hydromet.2009.04.004.

Jadhav, U., C. Su, M. Chakankar, and H. Hocheng. 2017. An ecofriendly approach for leaching of metals from waste foundry sand using *Acidithiobacillus thiooxidans* culture supernatant. *Journal of Environmental Management* (In press).

Jena, P. K., C. S. K. Mishra, D. K. Behera, S. Mishra, and L. B. Sukla. 2012. Does participatory forest management change household attitudes towards forest conservation and management? *African Journal of Environmental Science and Technology* 6(5): 208–213. doi:10.5897/AJEST11.362.

Ji, S., L. Wan, and Z. Fan. 2001. The toxic compounds and leaching characteristics of spent foundry sands. *Water Air and Soil Pollution* 132: 347–364.

Kaksonen, A. H., L. Lavonen, M. Kuusenaho, A. Kolli, H. Närhi, E. Vestola, J. A. Puhakka, and O. H. Tuovinen. 2011. Bioleaching and recovery of metals from final slag waste of the copper smelting industry. *Minerals Engineering* 24(11): 1113–1121. doi:10.1016/j.mineng.2011.02.011.

Kaksonen, A. H. and J. A. Puhakka. 2007. Sulfate reduction based bioprocesses for the treatment of acid mine drainage and the recovery of metals. *Engineering in Life Sciences* 7(6): 541–564. doi:10.1002/elsc.200720216.

Kaksonen, A. H., S. Silja, J. A. Puhakka, E. Peuraniemi, S. Junnikkala, and O. H. Tuovinen. 2016. Chemical and bacterial leaching of metals from a smelter slag in acid solutions. *Hydrometallurgy* 159: 46–53. doi:10.1016/j.hydromet.2015.10.032.

Khalezov, B. D., V. A. Nezhivykh, and L. A. Ovchinnikova. 2005. Semiindustrial trials of zinc recovery by a sodium hydrosulfide from a liquors of heap leaching. *Min. Inform. Anal. Bull. (Science Tech. J.)* 4: 278–279. (in Russian.)

Kinnunen, P. H.-M., S. Heimala, M.-L. Riekkola-Vanhanen, and J. A. Puhakka. 2006. Chalcopyrite concentrate leaching with biologically produced ferric sulphate. *Bioresource Technology* 97(14): 1727–1734. doi:10.1016/j.biortech.2005.07.016.

Liu, Y.-G., M. Zhou, G.-M. Zeng, X. Wang, X. Li, T. Fan, and W.-H. Xu. 2008. Bioleaching of heavy metals from mine tailings by indigenous sulfur-oxidizing bacteria: Effects of substrate concentration. *Bioresource Technology* 99(10): 4124–4129. doi:10.1016/j.biortech.2007.08.064.

Massinaie, M., M. Oliazadeh, and A. Seyed Bagheri. 2006. Biological copper extraction from melting furnaces dust of sarcheshmeh copper mine. *International Journal of Mineral Processing* 81(1): 58–62. doi:10.1016/j.minpro.2006.06.005.

Mehta, K. D., B. D. Pandey, and Premchand. 1999. Bio-assisted leaching of copper, nickel and cobalt from copper convertor slag. *Materials Transactions, JIM* 40: 214–221.

Miguel, R. E., J. A. Ippolito, A. B. Leytem, A. A. Porta, R. B. Banda Noriega, and R. S. Dungan. 2012. Analysis of total metals in waste molding and core sands from ferrous and non-ferrous foundries. *Journal of Environmental Management* 110: 77–81. doi:10.1016/j.jenvman.2012.05.025.

Mousavi, S. M., S. Yaghmaei, M. Vossoughi, A. Jafari, and S. A. Hoseini. 2005. Comparison of bioleaching ability of two native mesophilic and thermophilic bacteria on copper recovery from chalcopyrite concentrate in an airlift bioreactor. *Hydrometallurgy* 80: 139–144. doi:10.1016/j.hydromet.2005.08.001.

Müller, B., W. Burgstaller, H. Strasser, A. Zanella, and F. Schinner. 1995. Leaching of zinc from an industrial filter dust with *Penicillium, Pseudomonas* and *Corynebacterium*: Citric acid is the leaching agent rather than amino acids. *Journal of Industrial Microbiology and Biotechnology* 14: 208–212. doi:10.1007/BF01569929.

Muravyov, M. I., N. V. Fomchenko, and T. F. Kondrat'eva. 2011. Biohydrometallurgical technology of copper recovery from a complex copper concentrate. *Applied Biochemistry Microbiology (Prikl. Biokhim. Mikrobiol.)* 47: 607–614.

Muravyov, M. I., N. V. Fomchenko, A. V. Usoltsev, E. A. Vasilyev, and T. F. Kondrat'Eva. 2012. Leaching of copper and zinc from copper converter slag flotation tailings using H_2SO_4 and biologically generated $Fe_2(SO_4)_3$. *Hydrometallurgy* 119: 40–46. doi:10.1016/j.hydromet.2012.03.001.

Ojumu, T. V., G. E. Searby, and G. S. Hansford. 2006. A review of rate equations proposed for microbial ferrous-iron oxidation with a view to application to heap bioleaching. *Hydrometallurgy* 83(1): 21–28. doi:10.1016/j.hydromet.2006.03.033.

Oliazadeh, M., M. Massinaie, A. Seyed Bagheri, and A. R. Shahverdi. 2006. Recovery of copper from melting furnaces dust by microorganisms. *Minerals Engineering* 19(2): 209–210. doi:10.1016/j.mineng.2005.09.004.

Panda, S., S. Mishra, D. S. Rao, N. Pradhan, U. Mohapatra, S. Angadi, and B. K. Mishra. 2015. Extraction of copper from copper slag: Mineralogical insights, physical beneficiation and bioleaching studies. *Korean Journal of Chemical Engineering* 32(4): 667–676. doi:10.1007/s11814-014-0298-6.

Penkaitis, G. and J. B. Sígolo 2012. Waste foundry sand. Environmental implication and characterization. *Geologia., Série. cientifica USP* 12: 5–70.

Potysz, A., P. N. L. Lens, J. van de Vossenberg, E. R. Rene, M. Grybos, G. Guibaud, J. Kierczak, and E. D. van Hullebusch. 2016. Comparison of Cu, Zn and Fe bio-leaching from Cu-metallurgical slags in the presence of *Pseudomonas fluorescens* and *Acidithiobacillus thiooxidans*. *Applied Geochemistry* 68: 39–52. doi:10.1016/j.apgeochem.2016.03.006.

Riediker, S., S. Ruckstuhl, M. J.-F. Suter, A. M. Cook, and W. Giger. 2000. P -Toluenesulfonate in landfill leachates: Leachability from foundry sands and aerobic biodegradation. *Environmental Science & Technology* 34(11): 2156–2161. doi:10.1021/es9913434.

Romero, R., A. Mazuelos, I. Palencia, and F. Carranza. 2003. Copper recovery from chalcopy-rite concentrates by the BRISA process. *Hydrometallurgy* 70(1): 205–215. doi:10.1016/S0304-386X(03)00081-1.

Shen, H. and E. Forssberg. 2003. An overview of recovery of metals from slags. *Waste Management* 23(10): 933–949. doi:10.1016/S0956-053X(02)00164-2.

Siddique, R., G. de Schutter, and A. Noumowe. 2009. Effect of used-foundry sand on the mechanical properties of concrete. *Construction and Building Materials* 23(2): 976–980. doi:10.1016/j.conbuildmat.2008.05.005.

Siddique, R. and A. Noumowe. 2008. Utilization of spent foundry sand in controlled low-strength materials and concrete. *Resources, Conservation and Recycling* 53(1): 27–35. doi:10.1016/j.resconrec.2008.09.007.

Siddiquea, R., G. Kaur, and A. Rajor. 2010. Waste foundry sand and its leachate character-istics. *Resources, Conservation and Recycling* 54(12): 1027–1036. doi:10.1016/j.resconrec.2010.04.006.

Solisio, C., A. Lodi, and F. Veglio'. 2002. Bioleaching of zinc and aluminium from indus-trial waste sludges by means of *Thiobacillus ferrooxidans*. *Waste Management* 22(6): 667–675. doi:10.1016/S0956-053X(01)00052-6.

Sukla, L. B., R. N. Kar, and V. V. Panchanadikar. 1995. Bioleaching of copper converter slag using *Aspergillus niger* isolated from lateritic nickel ore. *International Journal of Environmental Studies* 47(2): 81–86. doi:10.1080/00207239508710947.

Vestola, E. A., M. K. Kuusenaho, H. M. Närhi et al. 2012. Biological treatment of solid waste materials from copper and steel industry. In *Materials Challenges and Testing for Supply of Energy and Resources*, T. Böllinghaus et al. (Eds.), pp. 287–296. Berlin, Germany: Springer-Verlag.

Wang, J., Q. Huang, T. Li, B. Xin, S. Chen, X. Guo, C. Liu, and Y. Li. 2015. Bioleaching mech-anism of Zn, Pb, In, Ag, Cd and As from Pb/Zn smelting slag by autotrophic bacteria. *Journal of Environmental Management* 159: 11–17. doi:10.1016/j.jenvman.2015.05.013.

Winkler, E. S. and A. A. Bolshakov. 2000. Characterization of foundry sand waste. Technical report, Chelsea Centre for Recycling and Economic Development, University of Massachusetts, Chelsea, MA.

Yang, X. W., Q. F. Shen, and Y. Y. Guo. 2003. *Biohydrometallurgy* Beijing, China: Press of Metallurgical Industry. (In Chinese.)

7 Recycling of Solar Electricity Waste

7.1 INTRODUCTION

The global increase in the use of fossil fuels for energy has resulted in potentially harmful climate changes and an increase in the generation of hazardous waste. Consequently, the interest in alternative and sustainable energy systems has increased, resulting in the innovation of solar energy as a future energy source. Today is the era of electricity generation using solar cells. Being a green technology, its use has dispersed in every corner of the world. Various types of solar cells are developed in order to convert the solar energy into electrical energy. In solar thin-film cells indium-tin-oxide film is used as semiconductor material (Angerer et al. 2009). In Cd-Te technology (IRENA 2015), the cadmium-telluride (CdTe) thin film is used as a semiconductor, whereas copper indium gallium selenide cells are based on a compound semiconductor made of copper, indium, gallium, and selenide (Kemell et al. 2005, Grandell and Höök 2015).

According to the European Photovoltaic Industry Association (2008), more than 9200 MW of solar photovoltaic systems were installed by the end of 2007, which reached to the cumulative value of more than 138 GW at the end of 2013 (Cucchiella et al. 2015). The amount of PV waste in 2017 is estimated to be 870 tons and the total amount of PV waste is estimated to be about 3,000,000 tons in 2035, of which about 800,000 tons will belong to the CdTe technology, and 45,000 tons to the copper indium gallium selenide technology (BIO Intelligence Service 2011).

The advantages of solar cells include no emission of any matter into the environment during operation; extremely long operation period (estimated average: 25 years), minimum maintenance, robust technique, and aesthetic aspects (Klugmann-Radziemska 2012). The low production costs and low energy and material requirement during the production of CdTe and copper indium disulfide/diselenide (CIS) thin film are increasing its popularity (Berger et al. 2010).

However, taking into consideration the escalating number of cells being installed every year, end-of-life management will become increasingly important. Nevertheless, they are creating a revolution, and the day will soon arrive when these huge quantities of solar cells will reach their end of life, and hazardous contents used in their panels will pose a threat to the environment, if not disposed of properly or recycled. To date, very few reports are available on the recycling of solar cells, that too using mechanical, pyro-, or hydrometallurgical techniques.

The company Solar World (Deutsche Solar) applies thermal and etching treatment to recycle solar modules (Wambach 2001). The First Solar company recycles CdTe thin-film modules using a combination of mechanical and chemical process steps (Fthenakis and Wang 2006).

The solar cell modules contain various metals like silver, copper, tin, aluminum, zinc, and the hazardous material lead (Kang et al. 2012). Moreover, these solar energy technologies are based on the use of rare metals like indium, gallium, and ruthenium, many of which are unavailable as primary ores and are found as by-products associated with primary base metal ores. The natural reserves of many of these ores are scarce and hence this will impact future production due to their limited availability (Andersson 2000, Feltrin and Freundlich 2008, Fthenakis 2009, Fthenakis et al. 2009). Initiating a technology for recycling and reusing spent photovoltaic panels is thus inevitable.

7.2 BIOLEACHING OF METALS FROM SOLAR ELECTRICITY WASTE

Very few reports are available on the recovery of metals from solar cells, but there are no reports available for the recovery of metals from solar electricity using biohydrometallurgy. Chakankar et al. (2017) utilized *Thiobacillus thiooxidans*, *Thiobacillus ferrooxidans*, and *Aspergillus niger* for recovery of metals from waste solar cells. The solar cells were ground and sieved and obtained powder was used in the bioleaching studies. A two-step bioleaching process was employed in the present study wherein solar cell powder was added to the culture supernatant after the growth of microorganisms, in order to avoid the toxic effect of heavy metals. The effect of various physicochemical parameters like temperature, shaking speed, and pulp density on bioleaching was also studied. *T. ferrooxidans* culture supernatant was found to be effective for bioleaching of metals from solar cell powder. B, Mg, Si, Ni, and Zn were removed with 100% efficiency within a reasonable time. *T. ferrooxidans* recovered 100% Cr, Mn, and Cu at 60°C; however, only 41.6% and 13.4% of Al and Te were extracted, respectively. These results suggest the higher temperature optimum for metal recovery from solar cells in the use of the organism. Bearing in mind the objective of recovering high-purity materials from the recycling process, bioleaching is the most appropriate and environmentally friendly method. Hence, application of this process for recovery of metals from solar cells will allow their reuse without affecting the environment.

7.3 CONCLUSION

The use of solar cells for energy production is the most promising and green technology. However, the end-of-life management of the widely installed solar cells needs to be given attention. They contain various base and precious metals that can be recovered and reused once their modules reach the end of their life cycle. Industrial processes to recycle both crystalline silicon cells and thin-film modules are already established and can retrieve substances like glass and aluminum, as well as semiconductor materials such as silicon, copper, indium, cadmium, and tellurium. However, they have their own disadvantages. Nevertheless, the biohydrometallurgical process will be an eco-friendly, efficient, and low-cost approach for the recovery of metals from waste solar cells. More research needs to be done in order to investigate various microorganisms having the ability to recover the metals from solar cells and develop the process so as to expand it at a commercial level.

REFERENCES

Andersson, B. A. 2000. Materials availability for large-scale thin-film photovoltaics. *Progress in Photovoltaics: Research and Applications* 8: 61–76.

Angerer, G., L. Erdmann, F. Marschneider-Weidemann et al. 2009. *Einfluss des branchenspezifischen Rohstoffbedarfs in rohstoffintensiven Zukunftstechnologien auf die zukünftige Rohstoffnachfrage.* Stuttgart, Germany: Fraunhofer IRB Verlag.

Berger, W., F. G. Simon, K. Weimann, and E. A. Alsema. 2010. A novel approach for the recycling of thin film photovoltaic modules. *Resources, Conservation and Recycling* 54(10): 711–718. doi:10.1016/j.resconrec.2009.12.001.

BIO Intelligence Service. 2011. Study on photovoltaic panels supplementing the impact assessment for a recast of the WEEE directive. Final report April 14, 2011.

Chakankar, M., C. Su, and H. Hocheng. 2017. Recovery of metals from solar cells by bioleaching. In *Proceedings of Fifth International Conference on Advanced Manufacturing Engineering and Technologies*, V. Majstorovic and Z. Jakovljevic (Eds.). NEWTECH 2017. Lecture Notes in Mechanical Engineering. Cham, Switzerland: Springer.

Cucchiella, F., I. D'Adamo, and P. Rosa. 2015. End-of-life of used photovoltaic modules: A financial analysis. *Renewable and Sustainable Energy Reviews* 47: 552–561. doi:10.1016/j.rser.2015.03.076.

EPIA, European Photovoltaic Industry Association. 2008. http://www.epia.org (accessed March 25, 2015).

Feltrin, A. and A. Freundlich. 2008. Material considerations for terawatt level deployment of photovoltaics. *Renewable Energy* 33(2): 180–185. doi:10.1016/j.renene.2007.05.024.

Fthenakis, V. 2009. Sustainability of photovoltaics: The case for thin-film solar cells. *Renewable and Sustainable Energy Reviews* 13(9): 2746–2750. doi:10.1016/j.rser.2009.05.001.

Fthenakis, V. M. and W. Wang. 2006. Extraction and separation of Cd and Te from cadmium telluride photovoltaic manufacturing scrap. *Progress in Photovoltaics: Research and Applications* 14(4): 363–371. doi:10.1002/pip.676.

Fthenakis, V., W. Wang, and H. C. Kim. 2009. Life cycle inventory analysis of the production of metals used in photovoltaics. *Renewable and Sustainable Energy Reviews* 13(3): 493–517. doi:10.1016/j.rser.2007.11.012.

Grandell, L. and M. Höök. 2015. Assessing rare metal availability challenges for solar energy technologies. *Sustainability* 7(9): 11818–11837. doi:10.3390/su70911818.

International Renewable Energy Agency (IRENA). *Renewable Energy Technologies: Cost Analysis Series.* Volume 1: Power Sector. Solar Photovoltaics. Available online: https://www.irena.org/DocumentDownloads/Publications/RE_Technologies_Cost_AnalysisSOLAR_PV.pdf (accessed March 25, 2015).

Kang, S., S. Yoo, J. Lee, B. Boo, and H. Ryu. 2012. Experimental investigations for recycling of silicon and glass from waste photovoltaic modules. *Renewable Energy* 47: 152–159. doi:10.1016/j.renene.2012.04.030.

Kemell, M., M. Ritala, and M. Leskelä. 2005. Thin film deposition methods for CuInSe 2 solar cells. *Critical Reviews in Solid State and Materials Sciences* 30(1): 1–31. doi:10.1080/10408430590918341.

Klugmann-Radziemska, E. 2012. Current trends in recycling of photovoltaic solar cells and modules waste. *Chemistry Didactics Ecology Metrology* 17(1–2): 89–95. doi:10.2478/cdem-2013-0008.

Wambach, K. 2001. Recycling von Solarmodulen. In *Gestalten mit Solarzellen*, S. Rexroth (Ed.), Heidelberg, Germany: C.F: Müller Verlag.

8 Recycling of Thermal Power Generation Wastes

8.1 INTRODUCTION

During the combustion of coal for energy production, fly ash is generated as an industrial by-product, which is an environmental pollutant. Extensive use of coal for power generation has augmented the generation of fly ash throughout the world, consequently creating a disposal problem and an environmental threat (Ahmaruzzaman 2010). A substantial amount of fly ash is disposed of in ash ponds and landfills, which may result in severe environmental degradation of land in future. The presence of Si, Al, Fe, Cd, Cr, Mn, B, As, Cu, Zn, and Mo in this by-product is hazardous to living organisms (Wong and Wong 1990, Rai et al. 2004). Thus, fly ash should be treated prior to its disposal. This chapter provides an insight into the biohydrometallurgical recovery of metals from fly ash.

8.2 BIOLEACHING OF THERMAL POWER GENERATION WASTE

8.2.1 Microorganisms for Metal Recovery from Fly Ash

Bosshard et al. (1996) introduced the use of *Aspergillus niger* for bioleaching of fly ash and compared it with chemical leaching. In a one-step process, fungus was grown in the presence of 50 g/L fly ash, while in a two-step process, after fungal growth in the absence of fly ash, the medium was filtered and fly ash was added to this filtrate for bioleaching studies. In addition to this, fungal leaching was compared with chemical leaching using commercial citric acid.

A. niger was found to grow on fly ash up to 10% (w/v). The higher concentration of fly ash resulted in the longer lag phase for fungus. Organic acid profiling suggested the production of citric acid in the absence of fly ash but the production of gluconic acid in the presence of fly ash. Citrate production is due presumably to a lack of mainly manganese (among other elements such as Zn, Co, Fe, and Ca) in the medium without fly ash (Karavaiko et al. 1977, Hahn et al. 1993). This leads to an inhibition of the enzymes of the TCA cycle except for the citrate synthase. In addition, a low pH has been reported to favor citrate production, whereas higher pH values (in the presence of fly ash) stimulate oxalic and gluconic acid production (Hahn et al. 1993, Hoffmann et al. 1993). The highest relative solubilization values (in percentage of ash added) were obtained in a 3% (w/v) fly ash suspension.

In the one-step process, 283 mM gluconic acid was produced in the presence of fly ash after 526 h and pH decreased from 9.3 to 4.8. In the two-step process, initially, the pH decreased from 5.5 to 2.3 due to the production of citric acid (110 mM) by fungus after 500 h. However, after filtration and the addition of fly ash pH increased from 2.3 to 6.4 due to the alkalinity of fly ash. In the one-step process, 57% of Cd, 52% of Pb and Zn, 41% of Mn, 33% of Cu, 30% of Al, 12% of Fe, and less than 10% of Cr and Ni were extracted, and the in two-step process, after 1 day, 81% of Cd, 66% of Zn, 57% of Cu, 52% of Pb, 32% of Mn, 27% of Al, and less than 10% of Cr, Fe, and Ni, respectively, were extracted at 6% (w/v) fly ash concentration. Leaching efficiency for Al, Fe, Mn, and Ni was significantly higher and for Cd, Cu, and Zn was significantly lower for the one-step process as compared to the two-step process. There was no difference in Cr leaching; however, Pb solubilization decreased in the two-step process from 52% to 23% after 23 days.

Chemical leaching with the commercial citric acid of equal molarity was found to be slightly more efficient than microbial leaching for all the metals, except for Cd, which was extracted to nearly 100%. Further, the authors prefer the use of the two-step process due to the ease of handling and enabling the process optimization for increasing bioleaching efficiency. Thus, this could be a methodical process to improve the environmental quality of the ash with respect to its reuse for construction (Bosshard et al. 1996).

In another study, Brombacher et al. (1998) performed the semicontinuous leaching of fly ash in a laboratory-scale leaching plant (LSLP) by using a mixed culture of *Thiobacillus thiooxidans* and *Thiobacillus ferrooxidans*. To reduce the capital and maintenance cost of the pilot plant for fly ash bioleaching, the treatment time for high pulp densities must be reduced without additional acidification. To achieve this, a semicontinuous three-stage leaching plant was designed, which comprised three serially connected reaction vessels, reservoirs for a fly ash suspension, a bacterial stock culture, and a vacuum filter unit. The LSLP was operated with an ash concentration of 50 g/L, and the mean residence time was 6 days (2 days in each reaction vessel). After RV-C, the pulp was transported by gravity flow into a 2 L vacuum glass filter unit, and the particle-free, metal-rich solution was further collected in a 5 L collecting vessel.

Economically valuable metals like Zn and Al were leached up to 81% and 52% respectively, whereas highly toxic Cd was completely solubilized (100%). Cu, Ni, and Cr were leached at 89%, 64%, and 12%, respectively. Metals can be biotically released from fly ash by mechanisms such as direct enzymatic reduction, the indirect action resulting from extracellular metabolic products, or acid formation (nonenzymatic dissolution), as previously shown in an anaerobic system (Francis and Dodge 1990). The role of *T. ferrooxidans* in metal mobilization was examined in a series of shake-flask experiments, which suggested that the release of copper present in the fly ash as chalcocite (Cu_2S) or cuprite (Cu_2O) was dependent on the metabolic activity of *T. ferrooxidans*, whereas other metals, such as Al, Cd, Cr, Ni, and Zn, were solubilized by biological sulfuric acid. Chemical (abiotic) leaching with 5 N H_2SO_4 showed significantly increased solubilization only for Zn. This developed LSLP could be the favorable step toward a high-capacity pilot plant for detoxification of fly ash and its further use in construction purposes and cost-effective metal recovery (Brombacher et al. 1998).

Next, Krebs et al. (2001) employed sulfur-oxidizing bacteria for metal bioleaching from municipal solid waste incineration fly ash. Sulfur-oxidizing *Thiobacilli* (pure or mixed culture) were cultivated with different amounts of the fly ash for several weeks with or without anoxic sewage sludge. In this study, anoxic sewage sludge was used as a nutrient source, as it contains many nutrients like ammonium and trace elements, is available cheaply in huge quantities, and also contains reduced S-compounds that can act as an energy source for *Thiobacilli*. Furthermore, it also contains indigenous *Thiobacilli* which could leach heavy metals and contribute to the acidification of solution. A consortium of *Thiobacilli* in sewage sludge grew faster as compared to the pure cultures or cultures with only sewage sludge as inoculum. The addition of sewage sludge not only enhanced the growth rate of *Thiobacilli* but also elevated the acidification of the suspension.

The addition of fly ash increases the pH of the solution, consequently slowing the acidification process. But the use of anoxic sewage sludge promoted the reduction of the time for fly ash by almost half. *Thiobacilli* were able to tolerate up to 8% (w/v) of fly ash. Thus, the mixed culture was able to extract 80% of Cd, Cu, and Zn, 60% of Al, and nearly 30% of Fe and Ni from the fly ash (Krebs et al. 2001).

The interactions between bacteria, metabolic products, fly ash particles, and leaching products during bioleaching of aluminum and iron from coal fly ash (CFA) by *Thiobacillus thiooxidans* were explored by Seidel et al. (2001). It was observed that the bacterial growth and the amount of metals leached from the CFA are coupled through biological and chemical interactions, which involve various components in this system. The increase in pH after the addition of CFA (up to 10% w/v) hampered the growth rate of *T. thiooxidans* culture. But after the adaptation of bacteria, the growth rate was similar to that of control culture, which indicated that the metals from CFA participated in the extension of the adaptation phase and did not affect the cell growth directly. Further, it was observed that in the presence of CFA, EPS production was 55–70 mg/L as compared to the increased control medium (25 mg/L polysaccharides).

SEM analyses showed the adherence of bacteria to CFA particles in fresh as well as 10-day-old cultures, which may be the reason for slow initial bacterial growth due to the dilution of the energy source (sulfur particles) by CFA particles. Precipitation of calcium sulfate on the cells and on the surface of sulfur particles may also interfere with cell growth. The extraction rate and level of aluminum and iron by sulfuric acid and *T. thiooxidans* were found to be similar, suggesting the role of sulfuric acid and at the same time showing the insignificant effect of other cellular metabolites (Seidel et al. 2001).

In another study, bioleaching of fly ash at various pulp densities was carried out using *Aspergillus niger* and compared with the chemical leaching using various organic and inorganic acids (Wu and Ting 2006). Both the one-step and two-step processes were employed. In one-step bioleaching, fungal growth was observed at 1% and 2% pulp density, whereas there was no growth at higher pulp densities. However, fungal growth was observed up to 4% pulp density in the two-step process. Even though the biomass concentration was the same at the end of the incubation period, the results demonstrated the long lag period during the one-step process and the early stationary phase in the two-step process. Both one-step and two-step bioleaching experiments showed similar metal extraction yield for 1% (w/v) of fly ash pulp density (80%–100% for Al, Mn, and Zn, 60%–70% for Cu and Pb, and about 30% for Fe). Optimum pulp density for bioleaching was observed at 1% (w/v), and an increase in pulp density decreased the leaching efficiency. A significant increase in the concentration of gluconic acid with an increase in fly ash pulp density in both processes clearly shows its role in metal extraction. However, the concentration of both citric and oxalic acids remained low as compared to the control. Gluconic acid, the main lixiviant, was produced only in the presence of fly ash. Enzyme glucose oxidase is reported to have optimum pH 4.5–6.5 and is inactivated at pH below 3.0 (Dellweg 1983). The increase in pH of the system due to fly ash activates the glucose oxidase, which converts glucose to hydrogen peroxide and, finally, hydrolyzes it to gluconic acid. Therefore, gluconate was only produced in the presence of fly ash (in one-step and two-step bioleaching) but not in the pure culture, as the pH of the latter decreased below 3 after 2 days of incubation.

The inhibition of citric acid production in the presence of fly ash was possibly due to the presence of metal ions (such as zinc, manganese, iron, and copper) in the ash (Burgstaller and Schinner 1993, Grewal and Kalra 1995). Among these metal ions, manganese is reported to strongly inhibit citric acid accumulation by stimulating the enzymes in the TCA cycle (Bowes and Mattey 1979, Grewal and Kalra 1995). In addition, the higher pH of the fly ash suspension also causes *A. niger* to accumulate gluconic acid by inhibiting citric acid secretion.

In the case of spent-medium leaching, an increase in alkalinity due to the addition of fly ash decreased the leaching yield. Also, the major leaching agent in the spent medium was citric acid.

When compared with the chemical leaching at 1% pulp density, Mn and Zn were extracted more efficiently by fungal leaching and the extraction yield for Al was similar for both. However, Cu extraction yield was lower in bioleaching than chemical leaching. Although bioleaching generally requires a longer period of operation compared to chemical leaching, it achieves a higher removal efficiency for some heavy metals. These results suggest that bioleaching by *A. niger* may be an alternative or adjunct to conventional physicochemical treatment of municipal solid waste (MSW) fly ash for the removal of hazardous heavy metals (Wu and Ting 2006).

Ishigaki et al. (2005) investigated the bioleaching of municipal solid waste incineration fly ash using individual and mixed cultures of a sulfur-oxidizing bacteria (SOB) and an iron-oxidizing bacteria (IOB). The IOB possessed high metal leachability but low tolerance to the addition of ash, which lowered the cell activity, resulting in a decrease in oxidation-reduction potential and increase in pH. In contrast, SOB showed more tolerance to increased ash concentration, though the metal leachability was limited. Mixed culture exhibited high metal leachability at 1% ash, leaching 67% Cu, 78% Zn, and 100% for Cr and Cd. Furthermore, comparably high leaching of Cu (42%) and Zn (78%) was observed even in the 3% ash cultures. This may be due to the combination of the high leaching potential of IOB and the high tolerability to ash addition of SOB. In mixed culture, the acidic and oxidizing condition remained stable throughout the experimental period, indicating the higher buffering ability of the mixed culture than that of pure cultures. Ferric iron remained at a high level in the mixed culture, and the metal leaching was enhanced by redox mechanisms coupling with the leaching by sulfate. Increased concentration of $FeSO_4$ in the mixed culture enhanced the Cr, Cu, and As leaching. The optimum sulfur concentration for As and Cr leaching was 5 g/L, and that for Cu leaching was 2 g/L. Further, the presence of the degradable and nondegradable organic compound had no significant impact on metal leaching by a mixed culture other than Zn. This is advantageous from the viewpoint of in situ application of the process (Ishigaki et al. 2005).

Yang et al. (2008) compared the one-step and two-step bioleaching process for heavy metal leaching from municipal solid waste incineration fly ash using *Aspergillus niger* (AS 3.879). In one-step bioleaching, *A. niger* was incubated with fly ash at the beginning of the bioleaching, and in two-step bioleaching, *A. niger* was precultured for a few days before adding the fly ash into the culture medium. The pH values, organic acid production, and metal extraction yield of the two methods were compared. The addition of fly ash decreased the pH of the medium. During the one-step experiments, the lag phase was observed as the result of the adaptive duration of *A. niger* spores to the culture medium containing fly ash. High fly ash concentrations resulted in a further increase in the lag phase. However, this lag phase was not observed in the two-step process as spores germinated into mycelium and secreted organic acids after preculturing of fungus for 2 days.

In a one-step bioleaching of 20 g/L fly ash, 98.7% of Mn, 87.6% of Cd, 69.7% of Cu, 68.5% of Zn, 42.1% of Cr, 31.7% of Fe, and 36.5% of Pb were extracted. The results demonstrated that one-step bioleaching was appropriate for the treatment of low concentrations of fly ash (10–20 g/L), whereas two-step bioleaching was suitable for the high fly ash concentrations (40–50 g/L).

In one-step bioleaching, fungus produced 36.6 mmol/L of gluconic acid after 336 h, subsequently decreasing the pH to 3.03. In two-step bioleaching, fungus was able to tolerate up to 50 g/L of fly ash and secreted 33.9 mmol/L of gluconic acid. Thus, the gluconic acid was found to be the major leaching agent in both processes. The TCLP results indicated the detoxification of fly ash which could be further safely disposed of or reused (Yang et al. 2008).

In continuation to this, Yang et al. (2009) studied the adaptation of highly metal-tolerant *Aspergillus niger* for bioleaching of fly ash. *A. niger* AS 3.40 and AS 3.879 strains were used for the same. One of the hurdles for large-scale application of fly ash bioleaching is the low tolerance of fly ash by organisms due to its toxicity. The main objective of the present study was the adaptation of *A. niger* to single- as well as multimetals solutions and their bioleachability thereafter, as multimetals are known to act synergistically and impose more toxicity than a single metal alone (Le et al. 2006). The Plackett–Burman design was used to select the significant toxic heavy metals in fly ash. When compared with the high pH, toxic heavy metals in the fly ash had a more inhibitory effect on fungal growth. The AS 3.879 strain showed a more multimetals tolerant nature than the AS 3.40 strain. The Plackett Burman design indicated that Al and Fe inhibited the growth of *A. niger* (AS 3.879 and AS 3.40) significantly, which may be attributed to the high concentration of those metals in fly ash. However, interestingly, Mg showed a positive effect on the growth of AS 3.879. The single-metal-adapted (Al and Fe) and multimetals-adapted AS 3.879 strain tolerated up to 3500 mg/L Al, 700 mg/L Fe, and 3208.1 mg/L multimetals, respectively.

The multimetals-adapted strain exhibited the highest fly ash tolerance up to 70 g/L as compared to single-metal-adapted or unadapted strains. It produced 37 mmol/L of citric acid, 1 mmol/L of oxalic acid, and 218 mmol/L of gluconic acid. After 288 h of bioleaching, 87.35% Cd, 21.5% Cr, 17.7% Fe, 64.8% Mn, 45.9% Pb, and 49.4% Zn were extracted. The TCLP tests revealed the successful detoxification of fly ash by the multimetals-adapted AS 3.879 strain (Yang et al. 2009).

In another study, Wang et al. (2009) demonstrated the effects of water-washing pretreatment on bioleaching of heavy metals from municipal solid waste incinerator fly ash using *A. niger*. The water-washing pretreatment prior to bioleaching of fly ash was introduced to reduce the bioleaching period and improve the metal extraction yield. The homogenized dry raw fly ash was washed with deionized water before the bioleaching process. The raw fly ash and water-washed fly ash were subjected to bioleaching by *A. niger* using the one-step and two-step processes at various fly ash concentrations.

The high concentration of alkali chloride and heavy metals in fly ash inhibited the fungal growth and resulted in the lag of the pH value drop (i.e., the increase of the lag phase) and a delay in the bioleaching process. Water-washing pretreatment extracted 50.6% of K, 41.1% of Na, 5.2% of Ca, and 1% of Cr from the fly ash. Due to the dissolution of alkali chlorides, which hold particles together, fly ash particles were smashed into smaller granules by the hydraulic flushing action caused by vibration. This reduced the lag phase and bioleaching period by 45% and 30%, respectively, in one-step bioleaching of 1% (w/v) fly ash. This step markedly increased the metal extraction yield in both the processes. For instance, 96% Cd, 91% Mn, 73% Pb, 68% Zn, 35% Cr, and 30% Fe were extracted from 1% water-washed fly ash, respectively, in two-step bioleaching (Wang et al. 2009).

Bankar et al. (2012) employed the tropical marine yeast *Yarrowia lipolytica* NCIM 3589 for bioleaching of fly ash. The effect of fly ash on the growth and morphology of metal-tolerant yeast was studied along with its leaching capacity. The yeast growth remains unaffected in the presence of 0.1%–0.3% fly ash; however, there was an evident decrease in the surface-to-volume ratio in response to toxic environmental stress factors. SEM analysis revealed the attachment of *Y. lipolytica* to the fly ash particles, which the authors suggest is mediated by hydrophobic interactions. Moreover, biofilms of *Y. lipolytica* cells were observed on the fly ash surfaces.

During the bioleaching experiments, the pH of the medium decreased to 2.7 as a result of organic acid production by yeast cells. Extensive foam formation in the presence of fly ash suggested the production of extracellular proteins that may be involved in bioleaching. Bioleaching of metals by *Y. lipolytica* was also compared with chemical leaching using citric acid. Yeast cells were found to be more efficient as compared to chemical leaching, and Cu was leached to a greater extent (59.41%) along with other metals (Zn, Ni, Cu, and Cr). TEM images (Figure 8.1) showed the accumulation of metals at the cell wall and cell membrane and in the cytoplasm of test cells. The metal accumulation efficiency of *Y. lipolytica* was Zn > Ni > Cu > Cr > Al > Si. Thus, the authors suggest the possible mechanisms of fly ash bioleaching by *Y. lipolytica* as: the production of citric acid as metal lixiviant, secretion of extracellular proteins in the presence of fly ash, and bioaccumulation (Bankar et al. 2012).

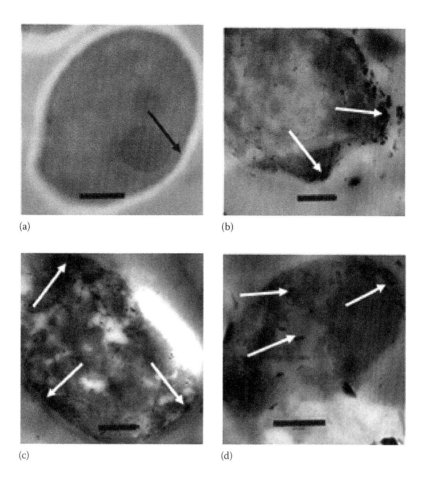

(a)   (b)

(c)   (d)

FIGURE 8.1   TEM images of *Y. lipolytica* NCIM 3589 (a) control cell. (b), (c), and (d) Presence of metal ions inside the cells (Magnifies × 25,000; *bar* equivalence, 0.50 μm). (From With kind permission from Springer Science + Business Media: *Appl. Biochem. Biotechnol.*, Bioleaching of fly ash by the tropical marine yeast, *Yarrowia lipolytica* NCIM 3589, 168, 2012, 2205, Bankar, A., Winey, M., Prakash, D., Kumar, A.R., Gosavi, S., Kapadnis, B., and Zinjarde, S.)

Various aspects like changes in fungal morphology, the fate of ash particles, and the precipitation of metallic salt crystals during bioleaching of municipal solid waste incineration fly ash by *Aspergillus niger* were studied by Xu et al. (2014). Bioleaching was conducted using both the one step and two step process. The fungi produce various organic acids like citric, oxalic, and gluconic acid in sugar media. However, citric acid was produced at a higher concentration in the absence of fly ash, while gluconic acid was produced at a higher concentration in the presence of fly ash and the concentration of oxalic acid was quite low.

EDX and XRD analyses confirmed the adsorption of calcium oxalate precipitates and fly ash particles on the surface of the fungi. The low concentration of oxalic acid in the bioleaching medium was possibly due to the immediate precipitation of insoluble metal oxalates. Precipitation of calcium oxalate apparently affected the bioleaching by weakening the fly ash, thus facilitating the release of other tightly bound metals in the matrix. Also, the released organic acid is available for complex formation with other metals as Ca concentration is decreased.

Although the mechanism of calcium oxalate hydrate precipitation was similar in both one-step and two-step bioleaching, the leaching rate of metals from fly ash was different. Bioleaching of fly ash was rapid in two-step as compared to one-step bioleaching due to the early formation of calcium oxalate hydrate. Also, the presence of organic acids prior to the addition of fly ash in two-step bioleaching rapidly decreased the pH of the medium. Further, the addition of fly ash after fungal germination reduced its toxicity toward fungal spore germination and fungal growth, thus speeding up the bioleaching process.

A remarkable effect of fly ash was observed on fungal growth and morphology by SEM analysis. In the absence of any stress factors, the fungi showed exuberant growth and intact morphology with linear hyphae having 2–4 µm diameter. However, in one-step bioleaching, fungal hyphae were found to be much larger and distorted with a 10 µm diameter and a swollen pellet structure. Some hyphae lost the linear structure and were abnormally short and swollen and showed highly branched distortion.

The presence of stress factors changes the network components, that is, internal and external signals at apical growth, affecting the shape and direction of growth. Excess metal ions in the growth environment cause tip swelling, an increase in branching, and thickness of transverse walls at subapical parts, as an adopted strategy to survive adverse conditions. Further, excess degradation or reduced synthesis of cell wall components may result in loosening of the cell wall, which in turn leads to swelling (Lanfranco et al. 2002). All these factors may be responsible for the observed morphological changes, which, however, have no impact on the observed morphology, although this may not have a significant impact on the organic acid production or leaching unless the enzymes involved in the mechanism are affected.

In two-step bioleaching, the diameter of hyphae was around 5 mm smaller. As the fungi had already grown and germinated before the addition of fly ash, the effect of fly ash on the fungus was not pronounced. However, the onset of distortion and the swollen structure of the hyphae occurred earlier in the presence of toxic metals in two-step bioleaching due to the earlier onset of growth. Notwithstanding this, the effect on bioleaching appears negligible, possibly due to the high production of organic acids before the addition of fly ash and exposure to toxic conditions (Xu et al. 2014).

Jadhav and Hocheng (2015) evaluated the use of *Aspergillus niger* culture supernatant for bioleaching of metals from thermal power plant fly ash. The water-washing treatment of alkaline fly ash reduced its pH to 8.5 from 10.5 along with the complete removal of sodium and 47%, 38.07%, 29.89%, and 11.8% removal of boron, calcium, magnesium, and potassium, respectively. The culture supernatant of *Aspergillus niger* 34770 was employed for the bioleaching process. One hundred percent removal of all the metals was obtained in 4 h, whereas 93%, 83%, 78%, and 70% of Cr, Ni, As, and Pb was removed, respectively. However, a further increase in incubation time did not increase the metal extraction. The authors suggest that the two-step process is advantageous as compared to direct bioleaching in terms of time required for bioleaching and also the toxicity of wastes to microbial cells. The results revealed the optimized process parameters as 4 h incubation time, 50 rpm shaking speed, 30°C temperature, and 1 g/100 mL solid/liquid ratio.

Soil polluted with fly ash exhibits phytotoxicity because of high heavy metal content. Phytotoxicity studies were carried out using mung bean (*Vigna radiata*) seeds in order to evaluate the effect of before and after bioleaching fly ash on its germination and growth. The results demonstrated the drastic effect of fly ash on the seed germination and root and shoot growth of *V. radiata*. At 20 g/100 mL fly ash concentration, no germination of *V. radiata* seeds was observed. With an increasing concentration of untreated fly ash, a gradual decrease in root/shoot length was observed. However, no such effect was observed during the use of bioleached fly ash as shown in Figure 8.2. After the bioleaching process, 78% (±0.19%) germination of *V. radiata* was observed with 20 g/100 mL fly ash. This indicates the effectiveness of the present bioleaching method for removal of phytotoxicity of fly ash (Jadhav and Hocheng 2015).

Ramanathan and Ting (2015) for the first time investigated and reported the copper leaching ability of indigenous alkaliphiles isolated from a fly ash landfill site with inherent pH tolerance and metal tolerance. The higher tolerance of the microorganisms for metal concentrations allows for bioleaching at a higher pulp density. Many studies report enhanced metal recovery from wastes by a mixed

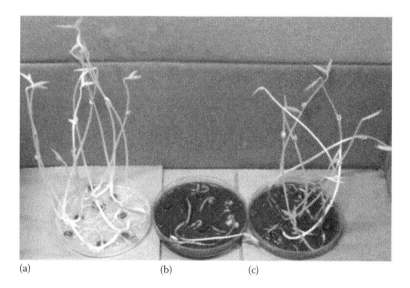

(a)   (b)   (c)

FIGURE 8.2 Phytotoxicity study: Effect of untreated and bioleached fly ash (10 g/100 mL) on *Vigna radiata*. (a) *Vigna radiata* grown in distilled water, (b) *Vigna radiata* grown in untrated fly ash (10 g/100 ml), (c) *Vigna radiata* grown in bioleached fly ash (10 g/100 ml). (From With kind permission from Springer Science + Business Media: *Appl. Biochem. Biotechnol.*, Analysis of metal bioleaching from thermal power plant fly ash by *Aspergillus niger* 34770 culture supernatant and reduction of phytotoxicity during the process, 175, 2015, 870, Jadhav, U.U. and Hocheng, H.)

consortium of bacteria (Ishigaki et al. 2005, Ilyas et al. 2007, Xiang et al. 2010). Four isolates, namely, *Agromyces aurantiacus* TRTYP3, *Alkalibacterium pelagium* TRTYP5, *Alkalibacterium* sp. TRTYP6, and *Bacillus foraminis* TRTYP17, were used as pure and mixed cultures to recover copper from fly ash at different pulp densities (1%–20% w/v).

The pH tolerance study suggests that TRTYP3, TRTYP5, and TRTYP6 are alkaliphilic, while TRTYP17 is alkali-tolerant, and hence these organisms are suitable for fly ash bioleaching. TRTYP5, TRTYP6, and TRTYP17 tolerated up to 10%, 20%, and 5% (w/v) of fly ash, respectively, while TRTYP3 showed poor fly ash tolerance (up to 1% (w/v)) and *Alkalibacterium* sp. TRTYP6 tolerated as high as 20% (w/v) fly ash. Each organism showed about 50% copper leaching from 1% (w/v) of fly ash. Glucose was the best carbon source, and the optimal pH was found to be 9.

Mixed culture of these bacteria resulted in higher leaching of copper. The optimal combination was TRTYP3, TRTYP5, TRTYP6, and TRTYP17 in the ratio 1:1:3:1, which leached 88%, 81%, 78%, 76%, 70%, and 55% Cu from 1%, 2.5%, 5%, 10%, 15%, and 20% (w/v) of fly ash. While Cu and Pb were bioleached into solution, Fe and Zn were precipitated. In a consortium, bacteria exist as a community and interact with each other to get acclimatized to the environmental conditions. This is advantageous for bioleaching as different mechanisms complement each other for efficient metal recovery. Moreover, increase in metabolite production and mineral adhesion due to intra- and intercellular communication in a microbial consortium *via* quorum sensing signals also plays an important role (Farah et al. 2005, Ruiz et al. 2008). High-performance liquid chromatography (HPLC) spectra exhibited the change in metabolite production, which is a sign of quorum sensing that may have played an important role in the growth and bioleaching by the mixed culture (Ramanathan and Ting 2015).

In another study, Ramanathan and Ting (2016) reported the isolation of indigenous bacteria from a local fly ash landfill site and evaluated their bioleaching ability. The incineration of MSW generated the fine residue called fly ash which rises with flue gasses and is collected using electrostatic precipitators. Several million tons of fly ash is generated annually around the world (Ramanathan and Ting 2014). The alkaline nature and high levels of toxic heavy metals have a detrimental effect on microbial growth as well as leaching ability. The use of acidophiles for fly ash bioleaching is not suitable because of its high pH and requires the acidification process. To avoid this hassle, the authors screened and isolated the indigenous autochthonous bacteria from the landfill site with inherent alkaline pH and fly ash tolerance and studied their capacity for bioleaching of fly ash. The bacteria were isolated from the eight fly ash samples collected from the Pulau Semakau Landfill site, located south of the main island of Singapore. Different media were selected for bacterial isolation based on relevant properties like pH and fly ash tolerance.

Thirty-eight total isolates were obtained from all the samples and different types of media. Eighteen isolates displayed notable metal leachability and pH and fly ash tolerance, which was further biochemically and genetically characterized. The 16s rRNA analysis revealed that 90% of the isolates belong to the phylum *Firmicutes*. Their dominance is attributed to their ability to grow in a metal-rich environment because of their capacity to activate mechanisms like strengthening the permeability barrier, activating metal efflux pumps, stimulating enzyme-based detoxification systems, promoting intra- and extracellular sequestration, and reducing the sensitivity of cellular targets to metal ions (Rathnayake et al. 2010). Bacteria from other phyla like *Bacteriodetes*, α-*Proteobacteria*, and *Actinobacteria* were also isolated. Further, the isolates were studied for their bioleaching potential, and Cu recovery and leaching kinetics were studied in detail. Six of the 18 isolates recovered about 20% Cu, whereas the others showed recovery around or less than 10%.

Authors also examined the isolates capable of growing at high fly ash concentrations. There are no reports of any microorganism, in municipal solid waste incineration fly ash bioleaching, that is capable of tolerating more than 10% w/v of fly ash. *Alkalibacterium* sp. TRTYP6 showed maximum fly ash tolerance at 20% w/v. The number of isolates found to exhibit maximum fly ash tolerance at 1%, 5%, 10%, and 20% w/v was 7, 5, 5, and 1, respectively. It also showed the growth over a pH range of 8–12.5.

Most of the microbes that are resistant to one metal are not resistant to another, and thus have the problem during bioleaching of fly ash which is invariable in its metal content. Comparatively, indigenous bacteria are naturally adapted to high metal content in waste and show harmonious tolerance to these metals (Errasquin and Vazquez 2003), with higher metal concentration in the waste leading to higher tolerance limits.

Shake-flask bioleaching experiments with fly ash showed nearly 52% Cu recovery by *Alkalibacterium* sp. TRTYP6. The results revealed the selective recovery of copper, which is beneficial in terms of further separation and purification of metal and reduces the problem of metal heterogeneity. Moreover, it also shows high fly ash tolerance and grows at alkaline pH, which is favorable for its large-scale bioleaching application as it obviates the need for acidification of fly ash. This will also help to improve the economics at a commercial level. Also, the after bioleaching TCLP analysis showed the metal values of Cd and Zn within regulated levels, suggesting the detoxification of fly ash (Ramanathan and Ting 2016).

The bioleaching ability of *Penicillium simplicissimum* for the recovery of metals from power plant residual ash (PPR ash) using different bioleaching methods such as one-step, two-step, and spent-medium bioleaching at 1% (w/v) pulp density was investigated by Rasoulnia et al. (2016). Additionally, the effects of thermal pretreatment on leaching of V, Ni, and Fe from PPR ash were studied. The power plant ash also acts as a source of economic value metals like V and Ni. *Penicillium simplicissimum* was employed for bioleaching of PPR ash owing to its high production of organic acids like citrate, gluconate, and oxalate. Thermal pretreatment of PPR ash at various temperatures was practiced due to its reported advantages like reducing the carbonaceous and volatile impurities of PPR ash, which are considered to be a growth-inhibitory factor to the fungus; increasing the vanadium content of PPR ash and thus having a liquor with a higher concentration of vanadium at the end of bioleaching; and lower acid consumption and consequently reaching a higher yield in the stage of precipitation of vanadium (Vitolo et al. 2001). This pretreatment was beneficial in terms of removal of the carbonaceous and volatile fraction of the ash, and also had a significant impact on fungal growth and metal leachability.

SEM analysis of PPR ash revealed that its smooth grains had undergone several microstructural changes in morphology after bioleaching and had a shriveled, rough surface with small grains and holes possibly due to the exposure to the acidic condition of the medium, the formation of metal–acid complexes, and the oxidation and reduction reactions. However, the thermal treatment resulted in agglomeration or crystallization of PPR ash grains and formation of an integrated matrix having a negative impact on metal leaching as it blocked the access of metabolites to the metal oxides.

XRD analysis showed the decrease in PPR ash mass and increase in the bulk concentration of metals after thermal treatment. Also, the oxidation of various metals at different temperatures and the volatilization of black carbon initially present in the ash resulted in a change in color of samples from gray to green and reddish brown.

The fungal growth in pure culture was investigated through measurement of produced organic acids *via* HPLC. The 15th day of the fungal growth was determined to be optimum for spent-medium bioleaching due to the maximum production of organic acids on that day. According to Amiri et al. (2012), the foremost fungal leaching mechanism is acidolysis, wherein under acidic conditions, the protons and the oxygen combine with water, consequently separating the metal from the surface. Complexolysis is an additional leaching mechanism of fungus that intensifies the metal ion solubility, which is formerly solubilized through acidolysis and facilitates the disengagement of metals from the surface *via* ligand exchange, polarizing the critical bonds. Nevertheless, the complexolysis mechanism is slower compared to acidolysis and becomes dominant under weakly acidic conditions.

The highest extraction yields of V and Ni were achieved for the original PPR ash, using spent-medium bioleaching in which nearly 100% of V and 40% of Ni were extracted. Thermal pretreatment had a negative effect on V and Ni leaching, whereas a positive effect was observed for Fe (maximum extraction yield 48.3%) at 400°C. The spent-medium bioleaching method showed the maximum recovery yields in all the experiments.

Furthermore, chemical leaching experiments were performed, using commercial organic acids at the same concentrations as those produced under the optimum condition of fungal growth (5237 ppm citric, 3666 ppm gluconic, 1287 ppm oxalic, and 188 ppm malic acid) for comparison with bioleaching. They showed that oxalic and citric acids play a more decisive role in the recovery of V due to their higher acidic nature. Fungal leaching was more efficient than chemical leaching. It was found that in comparison to chemical leaching, bioleaching improved V and Ni recovery up to 19% and 12%, respectively, suggesting the role of other fungal metabolites in bioleaching (Rasoulnia et al. 2016).

In continuation to the previous study (Rasoulnia and Mousavi 2016a,b) conducted the bubble column reactor studies to enhance the recovery of V and Ni from a PPR ash by optimizing the organic acid production by *Aspergillus niger*. Although the potential of a microorganism for metal recovery is investigated in shake-flask studies, the use of a bioreactor is the key step for developing a process from experimental surveys to industrial scale (Shahrabi-Farahani et al. 2014). Three influencing factors on the excretion of organic acids, aeration rate, sucrose concentration, and inoculum size, were optimized using central composite design (CCD) of response surface methodology (RSM).

A glassy bubble column bioreactor was used for the fermentation experiments. Water was pumped from a bath into the jacket around the bioreactor and circulated to maintain the temperature of the bioreactor at the desired temperature. Air flow was adjusted using an air compressor and a rotameter. Under the obtained optimized condition of aeration rate of 762.5 (mL/min), sucrose concentration of 101.9 (g/L), and inoculum size of 40 (mL/L), fermentation of the fungus was carried out for 7 days. 17, 185, 4539, 1042, and 502 (ppm) of oxalic, gluconic, citric, and malic acids were produced, respectively, at the end of the fermentation period. Further bioleaching tests were carried out using organic acid containing fungal spent medium with PPR ash pulp densities of 1, 2, 3, 5, 7, and 9 (%w/v) at 60°C and 130 rpm rotary shaking speed.

The results revealed that V (83%) was recovered more efficiently by produced organic acids than Ni (30%) even at a high pulp density of 9 (%w/v). Authors suggest that the lower Ni recovery was possibly due to the high concentration of oxalic acid in the spent medium obtained from bubble column bioreactor experiments. Reports suggest that some fungal metabolites such as oxalic acid react with nickel and form insoluble metal complexes like nickel oxalate, which in some cases act as an inhibitor, consequently slowing down the rate of metal dissolution (Amiri et al. 2012, Biswas et al. 2013).

The oxalic acid was a major fungal lixiviant in the bubble column bioreactor as a result of the shorter fermentation period, and thus the lower nickel recovery yield was obtained as compared to the authors' previous study in an Erlenmeyer flask, where citric acid was the main lixiviant due to the longer fermentation period of *Aspergillus niger* (Rasoulnia and Mousavi 2016a).

The results of the leaching kinetics of vanadium and nickel at a maximum pulp density of 9 (% w/v) during 7 days of incubation indicated that approximately 64% of vanadium and 28% of nickel were recovered on the first day of incubation. However, the prolonged leaching period slightly increased the nickel recovery to 34% up to the fifth day, which was again decreased to 30%. Precipitation of nickel in the presence of oxalic acid was the main reason for its low and almost constant leaching kinetics. In contrast, continuing leaching duration up to 7 days had a significant effect on vanadium recovery, which was enhanced up to 20%. Metal extraction kinetics was also investigated using the shrinking core model. The results indicated that the diffusion mechanism is the rate-controlling step. This is likely because of the relatively higher rate of acidolysis or low porosity of the PPR ash particles (Rasoulnia and Mousavi 2016b).

Funari et al. (2017) took the shake-flask bioleaching studies of fly ash to the reactor level and compared it with chemical leaching using H_2SO_4. A 2 L glass reactor equipped with a top-entered agitator and an aeration system, which supplied a continuous airflow from the bottom of the reactor, was constructed as shown in Figure 8.3. For bioleaching experiments, mixed acidophilic culture containing *At. ferrooxidans, At. thiooxidans, At. caldus, L. ferrooxidans, Sb. thermosulfidooxidans, Sb. Thermotolerans*, and some members of the *Alicyclobacillus* genus was employed. They produce sulfuric acid using metal sulfide phases found within fly ash and supplemented elemental sulfur as follows:

$$S^0 + 1.5O_2 + H_2O \rightarrow 2H^+ + SO_4^{2-} \tag{8.1}$$

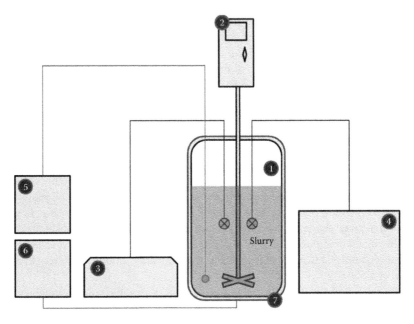

FIGURE 8.3 Experimental setup of leaching procedure in a 2 L glass reactor, equipped with titration system and benchtop meter for the continuous monitor of pH, redox and temperature: 1, reactor; 2, agitator; 3, benchtop meter; 4, titrator; 5, air supplier; 6, heating system; 7, hot water jacket. (Reprinted from *Waste Manage*. 60, Funari, V., Mäkinen, J., Salminen, J., Braga, R., Dinelli, E., and Revitzer, H., Metal removal from municipal solid waste incineration fly ash: A comparison between chemical leaching and bioleaching, 397–406, Copyright (2017). With permission from Elsevier.)

Prior to the reactor experiments, the mixed culture was adapted to increasing concentrations of fly ash. The reactor conditions for both chemical and bioleaching were similar. A rotation speed of 320 rpm and an airflow rate of 1.0 L/min were used. The reactor temperature was maintained at 30°C using a heated water jacket.

Prewashing of fly ash removed most of the water-soluble salts, which otherwise consume large amounts of acid and complicate (hamper) the separation of metals bonded to them during the leaching process. At the end of the experiments, both leaching methods resulted in comparable yields for Mg and Zn (>90%), Al and Mn (>85%), Cr (~65%), Ga (~60%), and Ce (~50%). Chemical leaching showed the best yields for Cu (95%), Fe (91%), and Ni (93%), whereas bioleaching was effective for Nd (76%), Pb (59%), and Co (55%). The bioleaching enhanced the removal of several hazardous elements such as As, Ba, Pb, and Sr, while chemical leaching is an option for the removal of Mo, Ni, and V.

When comparing the consumption of acid during both processes, the bioleaching process has an added value because of the finite consumption of sulfuric acid, which is nearly twice as low as chemical leaching, demonstrating the activity of sulfur-oxidizing bacteria and their ability to produce notable amounts of H_2SO_4. According to Equation 8.1 the biologic transformation of 1 ton of elemental sulfur produces approximately 3 tons of H_2SO_4. In U.S. dollars (2013), the price of elemental sulfur and imported H_2SO_4 product was 69 and 63 \$/ton, respectively (USGS, 2015). Therefore, chemical costs for bio-based H_2SO_4 are three times lower than the price of imported sulfuric acid.

The two leaching methods generated solids of different quality with respect to the original material as we removed and significantly reduced the amounts of metals and enriched solutions where metals can be recovered, for example, as mixed salts for further treatment. Compared to chemical leaching, the bioleaching halved the use of H_2SO_4, that is, a part of agent costs, as a likely consequence of bioproduced acid and improved metal solubility.

The H_2SO_4 production is a clear restraining factor, as it is required to supplement the sulfuric acid during the main leaching period (days 0–3) of bioleaching. Accordingly, the potential industrial application for fly ash bioleaching would consist of two reactors, the first one for improved biologic H_2SO_4 production from elemental sulfur (in the absence of fly ash) and the second reactor employing this bio-based lixiviant for fly ash chemical leaching. Further, authors have constructed the conceptual design of a hypothetical process for fly ash treatment as shown in Figure 8.4, which includes two optimized bioreactors ensuring closed circuits of washing water and an acidic solution, and suggests industrial uses of process by-products (Funari et al. 2017).

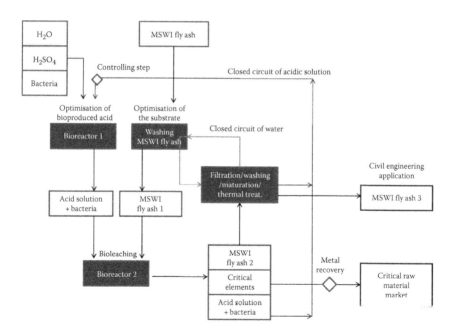

FIGURE 8.4 Design for a hypothetical process for the treatment of MSWI fly ash. (Reprinted from *Waste Manage.*, 60, Funari, V., Mäkinen, J., Salminen, J., Braga, R., Dinelli, E., and Revitzer, H., Metal removal from municipal solid waste incineration fly ash: A comparison between chemical leaching and bioleaching, 397–406, Copyright (2017). With permission from Elsevier.)

8.2.2 PROCESS OPTIMIZATION FOR METAL RECOVERY

Xu and Ting (2004) explored the impact of various factors affecting the bioleaching of fly ash in order to optimize the process conditions and improve the leaching efficiency. The four factors considered in this study were sucrose concentration, spore inoculum concentration, fly ash pulp density, and the time of the addition of fly ash to the *Aspergillus niger* culture.

The source of carbon and its concentration used for bioleaching are important as they not only support the microbial growth but also determine the yield of an organic acid which acts as a lixiviant for bioleaching. Inoculum size should be selected carefully as preculturing of microbes allows the attainment of high density and metabolite concentration before the addition of waste, which could be toxic to microbial growth. Further, the higher concentration of the solid pulp density results in a higher solid to liquid ratio, which no doubt will increase the concentration of heavy metals to be leached, but at the same time will hamper and inhibit the growth of microbes and subsequently acid production due to increased toxicity. Thus, the pulp density must be optimized.

The use of the "one-factor-at-a-time" technique does not reveal the interactions among all the factors, and hence the RSM, which simultaneously considers several factors at different levels, and may reveal corresponding interactions among these factors, is useful in this regard. Thus, the authors used the CCD for the optimization of this process. Thirty-one batch bioleaching tests were run under high and low values of the four abovementioned factors. Empirical models obtained through second-order (Taylor series approximation) regression provided the optimal bioleaching conditions. The sucrose concentration and pulp density were found to be more important factors than spore concentration and the time of the addition of the fly ash. The optimal sucrose concentration for metal leaching and gluconic acid production was about 150 g/L, a value 50% higher than Bosshard's medium, and the corresponding optimal pulp density was found to be 2.7%. Citric acid and gluconic acid attained the maximum concentrations of 51 and 281 mM, respectively. Al, Fe, and Zn leaching concentrations were 12.3, 12.3, and 77.6 ppm, respectively (Xu and Ting 2004).

Pangayao et al. (2015) studied the bioleaching of trace metals from coal ash (fly and bottom) using the mixed culture of *Acidithiobacillus albertensis* and *Acidithiobacillus thiooxidans*. Oxidation of sulfur and production of sulfuric acid during bacterial growth resulted in a decrease in pH of the solution from a range of 1.39 to 0.61. Nearly 56.69% Zn, 70.68% Mn, 79.86% Cr, 70.74% Fe, and 69.32% Cu were leached after 15 days at different variable parameters. Based on statistical analysis, the experimental data fit the quadratic model based on the Box–Behnken design of the experiment. The adjustment of initial solution pH favored microbial growth, thus showing the highest metal leachability at 5% pulp density. Further, based on the statistical analysis, initial inoculum was not preferred as the significant factor. At optimum parameters, 5% pulp density and 3 g sulfur, the metals leached were 56.88%, 70.88%, 85.01%, 74.44%, and 74.46% of Zn, Mn, Cr, Fe, and Cu, respectively.

In another study, Rasoulnia and Mousavi (2016a) investigated the use of *Aspergillus niger* and *Penicillium simplicissimum* for bioleaching of V and Ni from vanadium-rich power plant residue. The optimum fermentation period was decided on the basis of the organic acid production and pH changes by fungi. Before proceeding to the bioleaching experiments, the authors monitored the fungal growth and investigated different bioleaching methods, that is, one-step, two-step, and spent-medium bioleaching, the efficiency of which highly depends on the type of substrate and also the employed microorganism. Spent-medium leaching was selected as the most efficient method and was used further to optimize the conditions for fungal growth and organic acid production. The effect of various parameters like pulp density, leaching temperature, and duration as numeric factors and *Aspergillus niger* and *Penicillium simplicissimum* as categoric factors on the recovery of V and Ni was determined. Moreover, the interactions between these parameters were evaluated and optimized using the Box–Behnken design of RSM.

Both fungi were able to grow at a pulp density of 5 g/L, but the growth retarded at higher pulp density due to the increased toxicity of fly ash. Thus, the preliminary experimental data suggested the spent-medium leaching as an efficient method so as to reach high pulp density as well as higher metal recovery when compared with the other two methods. Depending on the results of HPLC profiling for organic acid production, the 14th day of fungal growth for *Aspergillus* and 15th day for *Penicillium* were selected as the optimum days for filtration of the cultures where the maximum organic acid production was observed. Leaching temperature of 60°C, leaching duration of 7 days, and pulp densities of 29.2 and 32.2 g/L, respectively, for *Aspergillus niger* and *Penicillium simplicissimum* were determined as the optimal conditions by the statistical models. Under these optimum conditions, 97% and 50% of V and Ni were extracted by *Aspergillus*, respectively, whereas 90% and 49% of V and Ni were recovered by *Penicillium*, respectively.

The FE-SEM photomicrographs reveal distinctive differences in the surface morphologies of fly ash before and after bioleaching. The significant increase in porosity of the surface of the bioleached sample and disintegration of some of the particles suggest the dissolution of metals from power plant residue due to bioleaching. According to TCLP test results, concentrations of V, Ni, and Fe were reduced below regulatory values, indicating the safe disposal of bioleached ash or its reuse for other purposes.

Further, chemical leaching tests were also conducted using commercial organic acids at equivalent values of the maximum amount produced by *Penicillium simplicissimum* (citric 5237 ppm, gluconic 3666 ppm, oxalic 1287 ppm, and malic acid 188 ppm) and *Aspergillus niger* (8080 ppm, gluconic 2126 ppm, oxalic 1170 ppm, and malic acid 1251 ppm) in pure cultures. When compared to chemical leaching, bioleaching using *Penicillium simplicissimum* increased V and Ni recovery from 79% and 37% (chemical leaching) to 90% and 49% (bioleaching), respectively. However, the recovery yields for *Aspergillus niger* were higher and it also improved chemical leaching of V and Ni up to 15% and 8%, respectively. According to some authors, bioleaching was significantly more effective as compared to chemical leaching due to the presence of other fungal metabolites in addition to citric, gluconic, oxalic, and malic acids in biomass-free spent medium, which play a remarkable role in the recovery of V and Ni. Nearly 145 different metabolites, such as galactonic, hydroxy pyruvic, fumaric, hexyl itaconic, and 4-hydroxymandelic acids, are produced by fungi

(Singh and Kumar 2007, Nielsen et al. 2009, Biswas et al. 2013). These unidentified fungal metabolites affect leaching efficiency by making more soluble complexes with the metals (Rasoulnia and Mousavi 2016a).

8.2.3 KINETICS OF METAL RECOVERY

In continuation to the previous study (Wu and Ting 2006), Xu and Ting (2009) investigated the bioleaching kinetics of fly ash by *Aspergillus* at various pulp densities (1%–6%) in a batch system. In the two-step bioleaching experiment, fly ash was added at various pulp densities to the fungal culture after 2 days. The modified Gompertz model was used to model *A. niger* growth and organic acid production. The experimental data fit the model very well over the entire range of the fly ash pulp density. The fungal growth rate gradually decreased with an increase in fly ash density due to the inhibitory effect of fly ash. Many factors responsible for toxicity to fungus during bioleaching are reported, like the blocking of essential functional groups of enzymes, the conformational changes of polymers in the cells, and the displacement of essential metals and the modification in membrane integrity and transport processes (Burgstaller and Schinner 1993). The increase in fly ash pulp density also resulted in an increase in lag phase (11.6 days at 5% pulp density) and decrease in the specific growth rate of fungus (0.033/day at 5% pulp density). The Monod growth kinetics model was used to evaluate the inhibitory effect. The critical inhibitor (i.e., fly ash) concentration (CI∗) above which no growth occurred was found to be 6.0%. The maximum specific growth rate (μmax) decreased to 0.115/day. Citric acid production was independent of pulp density, and a parallel increase was observed in citric acid concentration and biomass production at various pulp densities. The effect of pulp density on bioleaching revealed an increase in metal extraction (Al, Fe, and Zn) with an increase in pulp density up to 5%. However, at 6% pulp density metal extraction was lowest as there was no fungal growth. The simultaneous increase in the concentration of citric acid and the metals indicated that the acids produced by the fungus played a direct and important role in the bioleaching process (Xu and Ting 2009).

8.3 CONCLUSION

These studies conclusively suggest the significant role of microorganisms in the reduction, detoxification, and reuse of hazardous wastes along with metal recovery. As compared to their counterparts, bioleaching procedures are more efficient in high-yielding metal recoveries and can be used as an alternative to or in combination with other methods. Moreover, the use of a microbial consortium increases the efficiency of bioleaching. Of the three bioleaching methods reported, the two-step bioleaching and spent-medium bioleaching methods were more effective in contrast to the one-step bioleaching method. Pretreatment of ash by water washing prior to bioleaching reduces the time period of bioleaching and improves metal recovery. Very few reactor studies are reported in the case of fly ash and there is a need to develop more of these for industrial-scale implementation. Such scale-up studies will allow the detoxification of a huge amount of wastes for their further reuse and enhanced valuable metal recovery.

REFERENCES

Ahmaruzzaman, M. 2010. A review on the utilization of fly ash. *Progress in Energy and Combustion Science* 36(3): 327–363. doi:10.1016/j.pecs.2009.11.003.

Amiri, F., S. M. Mousavi, S. Yaghmaei, and M. Barati. 2012. Bioleaching kinetics of a spent refinery catalyst using *Aspergillus niger* at optimal conditions. *Biochemical Engineering Journal* 67: 208–217. doi:10.1016/j.bej.2012.06.011.

Bankar, A., M. Winey, D. Prakash, A. R. Kumar, S. Gosavi, B. Kapadnis, and S. Zinjarde. 2012. Bioleaching of fly ash by the tropical marine yeast, *Yarrowia lipolytica* NCIM 3589. *Applied Biochemistry and Biotechnology* 168(8): 2205–2217. doi:10.1007/s12010-012-9930-2.

Biswas, S., R. Dey, S. Mukherjee, and P. C. Banerjee. 2013. Bioleaching of nickel and cobalt from lateritic chromite overburden using the culture filtrate of *Aspergillus niger*. *Applied Biochemistry Biotechnology* 170: 1547–1559.

Bosshard, P. P., R. Bachofen, and H. Brandl. 1996. Metal leaching of fly ash from municipal waste incineration by *Aspergillus niger*. *Environmental Science & Technology* 30(10): 3066–3070. doi:10.1021/es960151v.

Bowes, I. and M. Mattey. 1979. The effect of manganese and magnesium ions on mitochondrial NADP+-dependent isocitrate dehydrogenase from *Aspergillus niger*. *FEMS Microbiology Letters* 6(4): 219–222. doi:10.1111/j.1574-6968.1979.tb03707.x.

Brombacher, C., R. Bachofen, and H. Brandl. 1998. Development of a laboratory-scale leaching plant for metal extraction from fly ash by *Thiobacillus strains*. *Applied and Environmental Microbiology* 64(4): 1237–1241.

Burgstaller, W. and F. Schinner. 1993. Minireview: Leaching of metals with fungi. *Journal of Biotechnology* 27: 91–116.

Dellweg, H. 1983. *Biotechnology*, vol. 3, pp. 415–465. Weinheim, Germany: Verlag Chemie.

Errasquin, E. L. and C. Vazquez. 2003. Tolerance and uptake of heavy metals by *Trichoderma atroviride* isolated from sludge. *Chemosphere* 50: 137–143.

Farah, C., M. Vera, D. Morin, D. Haras, C. A. Jerez, and N. Guiliani. 2005. Evidence for a functional quorum-sensing type AI-1 system in the extremophilic bacterium *Acidithiobacillus ferrooxidans*. *Applied and Environmental Microbiology* 71(11): 7033–7040.

Francis, A. J. and C. J. Dodge. 1990. Anaerobic microbial remobilization of toxic metals coprecipitated with iron oxide. *Environmental Science & Technology* 24(3): 373–378. doi:10.1021/es00073a013.

Funari, V., J. Mäkinen, J. Salminen, R. Braga, E. Dinelli, and H. Revitzer. 2017. Metal removal from municipal solid waste incineration fly ash: A comparison between chemical leaching and bioleaching. *Waste Management* 60: 397–406. doi:10.1016/j.wasman.2016.07.025.

Grewal, H. S. and K. L. Kalra. 1995. Fungal production of citric acid. *Biotechnology Advances* 13(2): 209–234. doi:10.1016/0734-9750(95)00002-8.

Hahn, M., S. Willscher, and G. Straube. 1993. Copper leaching from industrial wastes by heterotrophic microorganisms. In *Biohydrometallurgical Technologies*, A. E. Torma, J. E. Wey, and V. I. Lakshmanan (Eds.), pp. 673–683. Warrendale, PA: The Minerals, Metals & Materials Society.

Hoffmann, W., N. Katsikaros, and G. Davis. 1993. Design of a reactor bioleach process for refractory gold treatment. *FEMS Microbiology Reviews* 11(1–3): 221–229. doi:10.1111/j.1574-6976.1993.tb00288.x.

Ilyas, S., M. A. Anwar, S. B. Niazi, and M. Afzal Ghauri. 2007. Bioleaching of metals from electronic scrap by moderately thermophilic acidophilic bacteria. *Hydrometallurgy* 88(1–4): 180–188.

Ishigaki, T., A. Nakanishi, M. Tateda, M. Ike, and M. Fujita. 2005. Bioleaching of metal from municipal waste incineration fly ash using a mixed culture of sulfur-oxidizing and iron-oxidizing bacteria. *Chemosphere* 60(8): 1087–1094. doi:10.1016/j.chemosphere.2004.12.060.

Jadhav, U. U. and H. Hocheng. 2015. Analysis of metal bioleaching from thermal power plant fly ash by *Aspergillus niger* 34770 culture supernatant and reduction of phytotoxicity during the process. *Applied Biochemistry and Biotechnology* 175(2): 870–881. doi:10.1007/s12010-014-1323-2.

Karavaiko, G. I., S. I. Kuznetsov, and A. I. Golonizik. 1977. *The Bacterial Leaching of Metals from Ores*. Stonehouse, U.K.: Technicopy Ltd.

Krebs, W., R. Bachofen, and H. Brandl. 2001. Growth stimulation of sulfur oxidizing bacteria for optimization of metal leaching efficiency of fly ash from municipal solid waste incineration. *Hydrometallurgy* 59(2): 283–290. doi:10.1016/S0304-386X(00)00174-2.

Lanfranco, L., R. Balsamo, E. Martino, S. Perotto, and P. Bonfante. 2002. Zinc ions alter morphology and chitin deposition in an ericoid fungus. *European Journal of Histochemistry* 46: 341–350.

Le, L., J. Tang, D. Ryan, and M. Valix. 2006. Bioleaching nickel laterite ores using multimetal tolerant *Aspergillus foetidus* organism. *Minerals Engineering* 19(12): 1259–1265. doi:10.1016/j.mineng.2006.02.006.

Nielsen, K. F., J. M. Mogensen, M. Johansen, T. O. Larsen, and J. C. Frisvad. 2009. Review of secondary metabolites and mycotoxins from the *Aspergillus niger* group. *Anaytical and Bioanalytical Chemistry* 395: 1225–1242.

Pangayao, D. C., E. D. Van Hullebusch, S. M. Gallardo, and F. T. Bacani. 2015. Bioleaching of trace metals from coal ash using mixed culture of *Acidithiobacillus albertensis* and *Acidithiobacillus thiooxidans*. *Journal of Engineering Science and Technology* 10(Spec. issue 8): 36–45.

Rai, U. N., K. Pandey, S. Sinha, A. Singh, R. Saxena, and D. K. Gupta. 2004. Revegetating fly ash landfills with *Prosopis juliflora* L.: Impact of different amendments and rhizobium inoculation. *Environment International* 30(3): 293–300. doi:10.1016/S0160-4120(03)00179-X.

Ramanathan, T. and Y. P. Ting. 2014. Fly ash and the use of bioleaching for fly ash detoxification. In *Fly Ash Chemical Composition, Sources and Potential Environmental Impacts*, P. K. Sarker, (Ed.). New York: Nova Publishers.

Ramanathan, T. and Y. P. Ting. 2015. Selective copper bioleaching by pure and mixed cultures of alkaliphilic bacteria isolated from a fly ash landfill site. *Water, Air, and Soil Pollution* 226(11): 374–388. doi:10.1007/s11270-015-2641-x.

Ramanathan, T. and Y.-P. Ting. 2016. Corrigendum to 'alkaline bioleaching of municipal solid waste incineration fly ash by autochthonous extremophiles' [*Chemosphere* 160: 54–61]. *Chemosphere* 164: 692. doi:10.1016/j.chemosphere.2016.07.084.

Rasoulnia, P. and S. M. Mousavi. 2016a. Maximization of organic acids production by *Aspergillus niger* in a bubble column bioreactor for V and Ni recovery enhancement from power plant residual ash in spent-medium bioleaching experiments. *Bioresource Technology* 216: 729–736. doi:10.1016/j.biortech.2016.05.114.

Rasoulnia, P. and S. M. Mousavi. 2016b. RSC advances V and Ni recovery from a vanadium-rich power plant residual ash using acid producing fungi: *Aspergillus niger* and *Penicillium simplicissimum*. *RSC Advances* 6: 9139–9151. doi:10.1039/C5RA24870A.

Rasoulnia, P., S. M. Mousavi, S. O. Rastegar, and H. Azargoshasb. 2016. Fungal leaching of valuable metals from a power plant residual ash using *Penicillium simplicissimum*: Evaluation of thermal pretreatment and different bioleaching methods. *Waste Management* 52: 309–317. doi:10.1016/j.wasman.2016.04.004.

Rathnayake, I. V. N., M. Megharaj, N. Bolan, and R. Naidu. 2010. Tolerance of heavy metals by gram positive soil bacteria. *International Journal of Environmental Engineering* 2(5): 191–195.

Ruiz, L. M., S. Valenzuela, and M. Castro et al. 2008. AHL communication is a widespread phenomenon in biomining bacteria and seems to be involved in mineral-adhesion efficiency. *Hydrometallurgy* 94(1): 133–137.

Seidel, A., Y. Zimmels, and R. Armon. 2001. Mechanism of bioleaching of coal fly ash by *Thiobacillus thiooxidans*. *Chemical Engineering Journal* 83(2): 123–130. doi:10.1016/S1385-8947(00)00256-4.

Shahrabi-Farahani, M., S. Yaghmaei, S. M. Mousavi, and F. Amiri. 2014. Bioleaching of heavy metals from a petroleum spent catalyst using *Acidithiobacillus thiooxidans* in a slurry bubble column bioreactor. *Separation and Purification Technology* 132: 41–49. doi:10.1016/j.seppur.2014.04.039.

Singh, O. V. and R. Kumar. 2007. Biotechnological production of gluconic acid: Future implications. *Applied Microbiology and Biotechnology* 75: 713–722.

USGS. 2015. *2013 Minerals Yearbook*, Sulfur (Advance release). http://minerals.usgs.gov/minerals/pubs/commodity/sulfur/myb1-2013-sulfu.pdf (accessed November 30, 2016).

Vitolo, S., M. Seggiani, and F. Falaschi. 2001. Recovery of vanadium from a previously burned heavy oil fly ash. *Hydrometallurgy* 62(3): 145–150. doi:10.1016/S0304-386X(01)00193-1.

Wang, Q., J. Yang, Q. Wang, and T. Wu. 2009. Effects of water-washing pretreatment on bioleaching of heavy metals from municipal solid waste incinerator fly ash. *Journal of Hazardous Materials* 162(2–3): 812–818. doi:10.1016/j.jhazmat.2008.05.125.

Wong, J. W. C. and M. H. Wong. 1990. Effects of fly ash on yields and elemental composition of two vegetables, *Brassica parachinensis* and *B. chinensis*. *Agriculture, Ecosystems & Environment* 30(3): 251–264. doi:10.1016/0167-8809(90)90109-Q.

Wu, H. Y. and Y. P. Ting. 2006. Metal extraction from municipal solid waste (MSW) incinerator fly ash—Chemical leaching and fungal bioleaching. *Enzyme and Microbial Technology* 38(6): 839–847. doi:10.1016/j.enzmictec.2005.08.012.

Xiang, Y., P. Wu, N. Zhu et al. 2010. Bioleaching of copper from waste printed circuit boards by bacterial consortium enriched from acid mine drainage. *Journal of Hazardous Materials* 184(1–3): 812–818.

Xu, T. J., T. Ramanathan, and Y. P. Ting. 2014. Bioleaching of incineration fly ash by *Aspergillus niger*—Precipitation of metallic salt crystals and morphological alteration of the fungus. *Biotechnology Reports* 3: 8–14. doi:10.1016/j.btre.2014.05.009.

Xu, T. J. and Y. P. Ting. 2004. Optimisation on bioleaching of incinerator fly ash by *Aspergillus niger*—Use of central composite design. *Enzyme and Microbial Technology* 35(5): 444–454. doi:10.1016/j.enzmictec.2004.07.003.

Xu, T. J. and Y. P. Ting. 2009. Fungal bioleaching of incineration fly ash: Metal extraction and modeling growth kinetics. *Enzyme and Microbial Technology* 44(5): 323–328. doi:10.1016/j.enzmictec.2009.01.006.

Yang, J., Q. Wang, Q. Wang, and T. Wu. 2008. Comparisons of one-step and two-step bioleaching for heavy metals removal from municipal solid waste incineration fly ash. *Environmental Engineering Science* 25: 783–789.

Yang, J., Q. Wang, Q. Wang, and T. Wu. 2009. Heavy metals extraction from municipal solid waste incineration fly ash using adapted metal tolerant *Aspergillus niger*. *Bioresource Technology* 100(1): 254–260. doi:10.1016/j.biortech.2008.05.026.

Index